FACILITIES MANAGEMENT HANDBOOK

The world of facilities management has changed dramatically over the past 20 years. From relatively humble beginnings, the role of facilities manager now encompasses a wide range of complex and challenging roles, often across entire estates.

The *Facilities Management Handbook* gives a complete overview of these roles, demonstrating that facilities managers really are the stewards of the built environment. This highly practical book, now in its fourth edition, provides all the relevant legal compliance, strategic policies and best practice information needed to ensure the safe, effi cient and cost-effective running of any facilities function, and comes complete with checklists and sources of further information, offering easy-to-fi nd practical advice plus key chapters on the increasingly important subjects of risk management and business continuity planning.

The wide range of subjects covered in the *Facilities Management Handbook* includes:
- health and safety law for facilities managers
- the law regarding employed, contract and casual workers
- property law for facilities managers
- financial management
- transport policies
- outsourcing
- business continuity
- IT and communications
- workplace facilities and space planning
- access and security
- maintenance and risk management.

This up-to-date, thorough and comprehensive handbook will be your guide for the changing times ahead.

FACILITIES MANAGEMENT HANDBOOK

FOURTH EDITION

FRANK BOOTY

Routledge
Taylor & Francis Group

LONDON AND NEW YORK

First published by Butterworth-Heinemann

This edition published 2011 by Routledge
2 Park Square, Milton Park, Abingdon, Oxon OX14 4RN
711 Third Avenue, New York, NY 10017

Routledge is an imprint of Taylor & Francis Group, an Informa business.

First edition 2009

British Library Cataloguing in Publication Data
A catalogue record for this book is available from the British Library

Library of Congress Cataloging-in-Publication Data
A catalog record for this book is availabe from the Library of Congress

ISBN: 978-0-7506-8977-9

Printed and bound by CPI Group (UK) Ltd, Croydon, CR0 4YY

Contents

CHAPTER 2   Complying with the Law on Staff, Casual and Contract Workers

Part 2   Managing Your Business Effectively

CHAPTER 4   Financial Management

Part 3   The Built Environment

CHAPTER 10   Workplace Facilities

Foreword

As facilities management expands and makes its impact across the globe, it is essential that every stakeholder keeps abreast of the key developments and trends. This fourth edition of the *Facilities Management Handbook* is a testament to the maturity of the industry and the great thirst for knowledge exchange based on practical experience.

The contributors have covered the vast complexity of the industry in a logical and natural order and readers will benefit from this updated edition whether on the demand side of the industry representing client organisations, the supply side providing goods or services, or a consultant giving expert advice and guidance.

The role of the facilities manager has grown considerably over the past few years. As legislation places an ever-growing demand on the industry professional, this comprehensive book will help readers to understand any issue from risk management, environment and legal responsibilities, right through to expert service delivery. I am confident that whether readers are commencing a career in facilities management or are seasoned campaigners and practitioners, there is something of value to be gained in this new edition.

Ian R. Fielder
CEO
British Institute of Facilities Management (BIFM)

About the Authors

Frank Booty is former editor of *Facilities Management* and a contributor to, and editor of, other market-leading titles, books and web sites in the fields of business, IT and networking. He is an award-winning journalist, and has also devised and chaired many conferences in the facilities management market.

Connel Bottom is a facilities management consultant with the Business Advisory Services division of PricewaterhouseCoopers LLP, and was previously with Bernard Williams Associates, specialising in facilities cost and performance benchmarking/ modelling for corporate occupiers and service providers within the UK and Europe.

Nicholas Croft is assistant director, market and value advisory – real estate at PricewaterhouseCoopers LLP in London.

Nistha Jeram-Dave is a solicitor in the real estate group at Landwell, the associated law firm of PricewaterhouseCoopers, specialising in providing legal services to the commercial property industry.

Nicola Pecenicic is a solicitor at PricewaterhouseCoopers Legal LLP and is an Australian solicitor qualified to practice in Queensland.

Pat Perry is an Environmental Health Consultant and Managing Director of Perry Scott Nash Associates Ltd. She is a recognised expert in the fields of health and safety, food safety and CDM, and advises a wide range of clients on legal compliance and best practice.

Chris Taylor is a specialist in management, and the issues and concerns that affect business today, and is a former senior lecturer at the Business and Law faculty at the University of Lincoln.

Louis Wustemann is Managing Editor, Compliance, of all LexisNexis Butterworths Facilities Management, Health and Safety and Environment titles. He is former editor of *Flexible Working* magazine, with much experience as a writer and consultant on facilities and employment matters.

Introduction

Much has changed in the past few years since the previous editions of this book. A fresh raft of legislation has entered the facilities domain. Information technology (IT) and communications have been steadily advancing. Energy crises have impacted business dealings. The energy efficiency of a building is assuming a key and strategic importance. Sustainability is becoming ever more prominent. Security issues have leaped high up the corporate agenda. Many companies have started to review how they define 'core' activities in the wake of some high profile publicity over outsourcing contracts. Indeed there have been pundits questioning whether we are in fact seeing the start of the move to bring a lot of work back in-house. This is just what's happened in the past few years.

The world of facilities management has changed dramatically over the past two decades. Indeed, from relatively humble beginnings, the job of Facilities Manager now encompasses a wide range of complex and challenging roles, often across entire estates. In the CEN/SIC definition: "Facilities management is the integration of processes within an organization to maintain and develop the agreed services which support and improve the effectiveness of its primary activities". That is really just the starting point.

The UK SIC (Standard Industrial Classification) code is used to classify business establishments and other standard units by the type of economic activity in which they are engaged. It provides the Office of National Statistics (ONS) with a framework for the collection, presentation and analysis of data, and its use promotes uniformity. It can be used for administrative purposes as well, and by non-government bodies for classifying industrial activities into common structures. CEN (Comite Europeen de Normalization) is the European Committee for Standardization. CEN TC 348 is the facilities management standards committee operating across Europe, which works on European Standards Development.

Facilities Managers are the stewards of the built environment. They may well have to run a department responsible for everything from risk assessments to gas supply, from preparing tendering documents to managing IT systems and car parks. They are usually in the unenviable position of having to request funds for functions unconnected to the core business of the company, which are nevertheless essential. Given facilities managers' plethora of responsibilities, publications are a major requirement. There are many.

The goal of this book is to provide information and guidance for Facilities Managers, allowing readers quick and easy access to the answers to their most pressing questions. Its aim is not to offer a conclusive definition of the role of the facilities manager, which is constantly developing, but to help those working within the facilities function to develop strategies for succeeding in an increasingly high-tec and global market place, as most readers of the previous three editions have testified.

WHO WILL FIND THE BOOK HELPFUL?

Individual facilities managers and estate managers in all sizes of companies should benefit from this book. But so, also, will independent suppliers, architects, surveyors, lawyers, space planners, human resources staff and building services engineers who need an insight into the facilities universe.

KEEPING ABREAST OF THE LAW

Clearly, not falling foul of the law is a major concern for any facilities manager. The main areas of legislation in the sector are health and safety, employment/contract/casual workers and property law, each of which is covered by a separate chapter within this book. The laws and regulations regarding health, safety and employment in particular are fast moving and wide-ranging. Facilities managers have to cope with the needs of office workers for ergonomic workstations, access for the disabled and the demand of TUPE [Transfer of Undertakings (Protection of Employment)], to name just a few of the areas covered here in the form of practical and actionable advice: 2006 saw the beginnings of a major increase in legislative activity, with more occurring subsequently.

RUNNING A BUSINESS

Managing the facilities function can be like running a business in itself; setting and meeting budget targets, cutting costs and making savings have been challenging tasks for facilities managers for some time. The business issues section guides readers through the demands of contract management and the outsourcing process. It also offers up-to-date information on communications and IT, where the combination of a shrinking world and fast-developing technology mean facilities managers need to know exactly what they can – and can afford to – invest in. Private investments and partnerships, with huge implications for facilities managers, are on the increase and there is an ever-pressing need for business continuity arrangements in case of communications breakdown and other disasters, such as fire, flood and terrorist activity. These are also fully explored.

MANAGING THE BUILT ENVIRONMENT

Facilities professionals are responsible for an extraordinary array of workplace facilities. The procurement of utilities, the management of waste, the installing of integrated security systems, the purchase of office furniture, and the costing of cleaning and catering functions are just some of the challenges facing facilities managers. This section of the handbook offers best practice information and advice on these issues, pointing out the new legislative changes where relevant.

SECURITY IS A KEY ISSUE

Companies face potentially devastating litigation from their staff and the public if, in the aftermath of a terrorist atrocity, it could be demonstrated the company had failed to take measures to identify, assess and mitigate foreseeable risks to human life. It is all too easy to say, 'it won't happen to us', but if an incident occurs in the next road, an exclusion zone thrown around the site will mean no access to your buildings. Would your company be prepared for that?

FUTURE CHALLENGES

The future workplace will bring with it a number of distinct priorities. More attempts to make flexible working a reality are likely, requiring improved IT and communications infrastructures. The expansion of a 24/7 work culture will necessitate non-stop support facilities, with consequent outsourcing, contract and health and safety considerations needing to be taken into account. Public Private Partnership (PPP) and Private Finance Initiative (PFI) deals and economic pressures will push facilities professionals towards more benchmarking, more cost–benefit analysis and more performance measurement.

The tightening of building regulations and particularly the introduction of Part L [the manifestation of the European Union's Energy Performance of Buildings Directive (EPBD)] is having enormous ramifications for energy efficiency and the built environment. Buildings account for an estimated 40 per cent of all world resource consumption and over 40 per cent of all waste, including greenhouse gas emissions. EPBD is designed to increase the energy efficiency of all buildings by more than 20 per cent. A key point is whether anyone will want to rent or own an 'F'-rated building instead of an 'A' or 'B' one. All these developments should give facilities managers a fresh impetus to making their voices heard by senior management in the times ahead.

Facilities for businesses in the future will have to be more flexible and more streamlined than ever before. Managers will need to exploit growing communication opportunities and facilitate an ever-widening range of support services. Facilities managers need to become more professional and business oriented. They must develop a performance profile that is committed to the needs of their organization. Facilities managers need a variety of skills, as all organizations are different, but all should aim to gain the attention of the board and not just be visible when something goes wrong. It is one of the most challenging professions to conquer.

The key point to be made, touching on an earlier comment, is that facilities managers are indeed the stewards of the built environment.

Frank Booty
Editor

Complying with the law

Complying with Safety, Health, Fire and Environment Law

1

Pat Perry

Prosecutions under criminal law and compensation under civil law are the tip of the iceberg when it comes to working out what poor health and safety costs employers. Health and Safety Executive (HSE) studies suggest that for every pound in costs covered by employers' liability insurance, organisations must pay another £8 to £36 in uninsured costs. The message is that safe, healthy workplaces save money, not only in obvious ways such as reducing absenteeism, but also by making organisations more efficient and productive. By improving workplace design and ensuring that work-stations are ergonomically suited to the individual worker, for example, facilities managers play a crucial role in this process of improving staff morale and well-being.

The 2006/07 HSE survey of self-reported work-related illness estimated that 2.2 million employees were made ill by their work that year. Over one million people suffered musculoskeletal disorders, and over 400,000 suffered stress, anxiety or depression. Over 10.5 million working days were lost owing to stress, anxiety or depression in 2005/06. Overall, the total number of working days lost during 2005/06 was reputed to be over 30 million. The Confederation of British Industry (CBI) has estimated that the annual cost of absenteeism to UK plc is £10 billion, rising to £20 billion if indirect costs are counted.

As for accidents, 241 people were killed at work in 2006/07, and there were approximately 300,000 non-fatal injuries to employees. HSE estimates that injuries cost UK businesses £0.9 billion a year, illnesses £1.6 billion, and non-injury accidents between £1.4 billion and £4.5 billion. The total represents between 4 and 8 per cent of companies' gross trading profits.

On average, the cost of an injury or serious accident to an employer is approximately £8000 per incident.

Despite many health and safety awareness campaigns, and stricter law enforcement, accident rates have not significantly reduced. The construction industry in particular has much improvement to make.

COST OF POOR HEALTH AND SAFETY

Costs imposed by ill-health and accidents include:

- sick pay
- lost production

3

- damaged equipment and materials
- extra administration
- losing skilled staff
- training replacement staff
- increased insurance premiums
- loss of staff morale
- bad publicity (which may put off job candidates and investors)
- compensation, fines and court costs.

ENFORCEMENT

The cornerstone of health and safety legislation is the Health and Safety at Work etc. Act 1974 (HSW Act). It is an 'enabling' Act, under which detailed regulations on particular work-related risks are made.

These regulations are supported by Approved Codes of Practice (ACoPs) and guidance. ACoPs have a special legal status like the Highway Code. If a court proves that an organisation failed to comply with a relevant ACoP, it will be found guilty unless it demonstrates that it complied with the law in another way. Following guidance is not compulsory, and employers are free to take other action. But by following it, they will usually be doing enough to comply with the law.

The Health and Safety Commission and the Health and Safety Executive

The Health and Safety Commission (HSC) is the agency with overall responsibility for developing workplace health and safety and deciding on new legislation to lay before parliament. Its commissioners are drawn from employers' organisations, trade unions and local authorities. Its executive arm was the HSE, but in April 2008 the two divisions merged into one regulatory body which will be known as the Health and Safety Executive.

Responsibility for enforcing the law is shared between the HSE and local authorities. The HSE inspects factories, chemical firms and construction sites, while local authority environmental health officers (EHOs) inspect offices, shops, warehouses, hotels and leisure sites. Inspectors may issue an improvement notice if they find a breach of statutory duty, or a prohibition notice stopping an activity which involves a risk of serious injury. Immediate prosecutions can be instigated for severe or gross contravention of any health and safety legislation and may often be a route taken by inspectors after accident investigations.

REGULATIONS

Modern health and safety law in the UK is 'goal setting' rather than prescriptive. The responsibility is on employers actively to apply these broad legal principles to suit their particular organisation, rather than following fixed rules which may not be appropriate to their circumstances.

Goal-setting regulations can cause much anxiety among employers as they have no prescribed solutions giving a definitive approach to compliance. A common question raised by employers and others who are responsible for complying with the law is 'how do I know that I've done enough to comply with the law?' The answer is often in the ACoP, or in the term 'suitable and sufficient'. No one is expected to achieve perfection all of the time, but any compliance must be appropriate for the hazards and risks associated with the task or work environment.

The Health and Safety at Work etc. Act 1974

The HSW Act imposes a duty of care on employers, employees and suppliers.

Employers' duties (s. 2)

Employers should, 'so far as is reasonably practicable', ensure the health, safety and welfare at work of all employees by:

- providing safe equipment and working practices
- ensuring safe handling, storage and transportation of goods and substances
- providing information, training and supervision
- providing a safe place of work with safe access and egress
- providing a safe working environment and facilities.

Safety policy [s. 2(3)]

Organisations with five or more employees must:

- prepare a written health and safety policy, listing the hazards present in the workplace
- provide details of arrangements for carrying out the policy
- make employees aware of the policy and arrangements
- update the policy as necessary.

Duties to non-employees (s. 3)

Employers should ensure 'so far as is reasonably practicable' that they do not put anyone not in their employment (e.g. members of the public, contractors and agency staff) at risk from their activities. This duty can extend to trespassers, so facilities managers need to ensure that high-risk workplaces have adequate security and fencing.

'So far as is reasonably practicable' means that if the cost of removing a risk (in terms of time and trouble as well as financial cost) outweighs the possible benefit to employees and the public, an employer does not have to act. The burden of proof rests with the employer, meaning that, if prosecuted, it has to prove why a safety measure was not reasonably practicable.

Suppliers' duties (s. 6)

Those who design, manufacture, import or supply goods or substances for use at work must ensure that they are safe when used properly and provide instructions on how to use them safely.

Employees' duties (s. 7)

Employees must:

- take reasonable care for their own health and safety and that of anyone else affected by their work
- co-operate with their employer in ensuring it meets its legal obligations [by wearing personal protective equipment (PPE) or reporting accidents, for example].

Personal liability (s. 37)

Company directors and managers may be personally liable for a health and safety offence committed with their consent or connivance. They face being fined and disqualified from being a director for up to two years.

The Corporate Manslaughter and Corporate Homicide Act 2007

This Act came into force on 6 April 2008 and applies in:

- England
- Wales
- Northern Ireland
- Scotland, where the offence will be known as corporate homicide.

The new Act was needed because prosecutions of organisations (corporate bodies) were almost impossible under former legislation unless a single individual, identifiable as the controlling mind of the company, has been personally guilty of gross negligence or manslaughter. The new Act removes the need to identify the 'controlling mind' of an organisation.

The Act applies to:

- companies incorporated under companies legislation or overseas
- other corporations including:
  - public bodies incorporated by statute, for example, local authorities, NHS bodies and a wide range of non-departmental public bodies
  - organisations incorporated by Royal Charter
  - limited liability partnerships
- all other partnerships, trade unions and employers, associations, if the organisation concerned is the employer
- crown bodies such as government departments
- police forces.

The offence is concerned with the corporate liability of the organisation itself and does not apply to individual directors, senior managers or other individuals, but individuals can already be prosecuted for gross negligence manslaughter/culpable homicide and for health and safety offences.

The Act creates a new offence for convicting an organisation where a gross failure in the way activities were managed or organised results in a person's death. An organisation will be guilty of the new offence if the way in which its activities are managed or organised causes a death and amounts to a gross breach of a duty of care to the deceased.

Courts will look at management systems and practices across the organisation. Organisations must take the opportunity to review current management systems for health and safety to ensure that they are complying with current legal duties. The new Act does not create new legal duties. It provides the opportunity for organisations that disregard the safety of others at work, with fatal consequences, to face serious criminal charges.

Juries will consider how the fatal activity was managed or organised throughout the organisation, including any systems and processes for managing safety and how these were operated in practice.

A substantial part of any management failure must have been at **senior level**, which means the people who make significant decisions about the organisation or substantial parts of it. This includes both centralised headquarters' functions as well as those in operational management roles.

The organisation's conduct must have fallen far below what could have been reasonably expected; for example, breaches of specific health and safety duties would indicate gross breach, as would being advised of a safety hazard/risk and doing nothing about it.

Duty of care exists in respect of an employer's duty to provide a safe working environment for his or her employees and others, to provide safe and suitable equipment and to supply safe products and services to others. A duty of care can be imposed by civil law, that is, common law, or by statute law.

The Director of Public Prosecutions, Public Prosecution Service (Northern Ireland) or Procurator Fiscal (Scotland) must give consent before a case of corporate manslaughter can be taken to court.

Penalties on conviction may be:

- unlimited fines
- publicity orders
- remedial orders.

The sentencing authority has issued a consultation paper advising that fines should be between 2.5 and 10 per cent of company turnover.

The court can impose a publicity order on the company/organisation requiring them to publicise the details of its conviction in any media the court thinks appropriate, such as national newspapers, television and radio. Courts may require an organisation to take steps to address the failures behind the death; for example, remedy defects, carry out training or employ competent persons.

Work-related death will be investigated by the police and/or HSE or local authority (depending on type of business) or other regulatory body, for example, the Railway Inspectorate.

Under the Corporate Manslaughter and Corporate Homicide Act 2007, the 'corporate body' may be represented by its lawyers, that is, not an individual. Under health and safety laws and existing gross negligence manslaughter, it could be represented by any individual.

To have a successful defence for breach of the Corporate Manslaughter and Corporate Homicide Act 2007, you will have to prove that there was no gross breach of the duty of care by senior management to the person who was killed. Most work-related deaths will be investigated as corporate manslaughter cases. Facilities managers can play a key role in ensuring that both their clients and their own organisations are properly prepared by:

- determining who could be senior managers and assess their competency in respect of health and safety
- reviewing health and safety policies to ensure that standards set are achievable
- reviewing health and safety practices and procedures: undertake a health and safety management review
- reviewing risk assessments
- reviewing all employee training for health and safety
- increasing board level scrutiny of health and safety matters: include a board agenda item for health and safety
- introducing or reviewing an accident management protocol: investigate accidents, take remedial action
- developing a procedures manual for 'handling an investigation and preparing your defence'.

Leading Health and Safety at Work

To coincide with the introduction of the Corporate Manslaughter and Corporate Homicide Act 2007, the HSE and Institute of Directors have published a guidance document on leading health and safety within the workplace.

Research has shown that a company's or an organisation's attitude to good health and safety is influenced by the interest taken by senior managers and the more they lead from the top by instilling a safety culture, the more successful the organisation's health and safety management system will be. The guidance can be summarised in essential principles as follows:

- strong and active leadership from the top
- worker engagement and involvement
- assessment and review.

Directors (and other equivalent positions) should plan the direction of health and safety in the organisation, deliver health and safety, monitor systems and procedures, and formally review their performance.

The guidance document contains a useful health and safety management checklist which is reproduced below. It will be just as beneficial to facilities managers who have responsibilities for overseeing health and safety within the organisation and could provide the framework for a report to the client or others to raise the profile of health and safety in the organisation.

HEALTH AND SAFETY LEADERSHIP CHECKLIST

- How do you demonstrate the board's commitment to health and safety?
- What do you do to ensure appropriate board-level review of health and safety?
- What have you done to ensure your organisation, at all levels including the board, receives competent health and safety advice?
- How are you ensuring all staff, including the board, are sufficiently trained and competent in their health and safety responsibilities?
- How confident are you that your workforce, particularly safety representatives, are consulted properly on health and safety matters, and that their concerns are reaching the appropriate level including, as necessary, the board?
- What systems are in place to ensure your organisation's risks are assessed, and that sensible control measures are established and maintained?
- How well do you know what is happening on the ground, and what audits or assessments are undertaken to inform you about what your organisation and contractors actually do?
- What information does the board receive regularly about health and safety, for example, performance data and reports on injuries and work-related ill-health?
- What targets have you set to improve health and safety and do you benchmark your performance against others in your sector or beyond?
- Where changes in working arrangements have significant implications for health and safety, how are these brought to the attention of the board?

Source: HSE/IOD[1]

A successful health and safety policy

Under the HSW Act, employers must prepare in writing:

- a statement of their general policy on their employees' health and safety at work
- the organisation of the policy
- the arrangements for carrying out the policy.

Policies are normally divided into three sections to meet these three demands.

The statement of intent

This involves a general statement of good intent and a commitment to comply with relevant legislation. Many employers extend their policies to the health and safety of others affected by their activities, such as contractors and the public. To demonstrate commitment from the top, the statement should be signed by the chairperson or chief executive.

---

[1] INDG 417.

Organisational responsibilities

This section shows how the organisation will put its good intentions into practice and which responsibilities those at different levels in the management structure will hold. The section might cover:

- making adequate resources available to implement the policy
- setting health and safety objectives
- developing suitable procedures and safe systems
- delegating specific responsibilities
- monitoring people's effectiveness in carrying out their responsibilities
- monitoring workplace standards
- feeding concerns up through the organisation.

Arrangements

The policy need not contain all the organisation's health and safety arrangements but should state where they can be found, such as in a separate health and safety manual or within various procedural documents. Topics that may require detailed arrangements include:

- operational procedures
- training
- personal protective equipment
- inspection programmes
- accident and investigation arrangements
- fire and emergency procedures
- first aid
- occupational health
- control of contractors and visitors
- consultation with employees
- audits of health and safety arrangements.

Employees must be aware of the policy and their own responsibilities. They may be given their own copy, or the policy may be displayed around the workplace. Detailed briefings and induction training may be necessary for some arrangements.

Employers should revise their policies as often 'as may be appropriate'. Larger employers are likely to need to arrange for formal review and, where necessary, for revision to take place on a regular basis (by way of an ISO 9000 procedure, for example). Dating the policy document is an important part of this process.

Primary regulations relating to health and safety

Management of Health and Safety at Work Regulations 1999 (Management Regulations)

The Management Regulations make explicit many of the duties contained in the HSW Act. The scope of the Regulations and employers' duties are:

- Employers must make suitable and sufficient assessment of the risks to health and safety of their employees and to non-employees affected by their work. Each risk assessment must identify the measures necessary to comply with relevant statutory provisions.
- Risk assessments must be in writing where there are more than five employees and they must be reviewed regularly.
- Employers must introduce appropriate arrangements for effective planning, organisation, control, monitoring and review of the preventive and protective measures. These arrangements must be in writing where there are five or more employees. Where appropriate, employees must have health surveillance.
- Employers must establish and effect appropriate procedures to deal with emergencies, for example, evacuation or major chemical spillage explosion. Employees must stop work immediately and proceed to a place of safety, if exposed to serious, imminent and unavoidable danger.
- Employers must provide comprehensive and understandable information to all employees on risks identified in the risk assessments, emergency procedures, preventive and protective measures, and competent personnel.
- Employees must receive appropriate health and safety training on recruitment and throughout their employment.
- Employers must appoint a 'competent person' to assist in undertaking the measures necessary to comply with statutory provisions. Competent persons should be employees but external advisors/consultants are permissible if there are no suitable employees.
- Employers must consider the health and safety of young people at work, pregnant women and nursing mothers.
- Contact must be made by employers with external services, for example, fire, police and emergency services so that any necessary measures can be taken in the event of an emergency or rescue.
- Multioccupied sites must have a plan for co-operation and co-ordination, with one employer taking the lead role in respect of health and safety.
- Employers must provide information to any person employed to work on their premises regarding risks to their health and safety from the employer's undertaking, information on fire evacuation procedures and the fire risk assessment, etc.
- Employers must take into account the capabilities and training of all employees before assigning them tasks, etc.
- Employees are under a duty to use equipment, materials, etc., provided to them by their employer in accordance with safe systems of work, any training, etc.
- Employees must inform their employer of any matter relating to either their own or others' health and safety.
- Temporary workers are to be afforded health and safety protection, information on hazards and risks, health surveillance as appropriate, etc.

See later in this chapter, Risk assessment, for detailed information on conducting a risk assessment.

Workplace (Health, Safety and Welfare) Regulations 1992 (Workplace Regulations)

Employers and others with any control over the workplace (e.g. landlords) must ensure that the workplace, equipment and devices listed in the regulations are maintained in efficient working order and good repair. These regulations govern much of the responsibility that facilities managers have for ensuring that the work premises are clean, comfortable, well lit, well ventilated and well organised. Specifically, employers must ensure that:

- enclosed workplaces are ventilated by sufficient fresh or purified air
- a reasonable temperature is maintained
- there is suitable and sufficient lighting, especially natural light
- workplaces, furniture, furnishings and fittings are clean
- there is sufficient floor area and headroom
- workstations and seating are suitable for the person and the job
- floors are level, not slippery, effectively drained and kept free from obstruction, and staircases have handrails
- risks from falls or falling objects are removed or controlled, with fences and covers for tanks and pits
- glazing is protected against breakage and marked so that it can be seen
- windows and skylights can be cleaned, opened and left open safely
- traffic routes are marked and organised so that pedestrians and vehicles can circulate safely
- doors, gates, escalators and moving walkways are safely constructed and safe to use
- enough toilets and washing facilities are provided, which are accessible, ventilated, lit and clean
- an adequate supply of wholesome drinking water is provided
- facilities are provided for storing and changing clothes
- facilities for resting and eating meals are provided, with suitable arrangements for pregnant women and separate external facilities for smokers.

Health and Safety (Display Screen Equipment) Regulations 1992 (DSE Regulations)

Employers must assess the risks to 'habitual' users or operators of computer monitors and other display screens such as video monitors and reduce them to the lowest level reasonably practicable. 'Users' are employees, whereas operators are self-employed people or 'temps'. Habitual monitor users:

- use them continuously for an hour or more a day
- depend on them to do the job
- have no discretion over their use.

Employers must also ensure that:

- new workstations meet minimum legal requirements
- users take breaks from their monitor or change activities regularly (a 5–10 minute break every hour is better than 15 minutes every two hours)

- on request, employees receive free regular sight tests and corrective glasses
- employees receive adequate training and information.

Manual Handling Operations Regulations 1992

Manual handling is the transporting or supporting of a load by hand or bodily force, including lifting, putting down, pushing, pulling, carrying or moving. Employers' duties are to:

- avoid the manual handling operation if reasonably practicable, by redesigning the task to avoid moving the load or by automating the process
- conduct a further risk assessment, if eliminating the risks is not possible, and reduce the risk of injury to the lowest level reasonably practicable
- provide employees with indications and, where practicable, precise information on the weight of each load and its heaviest side.

Further,

- Risk assessments must be reviewed regularly, particularly if there is any reason to believe that there has been significant change in the handling operation.
- Manual handling controls are not limited just to heavy weights; consideration should be given to lighter loads which are handled repetitively.
- Employers should make available mechanical lifting aids whenever practicable, for example, sack trolleys, hoists.

Provision and Use of Work Equipment Regulations 1998 (PUWER)

Employers must ensure that all work equipment (e.g. tools, photocopiers, vehicles and manufacturing plant) is:

- suitable for its intended use
- safe to use, maintained in a safe condition and, in some cases, inspected for safety by a competent person
- used only by trained personnel
- accompanied by suitable safety devices (e.g. guards, warnings, stop buttons)
- used according to a safe system of work (e.g. maintaining machinery when it is shut down).

Some further points:

- If employees provide their own equipment, or work from home, the employer must ensure that it complies with PUWER.
- PUWER does not apply to equipment used by members of the public (e.g. compressed air equipment on garage forecourts), although the HSW Act will apply
- Mobile equipment (e.g. forklift trucks) meant to carry people must be suitable for the purpose, and employers must minimise the risk of it rolling over
- Power presses and their guard devices must be thoroughly examined at specified intervals, inspected daily for safety, and records kept.
- Warning notices must be displayed as necessary adjacent to equipment.

- Any second hand equipment brought in to a business must be classed as 'new' equipment and must meet the requirements of the Regulations.

Personal Protective Equipment at Work Regulations 1992 (PPE Regulations)

Employers must:

- assess whether PPE (face masks, fall-arresting equipment, gloves, etc.) is suitable and provide it to employees exposed to risks not adequately controlled by other means
- ensure that items of PPE worn together are compatible
- ensure that PPE is maintained in efficient working order and good repair
- provide storage for PPE that is not in use: if the PPE is contaminated, this must be separate from ordinary clothing
- provide information and training on the risks the PPE will limit, and how to use and maintain the PPE
- ensure that the PPE is properly used.

Other key regulations

Control of Asbestos Regulations 2006 (CAW Regulations)

These Regulations bring together the three previous sets of Regulations covering the prohibition of asbestos, the control of asbestos at work and asbestos licensing.

The Regulations prohibit the importation, supply and use of all forms of asbestos. They continue the ban introduced for blue and brown asbestos in 1985 and for white asbestos in 1999. They also continue to ban the second hand use of asbestos products such as asbestos cement sheets and asbestos boards and tiles, including panels which have been covered with paint or textured plaster containing asbestos.

Remember: The ban applies to new use of asbestos. If existing asbestos-containing materials are in good condition, they may be left in place, and their condition monitored and managed to ensure that they are not disturbed.

The Asbestos Regulations also include the 'duty to manage asbestos' in non-domestic premises.

The Regulations require mandatory training for anyone liable to be exposed to asbestos fibres at work (reg. 10). This includes maintenance workers and others who may come into contact with or who may disturb asbestos (e.g. cable installers) as well as those involved in asbestos removal work.

When work with asbestos or which may disturb asbestos is being carried out, the Asbestos Regulations require employers and the self-employed to prevent exposure to asbestos fibres. Where this is not reasonably practicable, they must make sure that exposure is kept as low as reasonably practicable by measures other than the use of respiratory protective equipment. The spread of asbestos must be prevented. The Regulations specify the work methods and controls that should be used to prevent exposure and spread.

Worker exposure must be below the airborne exposure limit (control limit). The Asbestos Regulations have a single control limit for all types of asbestos of 0.1 fibres/cm^3. A control limit is a maximum concentration of asbestos fibres in the air (averaged over any continuous four-hour period) that must not be exceeded.

In addition, short-term exposures must be strictly controlled and worker exposure should not exceed 0.6 fibres/cm^3 of air averaged over any continuous 10-minute period using respiratory protective equipment if exposure cannot be reduced sufficiently using other means.

Respiratory protective equipment is an important part of the control regime, but it must not be the sole measure used to reduce exposure and should only be used to supplement other measures. Work methods that control the release of fibres should be used. Respiratory protective equipment must be suitable, must fit properly and must ensure that worker exposure is reduced as low as is reasonably practicable.

Most asbestos removal work must be undertaken by a licensed contractor but any decision on whether particular work is licensable is based on the risk. Work is only exempt from licensing if:

- the exposure of employees to asbestos fibres is sporadic and of low intensity (but exposure cannot be considered to be sporadic and of low intensity if the concentration of asbestos in the air is liable to exceed 0.6 fibres/cm^3 measured over 10 minutes) and
- it is clear from the risk assessment that the exposure of any employee to asbestos will not exceed the control limit and
- the work involves:

  - short, non-continuous maintenance activities. Work can only be considered as short, non-continuous maintenance activities if any one person carries out work with these materials for less than one hour in a seven-day period. The total time spent by all workers on the work should not exceed a total of two hours,
  - removal of materials in which the asbestos fibres are firmly linked in a matrix. Such materials include: asbestos cement; textured decorative coatings and paints which contain asbestos; articles of bitumen, plastic, resin or rubber which contain asbestos where their thermal or acoustic properties are incidental to their main purpose (e.g. vinyl floor tiles, electric cables, roofing felt) and other insulation products which may be used at high temperatures but have no insulation purposes, for example, gaskets, washers, ropes and seals,
  - encapsulation or sealing of asbestos-containing materials which are in good condition, or
  - air monitoring and control, and the collection and analysis of samples to find out whether a specific material contains asbestos.

Under the Asbestos Regulations, anyone carrying out work on asbestos insulation, asbestos coating or asbestos insulating board (AIB) needs a licence issued by HSE unless they meet one of the exemptions above.

Remember: Although a licence may not be needed to carry out a particular job, it will be necessary to comply with the rest of the requirements of the Asbestos Regulations.

If the work is licensable there are a number of additional duties, including:

- Notify the enforcing authority responsible for the site where you are working (e.g. HSE or the local authority).

- Designate the work area (see reg. 18 for details).
- Prepare specific asbestos emergency procedures.
- Pay for your employees to undergo medical surveillance.

The Asbestos Regulations require any analysis of the concentration of asbestos in the air to be measured in accordance with the 1997 World Health Organisation (WHO) recommended method. A clearance certificate for reoccupation may only be issued by a body accredited to do so, such as the United Kingdom Accreditation Service (UKAS).

Confined Spaces Regulations 1997

The main duties under these regulations are:

- to avoid entry to confined spaces, for example, by having work done from the outside
- if entry to a confined space is unavoidable, to follow a safe system of work
- to put in place adequate emergency arrangements before the work starts.

Control of Substances Hazardous to Health Regulations 2002 (COSHH) (and various amendments)

Employers must:

- make a suitable and sufficient assessment of health risks to employees exposed to hazardous substances
- in order of preference, prevent exposure, control exposure, or provide PPE
- provide health surveillance for exposed employees
- keep exposed employees' health records for 40 years
- attach safety data sheets to COSHH assessments.

Registration, Evaluation, Authorisation and Restriction of Chemicals Regulations (REACH)

The REACH Regulations became law across Europe on 1 June 2007 and regulate the manufacture, supply and use of chemicals across the European Union (EU). All businesses can be affected by the Regulations if they rely in some way on chemicals, even if only as cleaning substances.

Chemicals include paints, metals, glues, solvents and cleaning materials, all of which may fall under a facilities manager's remit at some point.

Chemicals that are manufactured or imported into the EU will need to be registered with the new European Chemical Agency (ECHA), which is based in Helsinki. Registration will be phased in over the next few years.

REACH is designed to provide more information on chemicals for users and to increase confidence in the safe use of chemicals. It will be good for health and safety, and REACH may help to reduce the increasing number of people who face occupational ill-health due to their work activities, for example, by contracting industrial dermatitis or industrial asthma due to the chemicals to which they are exposed.

Safety data sheets will be improved for all chemicals and end-users will have better information on which to develop their COSHH assessments. Some chemicals may be ultimately banned by ECHA, and businesses need to be planning for any such event. Start by asking your chemical suppliers how REACH will affect them and whether you need to be aware of any consequences.

Electricity at Work Regulations 1989

Scope of regulations and employers' duties:

- All systems, plant and equipment to be designed to ensure the maximum practical level of safety.
- Installation and maintenance to reflect specific safety requirements.
- Access, light and working space to be adequate.
- Means of cutting off power and isolating equipment to be available.
- Precautions to be taken to prevent charging.
- No live working unless absolutely essential.
- Specific precautions to be taken where live working is essential.
- All persons to be effectively trained and supervised.
- Responsibility for observing safety policy to be clearly defined.
- All equipment and tools to be appropriate for safe working.
- All personnel working on electrical systems to be technically competent and have sufficient experience.
- Work activity shall be carried out so as not to give rise to danger.
- Electrical systems must be constructed and maintained to prevent danger. 'Danger' means the risk of injury.

Health and Safety (First Aid) Regulations 1981

Employers must provide:

- adequate and appropriate first aid facilities
- either qualified first aiders (the ACoP suggests one first aider for every 50 employees)
- or, in low-risk or small workplaces, an 'appointed person' to take charge in an emergency.

Health and Safety Information for Employees Regulations 1989

Employers must display an HSE information poster or provide each employee with an HSE leaflet.

Health and Safety (Consultation with Employees) Regulations 1996

Employers must:

- consult with employees on the introduction of any measure or new technology that substantially affects their health and safety
- consult employees either directly, that is, on an individual basis, or through elected representatives
- provide elected representatives with training, paid time off and other facilities reasonably required.

Safety Representatives and Safety Committees Regulations 1977

Trade unions may appoint employees as safety representatives with the right to:

- be consulted by the employer on health and safety
- carry out investigations and inspections
- have paid time off to perform their duties and attend training.

Information and Consultation of Employees Regulations 2004

Businesses with more than 50 employees have to consult with their workforce over a wide range of business issues, including health and safety. If at least 10 per cent of the workforce demand an Information and Consultation agreement you must begin negotiating one within three months.

Broadly, employees will need to be consulted about:

- the business activities
- the economic situation
- the employment situation within the business
- decisions likely to lead to changes in the business organisation or contractual relationships.

Health and Safety (Safety Signs and Signals) Regulations 1996

Employers must:

- provide warning signs if they cannot avoid a risk by other means
- train employees in the meaning of safety signs.

Signs must be either:

- hazard warning: yellow with black writing
- mandatory: blue with white writing
- prohibition: red with white writing.

Lifting Operations and Lifting Equipment Regulations 1998 (LOLER)

These regulations supplement PUWER and apply to equipment such as cranes, goods lifts, forklift trucks, mobile elevating work platforms, hoists and ropes. Employers must ensure that:

- lifting equipment has adequate strength and stability
- lifting equipment is positioned and installed to make the risk of the equipment or load falling or hitting a person as low as reasonably practicable
- lifting equipment is marked with its safe working load
- equipment for lifting people does not present a risk of them being crushed or trapped or of them falling
- any operation involving lifting equipment is properly planned by a competent person, and properly supervised and executed.

Noise at Work Regulations 2005

Employers must:

- reduce the risk of hearing damage to the lowest level reasonably practicable
- where noise cannot be further reduced, provide ear protection if personal daily noise exposures are 85 dB or more (or on request at levels of 80–85 dB)
- carry out noise assessments to determine whether employees are exposed to noise levels that could impair their hearing.

Reporting of Injuries, Diseases and Dangerous Occurrences Regulations 1995 (RIDDOR)

Employers must:

- notify the HSE or local authority of a death or serious injury immediately
- send a report form within 10 days if a person is absent following a work injury for more than three consecutive days or if there is a 'dangerous occurrence'
- notify the HSE or local authority if a person suffers a work-related disease (e.g. carpal tunnel syndrome, hand arm vibration syndrome, legionellosis, dermatitis, asthma or asbestos-related diseases)
- notify the HSE or local authority of specific dangerous occurrences.

Dangerous Substances and Explosive Atmospheres Regulations 2002 (DSEAR)

These regulations came into force in December 2002. They set minimum requirements for the protection of workers from fire and explosions arising from dangerous substances and potentially explosive atmospheres.

Employers must:

- carry out a risk assessment of any work activity involving dangerous substances
- provide technical and organisational measures to eliminate or reduce, as far as is reasonably practicable, the identified risks
- provide equipment and procedures to deal with accidents and emergencies
- provide information and training to employees
- classify places where explosive atmospheres may occur into zones and mark the zones where necessary.

DSEAR Regulations 2002 should complement existing health and safety management systems.

The Work at Height Regulations 2005

These Regulations apply to all working at height in all businesses and cover the safety of employed and self-employed persons. Where possible, work at height should be avoided but where that is not possible, the hierarchy of risk control should be followed, and the hazard and risk of falls and falling objects reduced as far as is reasonably practicable.

The Regulations repealed the '2.0 m' rule in which, previously, no edge protection was needed if the fall distance was 2.0 m or less.

RISK ASSESSMENT

HSE has produced an ACoP and guidance to the Management Regulations,[2] as well as a *Five steps to risk assessment* guide.[3] According to its own definitions:

- 'a **risk assessment** is nothing more than a careful examination of what, in your work, could cause harm to people so that you can weigh up whether you have taken enough precautions or should do more to prevent harm'
- 'a **hazard** means anything that can cause harm (e.g. chemicals, electricity, working from ladders)'
- '**risk** is the chance, high or low, that somebody will be harmed by the hazard'.

Spotting hazards

Ways of identifying hazards include asking employees or their safety representatives what risks they have noticed; consulting accident and ill-health records, suppliers' manuals, the trade press, relevant legislation and guidance; or seeking advice from consultants.

Typical hazards to watch for include:

- physical hazards (e.g. poorly guarded machinery, mezzanine floors, slipping/ tripping hazards, vehicles, poor electrical wiring, fire hazards)
- hazardous substances (e.g. chemicals, dust, fumes)
- a hazardous work environment (e.g. noise, poor ventilation, bad lighting, hot or cold workplaces)
- psychological hazards (e.g. stress, long hours, shift work)
- ergonomic hazards (e.g. repetitive work, lifting).

Who may be harmed?

Those at risk may be:

- employees (e.g. office, operational and maintenance staff)
- contractors (e.g. cleaners and security guards)
- members of the public and volunteers
- young and inexperienced workers
- new and expectant mothers
- staff with disabilities
- home, lone and mobile workers.

Risks may increase at certain times of day, such as after dark or during busy periods.

---

[2] ISBN 07176 24889.
[3] IND(G) 132L.

Controlling the risks

It is important to consider whether existing precautions:

- meet legal requirements
- comply with industry standards
- represent good practice
- reduce risks as far as reasonably practicable.

If not, an action plan will be necessary, categorising remaining risks as high, medium or low risk. Priority should go to those measures that will protect the whole workplace. The aim is to eliminate hazards altogether (e.g. by not using a hazardous substance) or, if this is not possible, to control risks, in order of preference, by:

- combating risks at source (e.g. if steps are slippery, treating them is better than displaying a warning sign)
- preventing access to the hazard (e.g. by installing machine guards, using permits to work which restrict access to authorised staff, isolating a dusty area)
- organising work to reduce exposure to the hazard (e.g. by rearranging work patterns to reduce stress)
- issuing PPE
- providing welfare facilities (e.g. washing facilities to remove contamination).

Taking advantage of technical advances can make work processes safer and more efficient. It is also crucial to provide relevant training and information to staff, and in shared workplaces, to swap information with other firms on site.

Recording the findings

Generally speaking, organisations with five or more employees must write down their risk assessment. The document can make cross-references to the health and safety policy and other relevant paperwork, rather than repeating everything. An inspector or a union safety representative may ask to see the risk assessment, or it may provide evidence in the event of a personal injury claim.

Reviewing and revising

Most health and safety legislation requires employers to review the risk assessment and revise it as necessary, if it is 'no longer valid or there has been a significant change'. Hazardous substance and asbestos assessments must, however, be reviewed 'regularly', and HSE guidance states that this is good practice for any risk assessment. Significant changes that may make the risk assessment out of date could include bringing in new machines, substances or procedures.

'Suitable and sufficient'

The Management Regulations require risk assessments to be 'suitable and sufficient'. This means:

- allocating appropriate resources
- making the level of detail proportionate to the risk: overcomplicated assessments of simple hazards are not required
- anticipating 'foreseeable' risks
- ensuring that consultants have sufficient understanding of particular work activities
- adapting any 'model' assessment to the actual work situation
- drawing up a timetable for implementing short, medium and long-term controls
- reviewing non-routine activities such as maintenance, cleaning, loading and unloading vehicles, changes in production cycles and emergencies
- reviewing off-site activities, such as home working
- complying with specific regulations; although repeating assessments is not necessary
- addressing what actually happens in the workplace, not what the works manual says should happen.

THE HSE'S FIVE STEPS TO RISK ASSESSMENT

1. Look for the hazards.
2. Decide who might be harmed, and how.
3. Weigh up the risks and decide whether existing precautions are adequate or more needs to be done.
4. Record your findings.
5. Review your assessment and revise it if necessary.

CRIMINAL SANCTIONS

Most breaches of health and safety law are heard in the Magistrates' Court (summary jurisdiction), although serious offences are more frequently being referred to the Crown Courts (indictment). The Health and Safety (Offences) Act 2008 is now in force and the maximum fine for breaching health and safety regulations is £20,000 per offence; whilst breaching the Health and Safety At Work Act 1974, Sections 2-6, also carries a maximum fine of £20,000. The levels of fines in the Crown Court are unlimited. Custodial sentences are now possible for nearly all health and safety offences, with the Magistrates' Courts able to impose Custodial sentences of up to 12 months and the Crown Courts up to two years.

Penalties in the Crown Court have, however, been increasing, especially since the Court of Appeal's judgment in the Howe case in 1998.[4] This laid out sentencing guidelines, including that:

- The level of fine should reflect the gravity of the offence, the degree of risk, and whether it was an isolated offence.

---

[4] R v Howe & Son (Engineers) Ltd, 6 November 1998.

- Aggravating factors include whether the defendant failed to heed warnings, deliberately flouted health and safety legislation for financial gain, and whether a fatality occurred.
- Mitigating factors include a prompt guilty plea, taking steps to remedy health and safety failures, and a good safety record.

Since Howe, there have been some well-publicised seven-figure fines, including:

- £1.5 million under s. 3(1) of the HSW Act against Great Western Trains for the Southall crash, plus £680,000 costs
- £10 million under s. 3(1) of the HSW Act against Balfour Beatty and £3 million under the same section for Network Rail for the Hatfield rail crash
- a total of £1.2 million against Balfour Beatty Civil Engineering and its tunnelling subcontractor, Geoconsult, following the Heathrow rail link tunnel collapse, plus £100,000 costs each, even though the incident was a 'near miss' and no one was injured
- a total of £1.7 million against Port Ramsgate, two construction and design firms and Lloyds Register of Shipping (which carried out inspection work), plus costs of nearly £250,000, following the collapse of a walkway which killed six people.

Facilities managers should note that even though Balfour Beatty and Port Ramsgate contracted out the work that led to the incidents, they were still held accountable under health and safety law, being fined £700,000 and £500,000 respectively. See elsewhere in this chapter, Managing contractors, for more on this issue.

The Crown Court can order imprisonment of up to two years for contravention of an improvement or a prohibition notice.

The Health and Safety (Offences) Act 2008[5] also makes particular offences which were only triable in the lower (Magistrates') courts, now triable by either the lower or higher courts. This gives Magistrates or prosecutors the opportunity to pass a case up to the higher court where fines could be much higher and the option for custodial sentences greater. Employees have duties to comply with health and safety requirements as imposed by Section 7 of the HSW Act 1974. Should anyone be prosecuted for such an offence, including facilities management personnel, they will be at risk of a £20, 000 fine and up to 12 months in prison if convicted in the lower courts, or an unlimited fine and up to two years if convicted in the Crown Courts.

CIVIL COMPENSATION

An employee, a contractor or a member of the public who has been injured or made ill as a result of a negligent employer's act or omission can bring a claim for compensation. This is based on the common law concept that every member of society owes a duty of care towards others. Employers may also be 'vicariously liable' for an accident caused by one of their employees during the course of their employment.

Civil actions must commence within three years of the claimant finding out about their injury or illness. Claimants can receive damages for a number of losses, including

---

[5] Health & Safety (Offences) Act 2008.

loss of earnings, pain and suffering, medical expenses and disfigurement. Employers, in turn, have a number of defences available. They can argue contributory negligence, which means the injured person contributed to the accident, for example, by ignoring safety rules, or they can argue that the injuries were not reasonably foreseeable.

The Civil Procedure Rules 1998 introduced measures to speed up claims, including a 'fast track' system for claims up to £15,000. In practice, most claims are settled out of court by the employer's insurers. Employers are required to have insurance against such claims under the Employers' Liability (Compulsory Insurance) Act 1969.

In 2000, Trades Union Congress (TUC) figures show that the unions secured £320 million in compensation for members injured or made ill at work, an average of £6000 for each of the 54,650 cases pursued.

PROMOTING OCCUPATIONAL HEALTH

Section 2 of the HSW Act provides that employers have a duty of care to ensure the health, safety and welfare at work of all employees. The following sections offer guidance to the facilities manager on how to comply with this duty.

Avoiding back pain

Back pain is the most common source of workplace ill-health. The causes include poorly designed workstations which encourage poor posture, and lifting and carrying of loads (manual handling).

More than one-third of serious workplace injuries reported each year are related to manual handling, and most of these are back injuries. Employers lose 10 million working days a year to musculoskeletal disorders, half of these to back pain. Many injuries are cumulative, in other words a number of apparently insignificant problems build up over a period of time until the effects become serious.

The best way to prevent injuries is to avoid the need for lifting and carrying in the first place by reorganising the task or automating the handling process. If mechanisation is chosen, facilities managers will need to ensure that new risks are not imported. For example, lifting equipment will need to be properly maintained and forklift trucks will need to be kept away from pedestrians.

Low-tech handling aids such as trolleys may provide a simple solution. But all aids should be readily accessible or they will not be used, and employees must receive training in their use.

Risk assessment

HSE's manual handling guidance advises employers to assess four problem areas when considering how to make lifting and carrying safer. These are:

- the way the task is done
- the working environment
- the load itself
- the people doing the lifting.

Facilities managers will play a key role in reducing risk in the first two of these areas (see below, Manual handling checklist). Changes to the task and the environment will often also increase efficiency and productivity.

The task

- Change the layout of the task, by storing loads at waist height, for example. Store only lighter loads, or those handled infrequently, higher or lower.
- Eliminate obstacles that the worker has to reach over or into, such as poorly placed pallets or excessively deep containers.
- Avoid employees having to lift loads from the floor while seated.
- Although a swivel-action seat will reduce the need to twist, bear in mind that a chair on castors may move accidentally.
- Ensure that the relative height of seats and work surfaces is well matched.
- Consider using two or more people to handle an awkward load.
- Maintain all handling equipment and PPE (e.g. gloves or safety footwear) properly.
- Improve the work routine, for example, through self-pacing or job rotation.

The environment

- Remove space constraints, for example, narrow gangways, doorways and working areas. Ensure that there is enough floor space and headroom, and think about the positioning of fixtures.
- Ensure that floors are flat, well maintained and properly drained. Clear spillages away promptly, and consider slip-resistant flooring.
- Avoid manual handling activities on more than one level. If this is unavoidable, ensure that there is a gentle slope or, failing that, well-positioned and properly maintained steps. Working surfaces should, where possible, be level with each other.
- Provide a comfortable environment, avoiding extremes of temperature, high humidity and poor ventilation.
- Provide sufficient well-directed light.

The load

- Consider making loads lighter, smaller, easier to grasp, more stable or less damaging to hold.

The individual

- Consider whether individuals are at particular risk of injury, because they are pregnant or have a history or back, knee or hip problems, for example.
- Provide information and training on manual handling injury risks and good handling techniques. But remember that training workers is not a substitute for well-designed systems and workplaces.

Preventing work-related upper limb disorders

Work-related upper limb disorders (WRULDs), sometimes known as repetitive strain injuries (RSIs) are injuries to muscles, tendons or nerves, especially in the hands, wrists, elbows or shoulders. Specific WRULDs include tenosynovitis, carpal tunnel syndrome and epicondylitis (tennis elbow). The courts have also awarded damages to employees suffering from diffuse RSI; that is, a collection of symptoms that do not constitute a medically recognised injury. Repetitive tasks such as keyboard work, assembly line work and packing are common causes.

As with avoiding manual handling injuries, facilities managers have a responsibility to improve equipment and the layout of work areas. But such measures will only be effective if carried out in conjunction with fundamental changes to the way in which work is organised, for example, by allowing workers to take a break from repetitive work.

MANUAL HANDLING CHECKLIST

1. The tasks
   Do they involve:

   - holding or manipulating loads at distance from trunk?
   - unsatisfactory bodily movement or posture, especially:
     - twisting the trunk?
     - stooping?
     - reaching upwards?
   - excessive movement of loads, especially:
     - excessive lifting or lowering distances?
     - excessive carrying distances?
     - excessive pushing or pulling of loads?
     - risk of sudden movement of loads?
     - frequent or prolonged physical effort?
     - insufficient rest or recovery periods?
     - a rate of work imposed by a process?

2. The loads
   Are they:

   - heavy?
   - bulky or unwieldy?
   - difficult to grasp?
   - unstable, or with contents likely to shift?
   - sharp, hot or otherwise potentially damaging?

3. The working environment
   Are there:

   - space constraints preventing good posture?
   - uneven, slippery or unstable floors?
   - variations in level of floors or work surfaces?
   - extremes of temperature or humidity?
   - poor lighting conditions?

4. Individual capability
   Does the job:

   - require unusual strength, height, etc.?
   - create a hazard to those who might reasonably be considered to be pregnant or to have a health problem?
   - require special information or training for its safe performance?

5. Other factors
   Is movement or posture hindered by personal protective equipment or by clothing?

Source: HSE. *Manual Handling Operations Regulations 1992 – Guidance on Regulations.*[6]

Risk assessment

Employees may be at risk of developing WRULDs if:

- the work involves:

  - awkward hand, arm, wrist or shoulder movements
  - rapid repetitive movements
  - prolonged physical pressure, such as gripping or squeezing
  - a prolonged uncomfortable position
  - few breaks
  - lack of variety
  - long hours
  - a fast work rate (e.g. to keep up with a conveyor)

- tools and equipment are:

  - too heavy
  - an uncomfortable shape
  - vibrating or noisy
  - designed for men but used by women

- workstations are:

  - the wrong height for the individuals using them
  - not adjustable
  - noisy, cold, or have other adverse conditions
  - badly lit.

Solutions

Ways to reduce the risk of WRULDs occurring include:

- redesigning workstations and equipment, by, for example:
  - reducing the reaching required
  - moving controls

---

[6] HSE. *Manual Handling Operations Regulations 1992 – Guidance on Regulations.* HSE Books.

- making conditions less cramped
- providing adjustable chairs, footrests, etc.
- improving lighting and cutting out glare, to prevent users sitting awkwardly.

■ redesigning tool handles, keeping tools sharp and lubricated, replacing hand tools with power versions, avoiding high-vibration tools
■ changing the method of work
■ allowing self-pacing
■ introducing rest breaks and variety of tasks
■ training workers, including new staff or those using new equipment or working methods, in warning symptoms of WRULDs, good posture and safe working practices.

Computer workstations

Computer workstations must meet a number of ergonomic standards laid out in the Schedule to the DSE Regulations to avoid user strain. In order to comply, HSE has developed a risk assessment checklist in its booklet *VDUs: an easy guide to the Regulations*.[7] This includes the following questions:

■ Is the display screen image clear?

- Are the characters readable?
- Is the image flicker free?
- Are the brightness and contrast adjustable?
- Does the screen swivel and tilt?
- Is the screen free from glare and reflections?

■ Is the keyboard comfortable?

- Is the keyboard tiltable?
- Is there enough space to rest hands in front of the keyboard?
- Is the keyboard glare free?
- Are the characters on the keys easily readable?

■ Does the furniture fit the work and the user?

- Is the work surface large enough?
- Is the work surface free of glare and reflections?
- Is the chair stable?
- Do the adjustment mechanisms work?
- Is the user comfortable?

■ Is the environment around the workstation risk free?

- Is there enough room to change position and vary movement?
- Are the levels of light, heat and noise comfortable?
- Does the air feel comfortable?
- Is the software user friendly?

---

[7] HSE. *VDUs: An easy guide to the Regulations*.

Seating

Adjustable chairs are essential to combat musculoskeletal disorders such as back pain and WRULDs. When choosing or assessing seating, it is important to consider the individual's needs, the type of work and the size of the workstation. HSE guidance, *Seating at work*,[8] suggests ensuring that:

- the chair is comfortable for the intended period of use
- the back is properly supported
- there is sufficient padding
- the seat and back height are adjustable and the backrest tilts
- armrests (if required) allow enough arm movement and allow the individual to get close enough to the desk
- footrests are supplied if necessary
- the chair meets special user requirements (pregnant women and disabled employees, for example, may have special needs)
- the chair meets special task requirements
- the adjustment mechanisms are well maintained.

Asbestos-related diseases

Asbestos kills over 3000 people in the UK every year, one-quarter of whom have been employed in the building industry. By 2020, 10,000 people a year are predicted to die from asbestos-related diseases, which would make asbestos the biggest single killer of men under 65 years old.

Although all three types of asbestos have now been banned, the substance remains in many buildings constructed between 1950 and 1980, especially those with steel frames or boilers with thermal insulation. Those most at risk of contracting mesothelioma, an incurable and fatal cancer, are those involved in demolition, maintenance and refurbishment work (including plumbers, electricians and carpenters), and other contractors working nearby when asbestos fibres are released into the air. Others who may disturb asbestos are contractors installing computers, fire alarms, window blinds and telecommunications systems. Employees present when the asbestos is disturbed could also be exposed, and they could carry fibres home on their clothes.

There is no safe level of exposure, and repeated low-level exposure can also cause asbestos-related diseases. It takes 15–60 years for the symptoms of mesothelioma, asbestosis of the lungs or asbestos-related lung cancer to appear.

The most common uses of asbestos, listed in order of how likely they are to release dust if disturbed, are:

- sprayed coatings and asbestos insulation used for fire protection, pipes and boilers, thermal and acoustic insulation
- insulating boards used for fire protection to doors, cladding on walls and ceilings, partitioning and ceiling tiles
- asbestos cement used for roofing, wall panels, pipes and water tanks.

---

[8] HSE. *Seating at work*.

Surveys and logs

The Control of Asbestos Regulations 2006 formalise existing good practice (see p. 16). In effect, facilities managers will need to ensure that their premises are surveyed for asbestos as a matter of course, rather than waiting until asbestos is accidentally exposed.

The duty to manage asbestos

The regulations include the 'duty to manage asbestos' and place responsibilities on the duty holder to identify, locate and manage asbestos-containing materials. HSE has issued several Approved Codes of Practice and Guidance documents.[9] The ACoP L127[10] covers the management of asbestos in non-domestic premises and sets out ways in which to comply with reg. 4. The duty under reg. 4 rests with the person in control of the maintenance activities and this is likely to be the facilities manager. The ACoP advises duty holders to:

- take reasonable steps to find asbestos material and check its location
- presume that materials contain asbestos unless there is strong evidence that they do not
- make a written record of the location and condition of asbestos and keep it up to date
- assess the risk of anyone being exposed to asbestos materials
- ensure that material is kept in good repair
- provide information to anyone at risk.

Guidance recommends consulting the leaseholder, architects or original building plans to find out if and where asbestos was used. It may be necessary to bring in accredited personnel to analyse samples. If asbestos is present, the risk of fibres being released must be assessed.

Leaving asbestos in place

If asbestos is in good condition and not likely to be either damaged or worked on, HSE recommends:

- leaving it in place and logging where it is (mark it on building plans and/or set up a register)
- warning contractors that some undiscovered asbestos could still be present elsewhere
- making a note of safe materials that could be mistaken for asbestos
- labelling asbestos materials with a warning sign
- ensuring that contractors know where asbestos is and its condition
- conducting regular reinspections.

---

[9] *Work with materials containing asbestos.* HSG L143.
[10] *Management of asbestos in non-domestic premises.* HSG L127.

Repair and removal

Depending on how poor its condition is, asbestos may need to be:

- repaired, sealed or enclosed, then logged following the above steps, or
- removed.

In either case, HSE-licensed contractors must carry out the work. They must double-bag the waste, clearly label it and dispose of it at a licensed site.

Duty to contractors and employees

HSE warns that during asbestos repair or removal work, employers should ensure that contractors

do:

- understand the health risks of asbestos exposure
- keep unnecessary personnel away
- take care not to create dust
- keep the material wet whenever possible
- wear a suitable respirator and protective clothing
- clean up with a vacuum cleaner complying with BS 5415 (type 'H').

do not:

- break up large pieces of asbestos materials
- use power tools: this creates dust
- expose workers who are not protected
- take protective clothing home to wash.

All persons likely to be exposed to asbestos fibres while at work must have mandatory asbestos awareness training. Training should cover, as a minimum, the following:

- the properties of asbestos, its health effects and the interaction of asbestos and smoking
- the type of materials likely to contain asbestos
- what work could cause asbestos exposure and the importance of preventing exposure
- how work can be done safely and what equipment is needed
- emergency procedures
- hygiene facilities and decontamination.

Training must be given at regular intervals and needs to be proportionate to the nature and degree of exposure and should contain the appropriate level of detail for the job or the role of the attendees.

Facilities managers should always check the level of asbestos awareness training which operatives from all contractors have unless they know that the buildings in which they operate are totally asbestos free.

Eight steps to asbestos risk management

HSE has provided eight key steps to consider before carrying out any repair, maintenance or refurbishment work to a building (Figure 1.1).[11]

1. Are you responsible for maintenance and repair activities for non-domestic buildings, either through a contract or tenancy agreement, or because you own the building? Yes: You are a duty holder, and have a responsibility to manage asbestos. GO TO STEP 2 | No: The person who is in charge should know where asbestos is. Ask them to show you a record.	**2.** Was the building built before 2000? Yes: Assume asbestos is present. GO TO STEP 3 | No: Asbestos unlikely to be present. NO ACTION	**3.** Do you have any information on asbestos in your building already? This may be previous asbestos surveys, building or insurance reports. Use this information as a starting point. GO TO STEP 4.
4. Walk round your building. Identify all materials that may contain asbestos, e.g. insulating board, ceiling tiles and insulation on pipework. You can presume that asbestos is present, but before any work starts samples should be checked for asbestos. If you do not check, full asbestos safety precautions will still have to be used to do the work. If the materials are showing signs of damage, GO TO STEP 5 If the materials are in good condition, GO TO STEP 6	**5.** Act on damage. Draw up a priority action list considering extent of damage and proximity of workers likely to disturb material. Repair or remove material with damage – a licensed contractor may be required – check the HSE website or call Infoline before starting. Record what you find and the action you take. GO TO STEP 6	**6.** Keep a written record. This needs to be easy to read and easily available. Record where the asbestos-containing material is and its condition. Also record roles and responsibilities with regard to managing asbestos. This record could be a plan or diagram of the building, a written list or a computer-based record. GO TO STEP 7
7. Tell people where the asbestos is. Consider who works on or near asbestos (maintenance workers/contractors) and tell them where the asbestos is before they start work. Anyone working on asbestos should be trained and use safe working methods. Remember, some work requires a licence. GO TO STEP 8	**8.** Keep your records up to date. Even after your action list is completed you need to continue to manage risk; this includes regularly checking materials to make sure they have not deteriorated. Action must be taken if deterioration has occurred. Walk around your building at least once a year and update your plan as needed.	

FIGURE 1.1

Eight steps to consider before repair, maintenance or refurbishment work.

[11] HSE. *Managing risk from asbestos: a basic guide to duty holders' legal responsibilities.*

Reducing noise

What do the Control of Noise Regulations 2005 require?

The Regulations place a duty on all employers whose employees are or may be exposed to noise while at work. The legislation sets out the action levels shown in Table 1.1 and states what action should be taken at each level by the employer to control noise levels. The action levels are the noise exposure levels at which employers are required to take certain steps to reduce the harmful effects of noise to their employees. The action levels shown in Table 1.1 are a daily or weekly average of noise exposure.

Two factors need to be taken into consideration when determining whether noise is a problem at work:

- how loud the noise is
- how long people are exposed to it.

As a simple guide you will need to do something about the noise if any of the following apply:

- The noise interferes with the day-to-day work activities for most of the day, for example, a busy street or a vacuum cleaner being used non-stop.
- Employees have to raise their voices to carry out a normal conversation.

Table 1.1 Control of Noise at Work Regulations 2005

Exposure Action Level	dB	Action Required if Level is Exceeded
Lower exposure action level	80 dB(A)	Carry out a risk assessment Make suitable ear protection available Implement a maintenance programme for the ear protection Must implement a training programme
Upper exposure action level	85 dB(A)	Reduce the noise at source Implement ear protection zones Ear protection must be provided and must be used by employees (the use of hearing protection is mandatory if the noise cannot be controlled by any other measure) Health surveillance provided for employees

- Noisy tools or machinery are used for more than half an hour throughout the working day.
- Employees work in a noisy industry, such as construction, road repair, engineering, canning, production, manufacture, foundry, paper or board making.
- There is noise in the workplace due to machinery impacts, such as hammering, pressing, forging, pneumatic equipment or explosive sources.

If any of the above apply then a risk assessment will need to be carried out to decide whether further action is needed. The aim of the risk assessment is to provide you with the information so that a decision can be made on what needs to be done to ensure the health and safety of employees who are exposed to noise. In some cases measurements of noise may not be necessary, but it is about collecting as much information as possible.

The noise risk assessment should contain the following information:

- Who is at risk from noise.
- Who may be affected.
- An estimate of the employee's exposure to noise compared with the lower and upper exposure action levels.
- What needs to be done to comply with the law: these are often called noise control measures and may include the provision of hearing protection, such as ear defenders or ear plugs. If noise control measures are required then the type should be included in the risk assessment.
- Details of any employees who need to be provided with health surveillance and whether any particular employees are at risk, because of the nature of their work.

There is no right or wrong way to complete a risk assessment. The law requires that it is 'suitable and sufficient'.

A risk assessment must contain suitable information to be useful to an employee to understand what hazards they may be exposed to when carrying out the task.

If noise cannot be controlled by other methods, such as new machinery, acoustic screening, provision of antivibration mounts to machinery or change in working patterns, then extra protection for employees will be needed. In addition, it may be needed as a short-term measure while other methods of control are being implemented.

Hearing protection should not be used as an alternative to controlling the noise by technical or organisational methods. It should only be used where there is no alternative way of protecting employees from noise exposure.

Employers may also need to provide health surveillance to employees exposed to continuous noise above the legal action levels. If you believe that you have serious noise issues in the workplace, it would be best to seek advice from an experienced and competent person or your solicitors.

- If employees ask for hearing protection, it should be provided for them.
- It must be made available for employees to use when the lower action level of 80 dB(A) is exceeded.
- It must be used by employees when the upper action level of 85 dB(A) is exceeded.

- Employers must provide training and information in the correct use of the hearing protection.
- Employers must ensure that any hearing protection that is provided is properly used and maintained.

Control measures

HSE recommends controlling noise exposure by:

- designing workplaces for low noise emission, for example, using absorption materials to limit reflected sound, segregating noisy equipment from occupied areas
- engineering controls, for example, cushioning impacts in a yard where unloading takes place by fitting a rubber surface, using damping methods, mounting machines to prevent vibration, fitting silencers, using 'active noise control' to cancel out one sound with another
- modifying noise transmission paths, for example, enclosing machines, providing screens, barriers, walls and noise refuges, increasing the distance between employees and the noise source
- substituting noisy equipment or processes with quieter alternatives, for example, replacing compressed air tools with hydraulic equipment
- maintaining equipment and replacing worn parts
- changing working methods, for example, reducing exposure times through job rotation.

For best practice advice on how to design noise transmission out of the workplace, see Workplace facilities: minimising noise through design.

Legionellosis

Legionellosis is a group of diseases caused by *Legionella* bacteria found in water. The bacteria occur in natural water supplies, but hot and cold water systems, air conditioning systems and cooling towers offer better breeding grounds. The most serious of the diseases is legionnaire's disease, a type of pneumonia. HSE figures reveal 200–250 cases of legionnaire's disease a year, but the HSE believes that there is underreporting. About half of these cases are linked to foreign travel.

Legionnaire's disease

Legionnaire's disease:

- kills around 12 per cent of sufferers
- mainly affects people aged between 40 and 70
- affects three times more men than women
- is most likely to affect smokers, alcoholics, people with diabetes, cancer, chronic respiratory or kidney disease, or those on renal dialysis or immunosuppressant drugs
- is caused by susceptible individuals inhaling water droplets contaminated with *Legionella* bacteria

- is not believed to be transmitted by drinking water containing *Legionella* bacteria, or by person-to-person transmission
- produces symptoms two to 10 days after infection, including chills, high fever, headache or muscle pain followed by a dry cough, and in extreme cases, pneumonia
- has broken out in or near large building complexes such as hotels, hospitals, offices and factories.

Legal duties

Legislation relevant to controlling legionellosis includes COSHH and the Notification of Cooling Towers and Evaporative Condensers Regulations 1992. The latter require those in control of work premises to notify the local authority or HSE in writing of any cooling towers and evaporative condensers, including any being decommissioned or dismantled. Other places that can harbour bacteria include showers, taps, water softeners, humidifiers, calorifiers and pipework.

HSE's Legionella ACoP[12] states that the duty holder 'is required to have access to competent help' to assess whether there is a risk of the disease occurring. The accompanying guidance adds that whether it uses its own personnel, or a consultancy or water treatment company, the duty holder should ensure that they are trained and have the necessary equipment. The duty holder should also appoint a manager or director with day-to-day responsibility for controlling the risk of legionellosis; in practice, this may be the facilities manager.

Designing out legionellosis

Ideally, health and safety in any environment is designed in at the start. Where facilities managers have input in the design of a new building or a refurbishment, they should ensure that the following design criteria are applied to hot and cold water systems:

- Ensure that storage cisterns and calorifiers are the correct size for their intended use.
- Do not site cold water tanks in a warm part of the building.
- Ensure that pipework is as short and direct as possible.
- Ensure adequate insulation of pipes and tanks.
- Use materials that do not encourage the growth of *Legionella*.
- Protect against contamination by, for instance, fitting storage tanks with lids.

Risk assessment and control

To minimise the risk of contamination in existing installations, HSE guidance recommends carrying out a site survey of complex premises, including an asset register of plant, pumps, strainers and so on, and an up-to-date diagram of the system. Some factors to consider include:

---

[12] *Control of Legionella bacteria in water systems.* ACoP L8 (2000).

- the water source (is it mains water?)
- possible sources of contamination before the water reaches the cooling tower or cold water tank
- the system's normal characteristics
- reasonably foreseeable but abnormal operating conditions (e.g. breakdowns).

Measures to control the risk of disease include:

- controlling the release of water spray (e.g. the aerosol created by a cooling tower or shower)
- avoiding water temperatures of 20–45 °C, which allow bacteria to proliferate
- avoiding water stagnation, which encourages biofilm to grow
- avoiding materials in the system such as sludge, scale, rust, algae and organic matter which harbour or provide nutrients for bacteria
- keeping the system clean to avoid sediments
- using water treatment programmes
- ensuring that the system operates correctly and is well maintained.

Routine monitoring of water systems is necessary, and the guidance suggests that this should take place at least weekly. Testing of water quality and monitoring numbers of general bacteria are also important. But HSE suggests that testing only for *Legionella* bacteria is technically difficult, and the results are difficult to interpret. The guidance states: 'A negative result is no guarantee that *Legionella* bacteria are not present. Conversely, a positive result may not indicate a failure of controls as *Legionella* are present in almost all natural water sources.'

If it is suspected that a user of the building has contracted legionellosis, HSE must be consulted immediately. Facilities managers should then involve senior management and take steps to identify the source of the infection and deal with any contamination as soon as possible.

Legionnaire's disease outbreak: Barrow in Furness

An outbreak of legionnaire's disease killed seven people and affected 180 others in the town of Barrow in Furness in 2002. The cause of the outbreak was traced to a council-owned sports centre and generally, through management failures, systems and procedures were not in place to manage the risk posed by *Legionella* bacteria.

The report of the public meetings[13] into the outbreak identified the key management failings as follows:

- poor lines of communication and unclear lines of responsibility with the council
- failure to act on advice and concerns raised
- failure to carry out risk assessments
- poor management of contractors and contract documentation
- inadequate training and resources
- individual failings.

---

[13] HSE. *Report of the Public Meetings into the Legionella Outbreak at Barrow in Furness 2002.*

The local authority and the individual senior manager responsible for managing the Council's leisure facilities were prosecuted. The Council was fined £125,000 and the senior manager £15,000. Both parties were cleared of manslaughter.

Accidents and incidents

There are several things you need to do if an accident happens, namely:

- Call an ambulance if necessary.
- Investigate quickly to find out what happened.
- Take steps to prevent a similar accident happening, that is, make repairs, close off the area, disconnect the equipment, etc.
- Enter the information into the accident book or onto the accident form.
- Telephone the enforcing authority and police if there has been a fatality and the enforcing authority if someone has had a major injury.
- Complete Form F2508 and send it to the incident centre or local authority.
- Notify your insurance company or brokers.

Remember, accidents that happen to customers or members of the public must be treated the same as accidents to employees. It is an offence not to report certain accidents and fines for non-compliance can be up to £5000 per offence.

Certain accidents must legally be reported to the local authority environmental health department under RIDDOR. Some may need to be reported to the HSE. These are accidents that cause:

- broken bones/fractures
- dislocation of limbs
- unconsciousness
- electrocution
- chemical burn or puncture to the eye
- admission to hospital for 24 hours or more
- death of an individual
- a member of the public being taken to hospital because of the premises or its staff
- injuries that cause employees to be off work for more than three days
- injuries caused by any violence from customers, for example, assault or robbery.

Accidents must be reported as soon as possible after they occur, with the exception of 'over three day injuries', which can be reported up to 10 days after they have happened.

Reporting of accidents is now quite straightforward and can be done in any of the following ways:

- directly to the local authority in whose area your premises are located (obtain this from the telephone directory) or directly to the local HSE office
- directly to the Incident Control Centre (a national call centre operated on behalf of the HSE) (Tel.: 0845 300 9923; Fax: 0845 300 9924; Post: Incident Contact Centre, Caerphilly Business Park, Caerphilly CF83 3GG)

- online by logging on to: www.riddor.gov.uk (use Form F2508; forms can also be filled in online at www.riddor.gov.uk).

You must keep adequate details of all accidents to employees and others. An accident book as produced by The Stationery Office and available at most good bookstores can be used or you can design your own forms. You must be mindful of the Data Protection Act.

Often, your accident records and what action you took at the time will be requested by the EHO if they investigate the accident. Once an accident has been reported to the enforcing authority, the enforcement officer may decide to investigate the accident to establish whether, as an employer, you have been taking your responsibilities for health and safety seriously.

If you have reported an accident there is a high probability you will have a visit from the council or HSE. If you have investigated the accident and taken remedial action you should record this as this may be valuable information for your defence if you are prosecuted. If you have amended your risk assessment because the accident showed you that your control measures were inadequate, then record this as well.

You may find that some customers will want to sue you for any injuries they have suffered. To launch a civil action case for compensation for injuries, shock, stress, etc., the person usually has to appoint a solicitor. If you receive a solicitor's letter it will probably be best to seek advice from your own solicitors.

Often you will be asked to release information such as:

- risk assessments
- training records
- maintenance records
- accident reports or records
- history of accidents
- copies of internal safety checks.

If you have been negligent and failed to address a hazard you will probably find that the court will find against you and you may have to pay compensation. (You could also be prosecuted or fined. In many incidents of civil claims, your insurance broker or company will be involved in the case, so report the matter to them as soon as possible.

It is extremely good practice to review all accidents, no matter how trivial. An accident usually occurs because a safety control has gone wrong, so your investigation allows you to see what has not been working. Perhaps maintenance repairs were not completed quickly enough or the equipment was faulty. Recording what you found and what you did to put it right will be invaluable information for any formal investigations.

Improving well-being

Sceptics may still doubt that 'sick' buildings really make people ill. But at the very least, problems such as poor air quality, passive smoking, flickering lights and fluctuating temperatures undoubtedly reduce employees' well-being and productivity. Moreover, the Workplace Regulations impose a legal duty to look after employees' welfare.

Sick building syndrome

The WHO has recognised sick building syndrome (SBS) since 1982. HSE estimates that 30–50 per cent of new or recently refurbished buildings cause some form of SBS. In the worst cases, 85 per cent of occupants may suffer symptoms.

Symptoms

Symptoms are flu-like and include skin problems, breathing problems (sore throats, coughs, blocked noses, sinusitis), muscle and joint aches (stiff shoulders, backache) and neurological disorders (tiredness, headaches, digestion problems). The symptoms will be much more common than average among workers in a particular building, and will reduce or disappear during weekends and holidays.

At-risk buildings

According to WHO, buildings most at risk of SBS are likely to date from the 1960s or later and have:

- open-plan offices
- a low level of user control over ventilation, heating and lighting
- lighting with high glare or flicker
- air conditioning with cooling capacity
- lots of soft furnishings
- lots of open storage
- synthetic furniture, carpets and paint
- poor maintenance and repair
- insufficient cleaning
- high temperatures or large temperature variations
- very low or high humidity
- indoor air pollutants (ozone from photocopiers and laser printers, chemicals released from carpet adhesive)
- airborne dust or fibres
- computers.

Solutions

Bucking the trend for open-plan offices, WHO recommends no more than 10 workstations in any room. Other problems may be resolved through maintaining air conditioning systems, replacing old photocopiers and laser printers, replacing fluorescent lights, and better cleaning. Since cleaning chemicals may themselves cause problems, however, the best solution is redesigning workplaces to be low maintenance.

Photocopiers

Well-maintained, modern photocopiers are rarely a health hazard. But if they are used frequently or there are a number of machines, there could be problems with noise, dry heat, and the release of ozone. The best solution is to place equipment in a separate room with separate ventilation. It is important to follow the manufacturer's recommendations for siting, cleaning and maintaining the machines, as well as following the precautions in the material safety data sheets for toner and other chemicals.

There should also be clear guidance on when it is safe for employees, with the appropriate training and information, to carry out repairs (e.g. clearing simple paper jams), and when a technician should be called. Modern equipment should turn off automatically when opened, but turning it off at the wall first is a simple safety precaution.

Lighting

Poor lighting can cause eyestrain, headaches, SBS and fatigue, as well as accidents (see elsewhere in this chapter, Towards a safe workplace: Preventing slips and trips).

LIGHTING RISK ASSESSMENT

- Is emergency lighting adequate?
- Are desk lamps provided?
- Are light fittings regularly cleaned?
- Are lamps replaced regularly (after about 7000 hours' use)?
- Are desks placed at right angles to windows to prevent glare?
- Are fluorescent light strips covered with a diffuser?
- Are windows fitted with blinds or curtains?
- Are walls, ceilings and furniture decorated in light, matt colours (to increase the level of light and reduce glare)?
- Is there a source of natural light?
- Are there any dark, unlit areas?

Source: UNISON. *Office Health and Safety – A Guide to Risk Prevention.*[14]

The ACoP to the Workplace Regulations states that:

- lights should be of a type and so positioned as to avoid glare
- lights should be repaired and cleaned to ensure effectiveness
- windows should be cleaned regularly
- local lighting should be provided as necessary.

The legislation also requires:

- an average illuminance for a whole work area of 200 lx with a minimum for any individual work position of 100 lx
- the location of lights and associated fittings should not cause a hazard, with employees able to find light switches easily.

In practice, generally accepted levels for offices are 400–500 lx at desk level, rising to 700 lx or more in specialist areas, such as architects' offices, where sharp focus is

---

[14] UNISON. *Office Health and Safety – A Guide to Risk Prevention.*

needed. Where each work task is individually lit and the area around the task is lit to a lower illuminance, as in local lighting in an office, the maximum ratio of illuminances of working area to adjacent area should be 5:1. Ambient lighting in common areas such as stairwells and lobbies may be considerably less powerful than for offices, down to 20 lx or less.

Natural light

Unlike some European countries, the UK has no legal requirement that workers should have access to natural light. But the tradition of placing offices for senior staff round the edges of a floor, making daylight access a privilege of managers and leaving those in open-plan cut off from natural light, is now widely challenged, and the recommended minimum ratio of artificial light to natural light at any long-term work setting is 1:5.

For best practice advice on workplace lighting, see Lighting, in Chapter 10.

Air quality

Temperature

The ACoP to the Workplace Regulations[15] says that work rooms should normally be at least 16 °C, or 13 °C for work involving 'severe physical efforts'. There is no legally enforceable maximum temperature, but the ACoP says 'all reasonable steps should be taken to achieve a comfortable temperature'. It is rare for temperatures to be so extreme that they cause serious illness (e.g. heat stroke or hypothermia), but less severe temperature problems can cause discomfort, loss of concentration, tiredness, irritability and SBS.

The Chartered Institution of Building Services Engineers (CIBSE) recommends the following temperatures:

- heavy work in factories: 13 °C
- light work in factories: 16 °C
- hospital wards and shops: 18 °C
- offices and dining rooms: 20 °C.

The international standard for office environments ISO 7730-1984 recommends an operative temperature range of 20–24 °C with a variation of less than 3 °C between head and ankle height. Radiators fitted with thermostats produce least complaint from office occupants of any heating system (probably because it allows for a degree of user control), but many larger office blocks (especially those built in the years following the energy crisis of the 1970s) have sealed windows and centrally controlled systems.

Uncomfortably hot workplaces can be tackled by:

- providing air conditioning or fans
- ensuring that windows can be opened
- shading windows with blinds
- siting workstations away from direct sunlight or hot areas

---

[15] ACoP: L24: Workplace (Health, Safety and Welfare).

- insulating hot pipes
- providing a free supply of drinking water
- giving employees breaks to cool down
- introducing more flexible hours to avoid the worst effects of working in exceptionally high temperatures.

Thermometers should also be provided for monitoring temperatures.

In intentionally cold workplaces, such as those preparing food, it will be necessary to provide a warm working station using localised heating, draught exclusion and so on. Catering areas could be particularly hot and humid, and fume extraction and extractor or circulation fans will be necessary. Air inlets must be carefully sited to allow air movement in all parts of the kitchen.

Ventilation

The Workplace Regulations require enclosed workplaces to be ventilated by 'a sufficient quantity of fresh or purified air'. The recommended minimum airflow is 5 litres per second, though 8 litres per second is a safer minimum to prevent complaints of airlessness. Care needs to be taken to ensure that ventilation does not cause draughts.

Humidity

The DSE Regulations, in s. 3(g) of the schedule, state that 'an adequate level of humidity shall be established and maintained' in offices where there are visual display unit (VDU) users. To comply with the legislation, employers need to monitor humidity constantly, to ascertain whether adequate levels are being provided. It is also important for employers to recognise any dry heat symptoms being experienced by staff, in order to identify whether any problems need to be addressed to ensure staff health and comfort. For advice on humidity levels and monitoring, see Maintaining adequate humidity, in Chapter 10.

Smoke-free premises

The Health Act 2006 and other Regulations[16] contain the legal requirement for all indoor premises and those that might be classed as substantially enclosed to be smoke free. Broadly, from July 2007 all premises that are enclosed and operated as business premises, public places or workplaces across the UK need to have introduced a smoking ban in all areas except for those that are in the open air. There are a few exceptions to the legal requirement, for example, care homes and hotel bedrooms. The new smoke-free requirements also apply to vehicles used by employees as a place of work.

Contravening the smoke-free legislation is a criminal offence and fines could be:

- Individuals prosecuted for smoking in an enclosed place could be fined up to £200 or, if a fixed penalty notice is served, £50.

---

[16] The Health Act 2006; The Smoke-Free (Premises and Enforcement) Regulations 2006; The Smoke-Free (Signs) Regulations 2007; The Smoke-Free (Penalties and Discounted Amounts) Regulations 2007; The Smoke-Free (Exemptions and Vehicles) Regulations 2007; The Smoke-Free (Exemptions and Vehicles) Regulations 2007.

- Any manager or person in control of premises who is prosecuted for allowing a person to smoke in enclosed premises could be fined up to £2500 per offence.
- Failing to display the appropriate notice on the premises could result in a fine of up to £1000 or a fixed penalty notice of £200.

The law is enforced by authorised officers of the local authority, not just EHOs and trading standards officers but also licensing officers, specially designated smoking compliance officers. Some authorities may authorise traffic wardens so that they can catch motorists who smoke in work vehicles.

The law applies 24 hours a day, seven days a week, and unless the premises are your home, you will be covered. Even your home, that is, your staff accommodation, could be covered by the legislation if it is used by fellow workers for rest breaks.

Anyone can call a hotline telephone number and report businesses or people flouting the law and these referrals will invariably lead to an inspection of the premises.

Smoking is banned in enclosed or substantially enclosed premises. To be enclosed, premises or buildings must have a ceiling or roof (either a fixed or movable device or structure) and, except for doors, windows and passageways, must be wholly enclosed on a permanent or temporary basis; substantially enclosed means having a ceiling or roof and an opening in the walls which is *less than half* the total area of the walls. Opening areas cannot include doors and windows.

In general, anything with a roof or ceiling and four walls is 'enclosed'. Any structure with at least two completely open sides and a roof is not enclosed. Any building or structure without a roof or ceiling is not enclosed. All buildings with a configuration different from the foregoing need to be individually assessed to determine whether they are enclosed, substantially enclosed or open.

The legislation was intended to be straightforward but the definition of enclosed places is complex and before deciding on new smoking shelters, for instance, you should seek legal advice as you could be making a costly mistake.

All premises must have suitable no smoking signs displayed.

Similar titled Regulations are enacted in Wales and Northern Ireland. Scotland has its own legislation which came into force in 2006. The provisions are similar to those in England.

Stress

What is stress?

In Japan there is a cause of death called karoshi. It claims 10,000 lives a year and is regarded as 'sudden death from overwork'. Stress-related absence is said to have increased by 500 per cent since the mid-1950s in the UK alone. It is the second biggest cause of work-related illness in Britain, and costs employers about £400 million per year. The wider cost to society has been calculated at around £4 billion a year.

Stress can mean different things to different people and is often a generic term for a set of symptoms that can be related to life, work and environmental situations. Much of the research carried out in relation to stress has shown that there is a link between poor work organisation and subsequent ill-health of employees.

The HSE defines stress as[17]: 'the adverse reaction people have to excessive pressure or other types of demand placed on them'. This definition makes an important distinction between pressure, which can be a positive state and improve performance if managed correctly, and stress, which can be detrimental to health. It is important to recognise that stress is not necessarily a bad thing. Within certain limits an individual's performance improves with increased levels of stress. When stress becomes excessive performance drops off.

Research carried out by the HSE has shown that up to five million people in the UK feel very or extremely stressed by their work and approximately half a million people believe they are experiencing stress to such an extent that they are being made ill by it. It should also be remembered that in some cases stress can be a sign of medical problems and employees should always be advised to seek medical advice.

There are various effects associated with stress. Each employee is individual and the reactions are likely to be different, both physically and mentally. Some of the common symptoms, which are often indicators that a person is suffering from stress, are as follows:

- lack of concentration
- inability to relax
- inability to think clearly
- distress and irritability
- lack of self-esteem
- not enjoying work
- being depressed and negative about everything
- tiredness and sleepiness
- reduction in effectiveness of immune system
- heart disease
- psychiatric illness
- anxiety
- headaches.

These are just some of the symptoms that may occur. Each individual is different and the type, severity and number of symptoms will vary accordingly.

Stressed individuals are no longer able to keep going with their job. Stress tends to occur among highly motivated individuals. It is seen to be linked with trying to achieve individual and organisational objectives and becoming frustrated in not being able to achieve the desired outcomes. This frustration causes a spiral of increasing stress that leaves the individual exhausted. Burnout often manifests itself as poor performance, poor decision making, negative attitudes and exhaustion.

---

[17] HSE. *Working together to reduce stress at work*. HSE Misc. 686.

Stress may also result from home pressures, such as money and family. Managers must deal with these situations sympathetically. If employees are suffering from stress while they are at work, the costs to the business and the performance will be affected. Work stress can result in:

- an increase in time off work due to ill-health
- an increase in accidents at work, due to tiredness, which in turn leads to lost time, stoppages and near misses; these may lead to an increase in civil claims, and increased compensation and insurance premiums
- an increase in staff turnover, which in turn adds to costs of recruitment and training
- an increase in customer complaints, if staff are customer facing and are tired and irritable
- a decrease in overall performance and productivity.

It is estimated that 30–40 per cent of all sickness absence from work is attributed to some form of mental or emotional disturbance (whether work related or not). People do not work effectively under stress and may induce stress in colleagues. Reducing stress is cost-effective and will lead to lower sickness rates, improved performance and less staff turnover.

The legal position

There is no specific law relating to stress at work, and it not mentioned specifically in any health and safety legislation. It is recognised by the HSE as a serious work hazard, and extensive guidance and standards have been issued by the authority, advising on how to deal with it. The HSE has set up a specific web site specifically for stress-related advice and information (www.hse.gov.uk/stress) and has introduced the Stress Management Standards.

The HSE argues that 'having a positive, satisfied and psychologically healthy workforce will produce economic benefits, through improved attendance, motivation and commitment'. The law requires employers to tackle work-related stress. Under the HSW Act employers have a general duty to ensure, so far as reasonably practicable, the health of their employees at work. This includes taking steps to make sure that they do not suffer stress-related illness as a result of their work.

Employers must take account of the risk of stress-related ill-health when meeting their obligation under the Management of Health and Safety at Work Regulations 1999. The main provisions of these Regulations, as far as stress is concerned, are a duty to assess, a duty to apply the principle of prevention, a duty to ensure employees' capability to provide training, and duties towards young people.

New landmark rulings were made by the Court of Appeal in the case of Sutherland v Hatton and others (2002). The court set out a number of practical propositions for the future claims concerning workplace stress:

- Employers are entitled to take what they are told by employees at face value unless they have good reason to think otherwise. They do not have a duty to make searching enquiries about employees' mental health.

- An employer will not be in breach of duty in allowing a willing employee to continue in a stressful job if the only alternative is to dismiss or demote them. The employee must decide whether to risk a breakdown in their health by staying in the job.
- Indications of impending harm to health at work must be clear enough to show an employer that action should be taken, in order for a duty on an employer to take action to arise.
- The employer is in breach of duty only if he fails to take steps which are reasonable, bearing in mind the size of the risk, gravity of harm, the cost of preventing it and any justification for taking the risk.
- No type of work may be regarded as intrinsically dangerous to mental health.
- Employers who offer a confidential counselling advice service, with access to treatment, are unlikely to be found in breach of duty.
- Employees must show that illness has been caused by a breach of duty, not merely occupational stress.
- Compensation will be reduced to take account of pre-existing conditions or the chance that the claimant would have fallen ill in any event.

Employers have a duty to ensure the health, safety and welfare of their employees. Every employer should be concerned about stress not only because of the legal responsibilities, but also because of the moral and common-law responsibilities on employers to have a duty of care to their employees.

All employers are increasingly at risk from being sued by their employees for stress and many cases have succeeded in the civil courts. It is the responsibility of the employer to show that he has discharged his duty of care to his employee. Civil claims for stress are increasingly successful and stress management is becoming a prerequisite for insurance cover under employers' liability insurance.

How to determine whether there is a problem with stress in the workplace

There are five major categories of work-related stress:

- factors intrinsic to the job
- role in the organisation
- relationships at work
- career development
- organisational structure and climate.

Factors intrinsic to the job

- Working conditions, for example, changes in noise, lighting, speed of work, physical exertion and repetition.
- Shift work: impacts on blood temperature, metabolic rate, blood sugar levels, mental efficiency, motivation, sleep patterns and social life.
- Long hours: it has been suggested that beyond 40 hours a week, time spent working is increasingly unproductive and can create ill-health.
- Risk and danger: the individual may be in a constant state of arousal, and the associated adrenaline rush, respiration changes and muscle tension may all be

threatening to long-term health. Note that appropriate training and equipment may support the individual to cope with the situation.

- New technology: the need to adapt to new ways of working may exert pressure.
- Work overload: quantitative overload refers simply to having too much to do. Qualitative workload refers to work that is too difficult for the individual.
- Work underload: not being sufficiently challenged by work, for example, repetitive, boring and unstimulating work.

Role in the organisation

- Role ambiguity: not having a clear idea about one's work objectives, responsibilities or others.
- Role conflict: job demands may be conflicting, the individual may have to perform things they do not want to do or things they do not believe are part of the job. The individuals may find themselves torn between different groups within the organisation. The ability to cope may vary with personality.
- Role incompatibility: the person and the job may not fit.
- Responsibility: responsibility in organisations usually relates to people and things, and this in itself can cause stress.

Relationships at work

- Relationships with superiors
- relationships with subordinates
- relationships with colleagues.

Career development

- Opportunities for promotion and reward and perceived inequality
- job security
- retirement
- performance review.

Organisational structure and climate

- Lack of participation
- not identifying with the organisation
- organisational structure of change/review.

Unemployment

- loss of identity
- loss of social contact
- loss of security
- loss of income
- loss of time structure.

In order to establish whether there is a problem of stress in the workplace employers need to consult with employees. This can be done via trade union representatives or employee representatives or at staff meetings. Remember to include everyone; it is not only management who can suffer from stress.

Questionnaires can be a good starting point, and should be anonymous so as not to intimidate people. Check records for levels of absenteeism: do certain departments or teams have a higher level of absenteeism than others?

Employers should involve employees at every stage of the fact finding; they are the ones who know what is going on and can provide a true picture of the situation within the workplace. In effect, this fact finding is like carrying out a risk assessment. Regulation 3 of the Management of Health and Safety Regulations 1999 requires employers to assess risks to health and safety from hazards at work. The five steps to risk assessment are as follows:

1. Identify the hazards.
2. Decide who might be harmed and how.
3. Evaluate the risk by: identifying what is being done; deciding whether it is enough; deciding what more needs to be done.
4. Record the significant findings of the assessment.
5. Review the assessment at appropriate intervals.

Step 1: Identify the hazards – determining whether there is a problem

- informal talks with staff
- performance appraisal
- focus groups
- managing and monitoring attendance, including return to work interviews
- sickness/absence data
- productivity data
- turnover
- questionnaires.

The HSE draws attention to seven broad categories of risk factors for work-related stress:

- **culture**: of the organisation and how it approaches work-related stress
- **demands**: such as workload and exposure to physical hazards
- **control**: how much say a person has in the way they do their work
- **relationships**: covering issues such as bullying and harassment
- **change**: how organisational change is managed and communicated in the organisation
- **role**: whether the person understands their role in the organisation and whether their organisation ensures that the person does not have conflicting roles
- **support and training**: factors unique to the individual.

Step 2: Decide who might be harmed

- Work-related stress can affect any member of your team, and those exposed to the risk factors identified by the HSE may be affected.
- At particular times in their lives people may be more vulnerable to work-related stress, for example, major life events.
- The ability to cope with stress varies from individual to individual.

Step 3: Evaluate the risk

For each of the hazards identified in step 1, three questions need to be asked:

1. What actions are you already taking?
2. Are these enough?
3. What more do you need to do?

Regulation 4 of the Management of Health and Safety at Work Regulations 1999 states in controlling risks, the principles below must be applied in the following order:

1. Avoid risks (e.g. make the work environment safer so staff are not anxious about the risk of violence).
2. Evaluate risks which cannot be avoided.
3. Combat risks at source (e.g. organise work appropriately and clarify roles).
4. Adapt the work to the individual.
5. Develop a coherent protection policy covering issues such as organisation of work, working conditions and relationships.
6. Give collective protective measures priority over individual protective measures (e.g. by tackling risk at source rather than just providing information and training to individuals, or access to employee assistance programmes).
7. Give appropriate instruction to employees.

Step 4: Record the significant findings of the assessment

The Management of Health and Safety at Work Regulations require employers to undertake a 'suitable and sufficient' risk assessment.

Regulation 3(6) requires that if an employer employs five or more employees they must record the significant findings of the risk assessment and any group of employees identified as being especially at risk.

Step 5: Review the assessment at regular intervals

Regulation 3(3) of the Management of Health and Safety at Work Regulations states that the risk assessment must be reviewed whenever there is reason to believe that it is no longer valid. The HSE suggests initially there should be a review of the assessment every six months. If after a year this period is seen to be too frequent (i.e. no significant changes) move to an annual review period.

Assessments should be revised if there are forthcoming changes that could affect employees. Changes in staff or individual employees' circumstances may affect the risk assessment.

Table 1.2 shows the risk factors and possible solutions associated with each risk. It should be used as a guide to the possible solutions that could be implemented in the workplace, but remember that each situation is different and a suitable solution should be agreed by all concerned.

All of the following set out to improve the overall situation in the workplace by implementing the actions revealed as necessary by the risk assessment process. This may include:

Table 1.2 Risk Factors and Possible Solutions

Risk Factor	Areas Included	Possible Solutions
Demands	Workload Working patterns	Regular meetings Discuss anticipated workload and concerns
	Physical environment	Develop personal work plans Adjust working patterns Ensure people are appropriately trained for the task
Control	Input by employees on the way they work	Implement systems where employees have a say Discussion forums Discuss employees' skills and the way they are used
Support	Level of support provided by managers and colleagues	Regular one-to-one meetings Team meetings to discuss pressures Ask how employees would like to have access to managerial support Develop training, including refresher training
Relationships	Promoting positive working to avoid conflict	Provide a written policy for dealing with unacceptable behaviour and communicate it
	Addressing unacceptable behaviour	Agree and implement procedures to prevent or quickly deal with conflict Encourage good communication
Role	Understanding of individual roles within the organisation	Hold team meetings so that employees can clarify their roles
	Conflicting roles within the organisation	Display targets and objectives Agree specific performance standards and review periodically

TABLE 1.2 *(Continued)*

Risk Factor	Areas Included	Possible Solutions
		Introduce personal work plans Develop suitable inductions
Change	Managing and communicating organisational change	Ensure staff are aware of why change is necessary Define and explain key steps of change Ensure staff are aware of the impact of the change on their jobs

- stress management training
- stress awareness training
- complementary training
- self-help staff booklets
- a stress management policy,

whereas the following deal with the treatment and rehabilitation of individuals who have already suffered ill-health as a result of stress:

- support schemes for employees who want to discuss their problems confidentially with someone outside the organisation
- personal counselling for staff either using an in-house team or by referral to external providers.

Other strategies that could be implemented include the following.

Positive organisational responses to stress

- Job design, for example, make the job interesting, ensure that it does not make unreasonable demands: are long hours and taking work home considered the norm?
- Involvement and communication: people may gain a greater understanding of the organisation and the rationale for actions, ambiguity may be reduced and sources of stress identified and acceptable strategies developed. This can be used as part of a strategy to cope with change.
- Awareness programs: provide people with information on the existence, manifestation and management of stress.
- Health programmes: boost the individual's capacity to cope with stress.
- Organisational design: the physical structuring of the organisation.
- Personal development: boost the individual's feelings of competence but also as a means of relaxation.
- Personnel policies: flexible working, crèche facilities.

- Procedural frameworks: provide guidance to staff on how to deal with situations and so reduce stress related to uncertainty.
- Improving the physical environment: ensure that factors such as heat, light and ventilation are appropriate.
- Improving the psychosocial environment, for example, bullying and harassment.

Conflict management

- Planning: plan workloads; provide clarity in relation to roles and responsibilities. Ensure that the workload is appropriate and that employees have the necessary capability and capacity. Remember that underload as well as overload can cause stress.
- Culture design: create a supportive culture, encourage participation, communication, support and mutual respect, recognise and deal with issues.
- Training and recruitment: ensure that there is an appropriate fit between individuals and the job.
- Provision of support services, for example, counselling.

Other considerations

Managers must be aware, and communicate that:

- stress-related problems are not a sign of weakness
- pressure of excess workloads can trigger illness
- stress and illness can be related.

Managers should:

- encourage group problem solving to discuss perceived causes of stress
- develop supporting culture and improve coping techniques.

Strategies for senior managers include:

- ensuring they are accessible to staff to discuss problems and anxieties
- taking lead in changing the view that being under stress is a reflection of personal vulnerability
- revising effective induction and introduction programmes for new staff
- encouraging staff to talk about their feelings and effects of stress
- encouraging, commending and recognising supportive behaviour of others
- developing cooperative rather than adverse management styles
- engendering team spirit.

Violence

Violence to employees encompasses everything from bank robbery, vandalism, hostage taking and bombing, to the aggression displayed daily by members of the public in places such as hospital waiting rooms, job centres and train stations. The risk to staff is twofold: they may be physically injured, or they may be affected psychologically either following an actual attack or after being threatened. For detailed information about violence in the workplace and measures to combat it, see Chapter 12.

TOWARDS A SAFE WORKPLACE

Preventing slips and trips

Slips and trips cause over one-third of major accidents at work, or about 11,000 major injuries a year. More than 27,000 accidents resulted in over three-day absences. The HSE has published guidance, *Slips and trips*,[18] which contains much useful advice for facilities managers. It recommends:

- getting workplace conditions right in the first place, for example, by selecting the right flooring and lighting, planning routes and avoiding overcrowding
- keeping areas free of obstructions
- checking flooring regularly for holes, cracks, loose mats, and so on
- using the correct cleaning methods to avoid slippery residues building up
- carrying out cleaning and maintenance after hours, or if this is not possible, using signs to warn people about wet floors, providing alternative routes and avoiding trailing cables
- organising work and machinery to avoid spills, and cleaning up or fencing off any that do occur immediately
- positioning electrical equipment to avoid trailing cables, and using covers to fix cables down
- providing safety footwear free of charge to employees if floors are unavoidably wet or dusty (there are no British Standards for slip resistance, but the HSE guidance includes advice on shoe sole materials)
- keeping lights working, clean and free of obstructions, ensuring they do not dazzle, and providing extra lighting at stairs and slopes
- providing handrails, treads and floor markings on slopes or changes of level, and doormats between wet and dry areas.

Flooring

Cost and appearance should not be the only concerns when choosing flooring. Suitability for the location and the demands that will be made on the surface are also important. The HSE guidance states that 'there is no need to put up with a slippery smooth floor because it is more hygienic or easier to clean'. There are devices for measuring slipperiness, but HSE warns that the resulting 'friction values' must be interpreted with caution because workplaces differ so widely.

There are also floor treatments to reduce slipping: concrete can be abraded or chemically treated, there are coatings containing abrasive particles, and adhesive strips or squares. Some treatments may need regular reapplication.

Flooring needs to be laid by competent people, and kept clean and in good condition to maintain slip resistance and prevent curling edges or lifting. Different flooring surfaces and certain spills may need different kinds of cleaning.

---

[18] HSE. *Preventing slips and trips at work*. HSE ING 225.

A considerable number of slips and trips are caused by poor cleaning practices; often the use of the wrong type of chemical or poor techniques, such as leaving floors wet or leaving cleaning equipment in walkways.

Facilities managers will have an important contribution to reducing the likelihood of slips and trips in their building by ensuring that good safety systems are in place for all cleaning and maintenance jobs and that operatives are trained in understanding how to clean, what materials to use, how to leave surfaces clean and dry, where and how to store equipment, and so on.

Safe cleaning of floors should not be taken for granted and a task review before work commences would constitute good management practice.

Equipment

Heavy plant or humble ladder, photocopier or forklift truck, the risk from all work equipment must be as low as possible. The HSE's guidance leaflet *Using work equipment safely*[19] advises employers to ensure that:

- new and temporary workers (who may lack experience), people who have changed departments, those with difficulties (e.g. impaired mobility), and cleaning and maintenance staff are properly trained and supervised
- guards and safety devices are convenient to use, not easily overridden, made from appropriate materials, and allow the machine to be cleaned safely
- control switches are clearly marked and carefully sited
- the type of power supply, (electric, hydraulic or pneumatic) is considered: each has different risks
- hand tools are not worn, the handles are not split, and there are enough tools to avoid employees improvising with something else
- routine checks are made, preventive maintenance is carried out where appropriate, and manufacturers' instructions are followed
- equipment can be safely maintained and repaired:

  - machines should be disconnected and not moving or hot
  - mobile equipment should have the engine off and brake on
  - flammable substances must be cleaned away
  - isolating valves must be locked off
  - equipment that could fall needs support.

Facilities managers should ensure that all work equipment is regularly checked and that the people due to use it are competent. Contractors who bring equipment in to the workplace should be asked to provide suitable certification for safety checks and evidence of operative training.

---

[19] *Using work equipment safely*. INDG 291.

Vehicles (other than motor vehicles used on public roads)

Accidents involving vehicles at the workplace kill around 70 people a year and maim 1000 more. Reversing vehicles are a particular hazard, causing one-quarter of such deaths. The accidents involve cars, vans, trucks and other mobile equipment. People can be injured by being knocked down, falling from vehicles, being struck by a falling load or vehicles overturning.

Under reg. 17(1) of the Workplace Regulations, every workplace must (so far as is reasonably practicable) be so organised that pedestrians and vehicles can circulate in a safe manner. Under reg. 17(2), traffic routes in a workplace must be suitable for the persons or vehicles using them, sufficient in number, in suitable positions and of sufficient size. Under reg. 17(3) and (4):

- Pedestrians or vehicles must be able to use traffic routes without endangering those at work.
- There must be sufficient separation of traffic routes from doors and gates and between vehicle and pedestrian traffic routes.
- Where vehicles and pedestrians do use the same traffic routes, there must be sufficient space between them.
- Where necessary, all traffic routes must be suitably indicated.

Work vehicles also count as work equipment for the purposes of PUWER. Regulation 5 says that they must:

- be constructed and adapted to suit their purpose
- be selected so as to avoid risks to the health and safety of persons where the vehicles are to be used
- only be used for the operations specified and under suitable conditions.

Regulation 6 says that work equipment, including vehicles, must be maintained in an efficient state, in efficient working order and in good repair. Vehicles must be provided with safety features (a reversing alarm, seat belts, lights, etc.). Under the Management Regulations, drivers must be trained for the job and supervised if necessary. Contractors also need to know the site rules so that they can carry out loading and unloading safely.

Safe traffic routes

Facilities managers have a particular responsibility to ensure that the layout of the workplace does not create risks, from either work vehicles or private cars. Safety measures could include:

- creating pedestrian areas, including safe crossings and barriers if necessary
- providing enough designated parking places to avoid stopping in unsafe locations
- ensuring that traffic routes are wide enough, well constructed and maintained, avoid blind bends and are free of obstructions
- increasing visibility for drivers and pedestrians
- providing road markings, signs, road humps and mirrors as required

- redesigning the workplace to remove the need for reversing, by introducing a one-way system, for example
- if reversing cannot be eliminated, ensuring that there is enough space for vehicles to reverse and minimising the distance they have to reverse.

The HSE guides on vehicle safety provide further information.[20]

FLEXIBLE WORKING

An employee's workplace is not always fixed. Some employees may work from home some or all of the time, or hotdesk around the office. Others may visit clients, spend a large amount of time travelling, work alone or work in a remote place. Employers have a duty to assess the risks associated with all these working conditions.

Homeworking

Employers may need to visit, and revisit, homeworkers to check their working conditions. It is important to remember that:

- Not only the employee, but also their children could be harmed, for example, by trailing cables or dangerous substances.
- The employer is responsible for maintaining any electrical equipment (normally a computer and printer) that it provides.
- The employee needs training in the safe use of equipment provided (e.g. the need to take regular breaks from computer work).
- The employer must ensure that the workstation and seating are correctly adjusted, the screen is positioned to avoid reflections, there is suitable lighting and there is enough space to work comfortably.
- First aid provisions may be necessary.
- New or expectant mothers may be particularly at risk from some types of work.
- Some employees find working alone stressful: regular meetings, support or social events may help.
- Employees may work from home only occasionally, using their home furniture and lighting; if so, they will need advice.

It could be useful to incorporate health and safety into the homeworker's employment contract, requiring them to maintain a separate, designated work area at home. Also, although the employer is responsible for assessing homeworking conditions, it is helpful to give homeworkers their own checklist to display as a good practice reminder.

Hotdesking

To avoid WRULDs, workstations used by more than one person (including part-timers, jobsharers and shiftworkers) will have to be easily adjustable. Employees will also need training on the importance of actually making the adjustments.

---

[20] *Workplace transport safety: an employer's guide.* HSG 136 (2005).

Employees may need a mobile cart to store their work, in which case the principles of good manual handling will apply (e.g. avoiding the need to wheel the cart further than necessary and avoiding obstacles). Finally, although a recent HSE-funded study found that hotdeskers were not more stressed than their 'traditional' colleagues, it found that it was important to consult employees before introducing new ways of working.

Laptops

Portable computers are covered by the DSE Regulations if they are used for more than an hour at a time. As well as the risks associated with any computer or monitor, there are problems associated with carrying around heavy, valuable equipment, and using cramped keyboards and screens in awkward locations (e.g. wedged against the car steering wheel or on the passenger seat during a traffic jam).

An HSE-commissioned study on the safety of portable computers found that fewer than half the organisations contacted had any guidance or policy at all on using laptops, and most of them had started to get complaints from laptop users about health and safety. The researchers, System Concepts, went on to recommend:

- buying laptops with as large a keyboard and screen as possible, while keeping the weight down to 3 kg at most
- supplying employees with rucksack-style laptop bags: they distribute the weight better and give away the contents less easily to would-be muggers
- checking the weight of transformers, cables, etc.
- choosing a machine with long battery life so that transformers and cables can be left behind
- ensuring that it is possible to attach a separate monitor, keyboard and/or mouse to a docking station for prolonged use
- encouraging employees to restrict the use of laptops in non-ideal locations, and use docking stations where possible.

Homeworkers' checklist

Eyesight

- Take frequent minibreaks by looking away from the screen and documents, preferably out of the window.
- Take a 10-minute break at least once an hour and leave the room: a short walk outside to lengthen the optical focus is ideal.
- Get your eyes tested regularly.
- If eyes regularly feel sore seek advice.
- Do not work in poor light or glare.

Strain injuries

- If you change your workstation set-up after it has been health and safety checked, either self-assess it within the company guidelines or ask your supervisor to do another check.

- Take frequent minibreaks, stretch and gently exercise fingers, wrists, arms, shoulders and neck.
- Take a 10-minute break at least once an hour, leave the room, and carry out another activity.
- Take fresh air and exercise at least once a day.
- Keep work surfaces uncluttered: do not let paperwork and files build up over time, and keep plenty of space for keyboard, mouse and documents.
- Do not store boxes under the desk where they may cramp legs and feet.
- Recognise warning signs such as tiredness, sore eyes, headache or soreness, tension or numbness in hands, arms, neck, back or shoulders. If in doubt seek advice.

Posture

- Always use a good, ergonomic office chair.
- Remember to sit up straight with your back supported by your chair.
- Ensure that your screen is close enough to see without leaning forward.
- If you need a footrest, use it.

Electricity

- Do not use multisocket plugs or extension leads without first consulting your manager.
- Do not put drinks or vases of flowers on computer equipment.

Fire

- Storage of documents and reference material may constitute a fire hazard. Keep a fire extinguisher in the work area and ensure that the smoke alarm is in working order.

Driving

It has been estimated that up to one-third of all road traffic accidents involve somebody who is at work at the time. This may equate to over 20 fatalities and over 250 serious injuries every week.

Health and safety laws apply to on-the-road work activities and while the Road Traffic Act remains the primary legislation governing road safety, accidents will increasingly be investigated under health and safety laws and, in particular, corporate manslaughter duties will apply.

Driving at work not only refers to driving vehicles, but also covers using motorcycles and pedal cycles.

Employers must manage the risks to employees' (and others') health, safety and welfare while they are driving on work-related business. As for all work activities, a thorough or suitable and sufficient risk assessment must be carried out covering all aspects of driving for work.

Road traffic accidents are often caused by lack of awareness to road conditions, distraction, tiredness and general anxiety about arriving at a destination in time for a meeting, for example.

An employer will be expected to consider training for employees in safe driving, must have policies in plan to enable employees to take rest breaks while on journeys and must not set unreasonable deadlines for destination times or expect the employee to spend unreasonably long hours driving.

Facilities managers may be required to manage their client's vehicle fleet and, if so, must be familiar with the health and safety issues of driving while at work.

A sensible approach would be to develop a safe driving policy document which sets out policy, risk assessments, organisation and responsibilities, monitoring and reviewing procedures. Vehicle-related accidents must be reported to senior management/executive board level and all incidents should be investigated.

The HSE has issued an excellent guide, *Driving at work: managing work related road safety.*[21]

Mobile phones

It is illegal to drive while using a handheld mobile phone. The police can bring prosecutions for causing death by dangerous driving, driving without due care and attention, or failing to have proper control of a vehicle, but the common penalty is a fixed penalty notice and fine (£60) and three endorsement points.

Evidence on whether mobile phone emissions are safe, even when using a hands-free kit, remains inconclusive. They do cause a slight warming of the brain, but the long-term effects of this are unknown. One option is to advise employees to make short, essential calls only, and to use land lines whenever possible.

The advantages of mobile phones should not be forgotten. For lone workers, those visiting clients at home or those on the move, they offer invaluable personal protection. The increasing use of mobile hands-free kits will reduce the direct exposure to radiation waves.

Lone working

Lone workers may include:

- those working outside normal office hours, such as maintenance, cleaning and security staff
- mobile workers, such as sales reps, estate agents, social workers, postal workers, painters and decorators
- homeworkers.

There is no express legislation prohibiting lone working, but the normal rules on risk assessment apply. The assessment may show that an activity is too dangerous to be carried out alone (e.g. lifting a heavy load or visiting a violent client at home) or

---

[21] HSE. *Driving at work: managing work related road safety.* INDG 382.

without supervision (e.g. working in a confined space). The risk assessment should consider not only normal working conditions but also emergencies.

Lone workers may need extra training or experience to cope in an emergency or to deal with aggression. Their employers need to set limits on what they can and cannot do on their own. There needs to be a system for them to report to base and for the employer to take action if they fail to do so. Lone workers need access to first aid facilities, and mobile workers will need their own first aid kit.

CONSTRUCTION WORK AND BUILDING MANAGEMENT

Health and safety and environmental laws apply to construction work no matter who has commissioned it and who is carrying it out. Construction activity is one of the highest health and safety risks and has a poor record for fatalities and major injuries. Fatalities are not just to workers but also to a significantly high proportion of the public. HSE statistics for 2006/07 for the construction industry are:

- 77 fatalities to workers/contractors
- seven fatalities to members of the public
- 3711 major injuries to employees
- over 7000 over three-day injuries reported.

The greatest cause of fatality and major injury in construction is falls from height.

In addition to physical injuries, there are large numbers of individuals who are affected by occupational diseases caused by working in the construction industry. As an example, it is predicted that over 4000 people will die from asbestos-related diseases every year until numbers peak in about 2018–2020.

Construction safety is enforced by the HSE and the construction division of the HSE is targeting all projects, large and small, to establish compliance with health and safety legislation.

Key health and safety issues in respect of construction activity

- Failure to assess risks adequately
- failure to recruit and retain skilled workers
- failure to ensure proper site management
- poor health and safety systems
- inadequate resourcing of site management
- poor leadership of front line managers/supervisors
- failure to plan work adequately
- inappropriate selection of equipment.

Construction (Design and Management) Regulations 2007 (CDM)

The CDM Regulations 2007 replaced in their entirety the 1994 CDM Regulations and incorporate the Construction (Health, Safety and Welfare) Regulations 1996, thereby giving one set of regulations to govern the construction industry. The

Regulations are supported with an ACoP: *Managing health and safety in construction.*[22]

CDM 2007 apply to all construction projects and there is therefore no longer any exemption for small construction projects employing fewer than five operatives (as under the 1994 regulations). Those with responsibility for facilities management in buildings will need to make sure that procedures are in place for all refurbishment works, new build, maintenance and repair works regardless of the number of operatives involved.

If the construction phase of a project lasts for more than 30 days or 500 person-days it will be notifiable to the HSE on Form F10. Such notifications enable the HSE to programme its inspection visits, although it does have legal jurisdiction over non-notifiable projects.

The application of CDM 2007 to all construction projects also means that local authority environmental health practitioners may also enforce the Regulations, especially if the construction site area is not fully segregated from any work area; for example, during internal alterations and redecoration works.

Notifiable projects

When a project is notifiable, the client must appoint a CDM coordinator and a principal contractor. The CDM coordinator replaces the planning supervisor found in the 1994 Regulations and carries enhanced duties regarding advice to clients, assistance with preparation and planning time, and cooperation and coordination duties. The principal contractor role stays very much the same as in the earlier Regulations.

Notifiable projects no longer need a pre-tender health and safety plan as this is replaced with 'preconstruction information'. This information can be in any format or location and the role of the CDM coordinator is to coordinate the information and to ensure that it is available to all those who may need it in order to deliver a safe construction project and building for future use.

The HSE is strongly advocating 'the right information to the right people at the right time'.

While notifiable projects will have preconstruction information prepared or coordinated by the CDM coordinator, non-notifiable projects will need to have preconstruction information collated and provided by the client.

Proper planning and preparation, including the assimilation of information, will ensure a safer construction project.

Duty holders identified within the Regulations are:

- the client
- designers
- the health and safety or CDM coordinator (formerly the planning supervisor)
- principal contractor
- contractors.

---

[22] *Managing health and safety in construction.* ACoP: L144 (2007).

What are the duties under the Regulations?

All duty holders

No person shall:

- appoint a duty holder without ensuring that they are competent to perform the duties
- accept an appointment unless competent
- instruct a worker unless the worker is competent or supervised

All persons shall:

- seek and provide cooperation
- coordinate their activities with others in the interest of health and safety.

Clients

- Clients will not be able to appoint a client's agent.

Clients must:

- ensure that suitable arrangements are made to manage the project safely
- appoint a CDM coordinator and principal contractor for all notifiable construction projects
- ensure that suitable welfare arrangements are in place during construction
- ensure that designers and contractors are promptly supplied with information relevant to their purposes
- ensure that contractors (principal contractor on notifiable projects) are informed of the minimum time to be allowed for planning and preparation before construction commences
- ensure that construction work does not start on notifiable projects before a construction phase health and safety plan is in place
- ensure that suitable arrangements are made to protect the health and safety of users of any structure designed as a workplace, as well as of construction workers, cleaners and maintenance workers.

The CDM coordinator

This role replaces the planning supervisor, although many of the duties are re-enacted in the coordinator's role. The coordinator must:

- demonstrate that they are competent to carry out the role and that they have the resources for the commission
- advise and assist the client on arrangements for managing the project
- advise and assist the client on appointments
- advise and assist the client on determining if the project is notifiable, and if so, notify the HSE
- advise and assist the client on the competence, capability and resources of others

- provide information, coordination and planning and preparations for project construction work
- work with designers on risk reduction and health and safety management
- advise and assist clients on the suitability of the start on site date
- deal with design work during the construction phase
- deliver a suitable health and safety file to the client at the end of the project.

The designer

Anyone who specifies materials or type of design will be classed as a designer, including clients.

- Designers' duties apply to all projects.

Designers must:

- not start work on a notifiable project unless a CDM coordinator has been appointed
- be competent and have adequate resources
- cooperate with and seek the cooperation of others, whether on the project, another project, adjacent site and anywhere as necessary
- coordinate their activities with one another so as to ensure, as far as is practicable, the health and safety of all persons
- avoid risks to construction workers, cleaners, maintenance workers and anyone affected by their designs or activities, together with anyone using the structure if it is designed as a workplace
- eliminate hazards and reduce the risk from remaining hazards, giving priority to collective measures
- provide sufficient information regarding the design to assist the client, the coordinator and other designers or contractors
- provide information promptly to the coordinator so that the health and safety file can be prepared and issued on completion of the project.

Principal contractors

Duties imposed by CDM 2007 on the Principle Contractor include:

- planning, managing and monitoring the construction phase to ensure that it is carried out without risks to health or safety
- ensuring that adequate welfare facilities are provided throughout the project for those working on the site
- drawing up and implementing site rules
- drawing up and implementing the construction phase health and safety plan
- preventing unauthorised access to the site
- ensuring co-operation and co-ordination between all persons working on the site
- ensuring suitable arrangements are in place for communicating health and safety matters and other information to the workplace.

CDM 2007 enhances clients' duties by making their current management responsibilities in generic health and safety law more explicit in relation to CDM. They will be more accountable for the standards of health and safety applied to a project.

Regulation 9 requires clients to take reasonable steps to ensure that there are suitable management arrangements in place throughout the project. These arrangements must ensure that the construction work can be carried out without risks to health and safety, that welfare arrangements are in place before construction works start on site, and that the structure designed for use as a workplace complies with the Workplace (Health, Safety and Welfare) Regulations 1992.

Clients must also ensure that the arrangements for health and safety include allocation of sufficient time and resources to achieve safety on site and they must ensure that contractors are advised of the time allowed for planning and preparation before construction works start.

Clients will be expected to take advice from the health and safety coordinator, especially if they have little experience in overseeing construction projects.

CDM 2007 will remove the option for a client to appoint a client's agent: clients will always have accountability under CDM 2007.

The new role of 'CDM coordinator' will be a more involved role than that of the former planning supervisor and a key new duty will be to advise and assist the client in discharging their duties. A particular duty on the coordinator will be to assist the client by advising on the adequacy of other duty holders' arrangements for controlling risks arising from the project.

CDM coordinators must have sufficient independence from the project team so as to carry out their duties effectively.

The CDM coordinator will need a sound understanding of:

- health and safety in construction work
- the design process
- the importance of coordination of the design process and an ability to identify information that others will need to know about the design in order to carry out their work safely.

It is intended that the CDM coordinator will be more involved in design meetings than perhaps the planning supervisor was, and this may increase time and fees for the appointment. If the coordinator is to discharge their legal duties effectively then increased involvement in the project is essential.

The HSE sees the CDM coordinator as the 'power behind the throne' in ensuring that the client complies with duties under CDM, and also, that all other duty holders comply with their duties. Clients and the CDM coordinator will be accountable for compliance with CDM.

If enforcement action is needed on a project, the HSE will judge both the client and the coordinator according to their level of construction experience and competence, whether their actions were reasonable according to the facts and circumstances of the case.

The client should remember that they are entitled to rely upon the professional advice given by the coordinator and others, and if this advice was negligent or incompetent then the client may have a defence against proceedings, but if the advice was good but the client failed to follow it, then the client would have no defence, but the coordinator would have.

Designers' duties have not changed dramatically but CDM 2007 clarifies their duties for minimising risks in respect of their design.

Unnecessary paperwork should reduce under CDM 2007 and the emphasis will be on effectively planned projects with the right information being given to the right people at the right time.

Facilities managers may act in the client role or may appoint themselves as CDM coordinators. It is vital that they are clear as to which duty they are performing and must ensure that clear procedures are in place. Preparation and planning time is key to ensuring a safe project and the client and CDM coordinator (where appointed) set the agenda in this respect. The time allowed for preparation and planning before works start on site has to be notified to the HSE on Form F10 and they are likely to target short lead-in projects.

The preconstruction information pack

Any information that could be important in assisting designers or contractors in carrying out their duties under CDM 2007 could be classed as preconstruction information. The information required under CDM 2007 is really only that which is relevant to health and safety so that hazards and risks can be identified and subsequently addressed.

The client is responsible for deciding what information is available and what can be provided to designers and contractors. Where projects are notifiable, the CDM coordinator can advise the client as to what information will be needed and assist the client in obtaining the information.

The client has the duty to provide information to every person designing the structure and to every contractor who has been or may be appointed by the client.

When a project is notifiable, the client has to provide the information to the CDM coordinator and the CDM coordinator has to pass it on to every person designing the structure and to any contractor, including the principal contractor, who has been or may be appointed by the client to the project.

Designers have to provide information to the client, other designers, the CDM coordinator and other contractors as is necessary for them to carry out their duties. Some of this information will need to be included in the preconstruction information pack.

The CDM Regulations do not specify the format that the preconstruction information should be in and, indeed, do not really refer to a 'pack' as such.

Information could be available in a wide variety of locations and there is no requirement to duplicate the information. The HSE is keen to see the amount of unnecessary paperwork on CDM projects reduce significantly. Whatever information is available and wherever that information is located, it is important that it is clearly identified and its availability made clear to those who may need it. An information 'road map' could be created, listing what is available, where, and who holds it.

Where projects are notifiable, the CDM coordinator could advise the client early in the project on the format and content of the preconstruction information pack. The preconstruction information shall consist of all the information in the client's possession, or which is reasonably obtainable, including:

- any information about or affecting the site or the construction work
- any information concerning the proposed use of the structure as a workplace
- any information about the existing building.

The client must ensure that he has obtained anything about the site, including environmental issues, local traffic, use of the site, previous use of the site, use of adjoining premises/land, overhead power cables, public transport, etc.

Designers and contractors will need to know what the proposed use of the building is, especially if it will be a workplace, because designers, in particular, will need to ensure that they design the structure in accordance with the Workplace (Health, Safety and Welfare) Regulations 1992.

Very broadly, every person designing the structure and every contractor who has been or may be appointed by the client will need to have relevant information.

Designers will be:

- architects
- quantity surveyors
- structural engineers
- surveyors
- civil engineers
- interior designers
- landscape architects
- specialist contractors, for example, temporary works
- design and build contractors.

Contractors will be:

- any of those who are tendering
- specialist contractors
- demolition contractors
- mechanical contractors
- electrical contractors
- groundwork contractors
- civil and structural contractors.

The list of contractors who may be appointed could be considerable and the requirement for information could generate vast amounts of paperwork.

The client, and on notifiable projects, the CDM coordinator, could identify key information that contractors will need to quote adequately for the work required.

Once a shortlist of tenderers has been drawn up, or one contractor chosen, the client could discuss further details of the information that would be specific to the execution of the works.

A new requirement of CDM 2007 is that the client must advise the contractor, or any contractor, the minimum amount of time before the construction phase which will be allowed to the contractors appointed by the client for planning and preparation for construction work.

Contractors will need to determine the resources they need to carry out the project safely and they therefore need to have an idea of the time available for further research and mobilisation for the project.

Preparation and planning is not defined in the Regulations but is taken to mean the time period in which the contractor considers exactly how to do the job, exactly what equipment will be needed, what specialist contractors, etc., need to be procured, what welfare facilities are required and how they will be installed on the site, the number and skill base of operatives, etc.

Unrealistic deadlines and failure to allocate sufficient resources, both financial and human, are two of the main causes of construction site accidents and poor risk management. Clients should consult with the CDM coordinator on notifiable projects and also with all contractors and discuss the timescales needed, expected and offered. Naturally, clients will often believe that contractors are prolonging the period they need for mobilisation and for the construction phase, and they will be keen to ensure a shorter timescale if possible. Contractors often say that clients, or indeed their advisers, do not understand the full complexities of what they have to do to mobilise and complete a project. If a contractor says the project can be completed in 24 weeks, the client will often counter that it has to be finished in 20, believing that the contractor will be 'dragging his heels for four weeks'. Sensible dialogue is necessary and a true consensus and expectation need to be reached. Again, the CDM coordinator could assist in these discussions, and if the project is not notifiable, and the client is unsure of what timescales would be reasonable, the client could consult the competent person appointed under the Management of Health and Safety at Work Regulations 1999.

The preconstruction information pack should include the following:

- a general description of the construction works comprised in the project
- details of the time within which it is intended that the project, and any intermediate stages, will be completed
- details of the project team, including CDM coordinator, designers and other consultants
- details of existing plans, any health and safety file, etc.
- details of risks to health and safety of any person carrying out the construction work so far as such risks are known or are reasonably foreseeable or any such information as has been provided by the designers, including design risk assessments
- details of any client requirements in relation to health and safety, for example, safety goals for the project, site rules, permits to work, emergency procedures, management requirements
- such information as the client or CDM coordinator knows or could ascertain by making reasonable enquiries regarding environmental considerations, on-site residual hazards, hazardous buildings, overlap with the client's business operation, site restrictions, etc.

- such information as the client or CDM coordinator knows or could ascertain by making reasonable enquiries and which it would be reasonable for any contractor to know in order to understand how he can comply with any requirements placed on him in respect of welfare by or under the relevant statutory provisions, for example, availability of services, number of facilities required to be provided, details of any shared occupancy of the site
- content and format of the expected health and safety file.

The ACoP gives guidance as to what to include in the document and this can be summarised as:

- nature and description of project
- client considerations and management requirements
- existing environment and residual on-site risks
- existing drawings
- significant design principles and residual hazards and risks
- significant construction hazards.

The information pack is expected to include only that information which it is reasonable to expect or which could reasonably be ascertained by making enquiries. The Regulations do not expect every aspect of potential hazard and risk to be known at the outset of the project, but expects reasonable enquiries to be made.

The construction phase health and safety plan

The CDM Regulations require that a construction phase health and safety plan is produced by the principal contractor before works on site commence. The construction phase health and safety plan must comply with reg. 23 and must include:

- arrangements for the project which will ensure the health and safety of all persons at work carrying out construction work
- risks involved in the construction work
- site rules
- information about welfare arrangements on the site.

Copies of the health and safety plan must be made available to those who need to know about site safety. The principal contractor must manage contractors on the site and must stipulate the site rules to be followed. Copies of the F10 notification must be displayed on site.

The facilities manager may need to approve the construction phase health and safety plan on behalf of the client even if the CDM coordinator has given advice on it. The ultimate responsibility for ensuring that safety has been preplanned rests with the client and it is the client who would be in breach of reg. 16, facing criminal prosecution.

The health and safety file

An important requirement of the CDM Regulations is the preparation of a document known as the health and safety file. This is a collection of all relevant health and safety information relating to the construction project, covering in particular the residual hazards and risks associated with the building from the occupier's point of view.

The health and safety file should contain information on the design intent of the building and should explain the rationale behind the safe systems of work for maintenance and cleaning of the building. The CDM coordinator must prepare the file or review and update any existing document. The principal contractor provides the information. The health and safety file must be handed to the client at the end of the project and it must be available to whoever needs to access the information, for example, contractors for future maintenance and new owners.

Design risk assessments

These are a contentious issue within the CDM Regulations and should be completed, in some form or other, by designers undertaking design works. Designers have a significant contribution to make to reducing the hazards and risks associated with construction activity, and the intent of CDM is that they address these issues and design out, or reduce, significant hazards.

If working at heights contributes to the greatest fatality rate, it is incumbent on a designer to attempt to design out the need to work at height by sequencing the construction work differently, locating plant elsewhere than on the roof, designing windows that can be cleaned from the inside, etc.

There is no legal format for a design risk assessment: the designer needs to be able to demonstrate that they have:

- taken reasonable steps to ensure that clients are aware of their duties under CDM before starting design work
- prepared designs with adequate regard to health and safety, and to the information supplied by the client
- where practicable, designed structures to avoid foreseeable risks
- provided adequate information in, or with, the design
- designed in accordance with the Workplace (Health, Safety and Welfare) Regulations 1992
- cooperated with the CDM coordinator and other designers so that everyone can comply with their legal duties.

Design risk assessments can be separate risk assessment forms, can include information annotated onto drawings or can be schedules of information. As long as it is clear what residual risk remains from the design then responsibilities under the Regulations will have been met.

Managing refurbishment works

Refurbishment works will often fall within the ambit of CDM and should be considered no differently from other construction projects. It could even be argued that refurbishment works are potentially more hazardous because the building will often remain occupied, works may be in restricted areas, other processes may be happening, employees and public may have access, and so on.

Facilities managers must pay careful attention to refurbishment works, planned and term maintenance projects, and must establish that CDM is applied to all projects. If the facilities management company is competent, it can appoint itself as CDM coordinator for notifiable projects, and if it can be proved that it carries on the business of construction, it can appoint itself principal contractor as well.

It is vital to integrate the employer's health and safety policy into any construction phase health and safety plan.

CDM 2007 applies to all construction works, not just to those that are notifiable, that is, they last over 30 days or more than 500 person-days.

Non-notifiable projects still need to have relevant information provided to those who need it, need to have welfare facilities provided, need adequate preparation and planning time and must have suitable arrangements in place for managing health and safety. The client is responsible for ensuring that all elements of CDM 2007 are met.

Building management

In addition to the CDM Regulations, general health and safety applies to all projects undertaken regarding maintenance, refurbishment, repair and development of buildings and structures.

Clients, employers and persons in control of premises are expected to ensure that health and safety management systems are in place and that contractors work safely. It is no longer acceptable for clients/employers to leave contractors to 'fend for themselves', and anyone who engages a contractor of any sort is expected to guide them on standards of safety. All contractors must be competent to undertake the tasks allocated to them.

Employers need to develop robust Contractor Safety Codes of Practice, often known as employers' requirements, and ensure that these are followed during all works.

In particular, access to work areas at heights must be especially controlled, as must access into confined spaces, on to fragile materials, using hot works, etc. Safe systems of work must be determined with the contractor and employers' requirements for permit to work procedures must be clearly stipulated.

Structure and access

The HSW Act requires there to be a safe means of access to and egress from a place of work.

- A place of work is anywhere where an employee would carry out job tasks expected by the employer.
- Plant rooms, roofing areas, working platforms, gantries and mezzanine floors are all places of work. Working up a ladder, on a scaffold or tower scaffold is a 'place of work'.

- Access to a place of work must be safe and without risks to health. There must be a 'safe system of work' for the cleaning and maintenance tasks.
- The Work at Height Regulations 2005 require risk assessments to be completed and safe systems of work to be followed irrespective of the distance of falling.
- The Management of Health and Safety at Work Regulations 1999 require risk assessments to be completed for all job activities. This will include how you expect someone to get to their place of work.
- The Provision and Use of Work Equipment Regulations 1998 require that any equipment provided for access to the place of work is safe and properly maintained.
- The Lifting Operations and Lifting Equipment Regulations 1998 require that any lifting equipment is inspected and tested and maintained in a safe condition.
- The facilities manager must consider the safety of all access routes to roofs, plant rooms, gantries, etc. Anywhere that equipment is located will need at some stage to be accessed so that the equipment or plant can be cleaned or maintained. Also, parts of the structure may need cleaning, for example, gutters need cleaning out, windows need cleaning, roof lights need cleaning.
- The facilities manager may be the employer of the persons needing access, or they may be the person in control of the premises. In any event, health and safety responsibilities are mandatory.

Cleaning windows

Safe means of access for cleaning windows both inside and out is essential. Usually, the common practice is to use a ladder, but a risk assessment should be conducted to establish whether a ladder is best. Would a tower scaffold, mobile elevating work platform or 'cherry picker' be better? The Work at Height Regulations 2005 must be applied.

How will the window cleaner carry the bucket and cleaning cloths? Will he have to overreach? What happens in poor weather? Could the ladder be knocked into?

Perhaps a safer method of access to the outside windows is from the inside. Could windows be 360-degree swivel? (This would be good design under the CDM Regulations.) For all new buildings, where the cleaning of any window, transparent or translucent wall, ceiling or roof involves the potential for falling, the designer must complete a design risk assessment.

The facilities manager cannot leave the contractor to it and turn a blind eye to the way that windows are cleaned. The window cleaning contractor must provide risk assessments and method statements. As the facilities manager, you must provide adequate information on the hazards in your building, such as falls from heights, so as to enable the contractor to plan his job safely.

Using ladders

More accidents are caused by poor use of ladders than any other access equipment. Ladders should be properly maintained and regularly inspected as they are 'work equipment' (PUWER 1998). They should be suitable for the purpose; for example, wooden ladders may not be suitable in some environments. The Work at Height Regulations 2005 apply to the use of ladders. The Regulations do not ban ladders or stepladders but require a specific risk assessment to be completed.

Ladder safety tips:

- Secure them to prevent slipping, especially at the top.
- Have a second person to 'foot' them (but only if the ladder is less than 5 m).
- Extend the top of the ladder to at least 1.0 m above the landing space.
- Do not place ladder against fragile surfaces, for example, gutters.
- Place ladders on firm and level ground.
- Set ladders to have an incline of one in four.
- Check all rungs for damage.
- Regularly inspect ladders and record findings.

Access for cleaning and maintenance

- Access routes and passageways must be well lit and properly ventilated.
- There must be no obstructions, disrepair to floors, etc.
- Adequate room for access must be provided; and remember, materials, spare parts and plant may need to be brought to the place of work.
- Is the place of work a 'confined space'? If so, a risk assessment must be carried out and the Confined Spaces Regulations 1997 complied with.
- Consider egress from the place of work: can operatives get out the way they got in?
- What emergency procedures have been put in place: fire evacuation, accidents and emergencies, etc.? How might people be got out if they suffer a severe injury and are not mobile?
- Does anyone have to traverse fragile materials with a risk of falling through them? This would be unacceptable, so an alternative method will have to be found or crawling boards used, strengthening the material. As a last resort, fall arrest systems or fall arrest netting could be used.

Consequences of failure to have safe access and egress

Consequences may include:

- serious injury to someone
- fatalities
- prosecution: fines of up to £20,000 for contravening the Work at Height Regulations 2005
- prosecution: unlimited fines in the Crown Court
- Statutory notices: either improvement or prohibition notices under the HSW Act
- corporate manslaughter.

Top tips for access and safety

- Assess access routes to all work areas, plant, equipment, etc.
- Complete risk assessments.

- Devise safe systems of work.
- Work on any level where there is a likelihood of a fall.
- Use ladders only as a means of access, not as working locations.
- Ask architects and designers what safety provisions they have allowed for access for cleaning and maintenance.
- Check any access equipment, fall arrest systems, eye bolts and lifting equipment regularly and inspect lifting equipment every six months.
- Assess the competency of contractors.

Permit to work procedures/safe systems of work

There are many instances where a residual hazard exists within work premises and the employer has to put in place procedures and methods for controlling exposure to the risks.

Hazardous environments, such as plant rooms, roof areas, confined spaces, tanks and vats, are all areas where significant hazards can be present and the preferred safety procedure for managing those risks is a permit to work or enter system.

Permits to work set out the safety procedures to be followed by those undertaking the work and should:

- describe the work being undertaken

  - who is doing it
  - how long the work will last
  - where is it being undertaken

- what are the hazards

  - how will these be controlled

- emergency procedures
- requirement for training or competency
- site/employer safety rules
- handback/sign-off procedure
- PPE to be used.

A permit to work system is an invaluable safety tool and should be used for all works carried out by external contractors in areas in which they are unfamiliar. A permit process gives the facilities manager the opportunity to check the safe system of work of the contractor or employee and brings into the open any site safety hazards and their expected controls.

A permit to work should be signed by both parties and must be completed at the end of the day or period specified.

If an accident were to ensue despite the permit, the permit would be used as 'due diligence defence', provided of course it had been drawn up by a competent person and was valid and relevant to the job.

Hot works permits are essential for managing fire safety risks.

Method statements

These are really written safe systems of work and should explain the process, procedures and safe controls to be taken when a job is tackled. Method statements are not legally required but are good practice and a sound safety discipline.

The Building Regulations

The current edition of the Building Regulations is 2002, as amended, and the majority of building projects are required to comply with them.

The Building Regulations exist to ensure the health and safety of people in and around all types of buildings. They also cover energy conservation and access and facilities for people with disabilities.

Anyone wanting to carry out building work (as defined in the regulations) which is subject to the regulations is required by law to make sure that it complies with the regulations and to use one of two types of Building Control service.

A Building Control service is provided so that all building works can be screened and checked so as to ensure that the works comply. Building Control services are provided either by a local authority or by an approved inspector or firm. There is a charge for either service. The responsibility for complying with the Building Regulations rests with the person carrying out building works: the facilities management company if you are using your own employees, or the contractor or builder if using external resources. It is vital to establish at the outset of any contract who is making the Building Regulations application.

The owner of the building will be the person served with an enforcement notice if the buildings works fail to comply with the requirements.

The Building Regulations deal with 13 parts, each dealing with individual aspects of building design and construction: structural matters, fire safety, conservation of energy, hygiene, sound insulation, access and facilities for disabled people, ventilation, drainage, resistance to moisture, combustion appliances, glazing, protection from falling and use of toxic substances.

Each part sets out the broad objectives that the individual aspects of building design must achieve. The 'parts' are accompanied by 'approved documents' which contain practical ways on which to comply with the requirements.

The information and guidance contained in the approved documents is not a set of statutory requirements and is not mandatory. However, anyone carrying out building works must be able to demonstrate that they have built or are going to build, a structure that meets the requirements in other ways.

Building work is defined as:

- the erection or extension of a building
- the installation or extension of a service or fitting which is controlled under the regulations
- an alteration project which will be relevant to the continued compliance of the building
- the insertion of cavity wall insulation
- underpinning of the foundations.

With a few exceptions, the person undertaking the building works can decide whether to appoint the local authority Building Control Service or an approved inspector as the 'checking authority'.

Before works can commence either 'full plans' have to be submitted or a building notice has to be lodged.

If plans are submitted and they comply with the regulations, a notice is issued to that effect. Building works can then start. If the plans do not comply, modifications can be made and the local authority will advise.

Building notices apply predominantly to small works and avoid the need to submit full plans. Building notices cannot be used for fire safety works under Part B. Once a building notice has been issued, works can commence and the local authority will inspect works as they progress.

If Building Regulations approval is sought under the 'full plans' principle, a building completion certificate is issued. No such certificate is issued under a building notice approval.

An approved inspector will check plans, issue a plans certificate, inspect the work as it progresses and, if everything complies, issue a final certificate. The approved inspector will notify the local authority of the intended building work on an 'initial notice'.

Contraventions of the Building Regulations are usually dealt with by way of fines, of up to £5000 in the Magistrates' Court and £50 per day for each day the offence continues, or by way of an enforcement notice requiring the owner to alter, remove or amend works which contravene the Regulations. Should an owner not comply, the local authority has the power to 'act in default' and undertake the works themselves and recover the costs. Enforcement notices normally give 28 days in which to rectify the defects. Appeals can be made against enforcement notices to the Magistrates' Court.

Maintenance and repair

Facilities managers will generally delegate maintenance and repair work on the premises to outside contractors. They cannot, however, delegate their health and safety duties in the event of death, injury or accidental exposure to asbestos or another hazardous substance.

Contractors have duties too, but if their employees are injured on your premises, or using your equipment, you could be liable. Equally, contractors may import risks onto your normally safe premises, putting your employees at risk and exposing you to possible legal action if you fail to take preventive measures.

Managing contractors

Many large firms keep a list of contractors who they prefer to use, and who meet their safety criteria. Those who want to make it onto the list may have to undergo an audit on their own premises or while doing work on another site. The qualifications of their subcontractors are also likely to be checked. Successful bidders for a contract will be asked to draw up a safety plan for the work, which may include targets and a contractual requirement for continuous improvement.

During the work, the employer should monitor the contractors' performance and investigate incidents and near misses. Typically, large firms will expect their contractors to meet the same safety standards as their own employees, to avoid differences across their site. They may set up safety competitions for the various contractors on site, or circulate details of each contractor's performance. Poor performance will, at best, lead to the contractor's removal from the 'preferred' list. At worst, the contractor may be dismissed before the end of the contract and have to pay for a replacement.

FIVE STEPS TO MANAGING CONTRACTORS' HEALTH AND SAFETY

1. Planning
   - Define the job.
   - Identify hazards.
   - Assess risks.
   - Eliminate and reduce risks.
   - Specify health and safety conditions.
   - Discuss with contractor.

2. Choosing a contractor
   - Check contractor is competent for the job (ask questions, get evidence).
   - Discuss the job, the site and site rules.
   - Obtain a safety method statement.
   - Decide if subcontracting is acceptable and will be safe.

3. Managing contractors on site
   - Ensure that contractors sign in and out.
   - Name a site contact.
   - Reinforce health and safety information and site rules.

4. Keeping a check
   - Assess how much contact with contractors is needed.
   - Is the job going as planned?
   - Is the contractor working safely and as agreed?
   - Have there been any incidents?
   - Have there been changes in personnel?
   - Are any special arrangements required?

5. Reviewing the work
   - How effective was the planning?
   - How did the contractor perform?
   - Record the findings.

Source: HSE. Managing contractors.[23]

---

[23] HSE. *Managing contractors*. ISBN 07176 11965.

Legal duty to contractors

The principle that organisations retain responsibility for the safety of contractors working on their premises was established in the Associated Octel case, heard in the House of Lords in November 1996. There have been much higher fines, and much worse incidents, as the figures at the start of this chapter showed, but for facilities managers, the lessons are salutary.

The case involved a maintenance job in a confined space, in which a contractor's employee was injured because he used the wrong equipment. The man, Mr Cuthbert, was repairing a tank lining at Octel's Ellesmere Port chemical plant, which was shut for maintenance. He was working by an electric bulb inside the tank, cleaning the lining with acetone. The acetone was in an old paint bucket which he had retrieved from a rubbish bin. The open container allowed the acetone to give off highly flammable fumes. The light bulb broke, and there was a flash fire in which Mr Cuthbert was badly burned.

Octel was prosecuted under s. 3(1) of the HSW Act. This states:

> '*It shall be the duty of every employer to conduct his undertaking in such a way as to ensure, so far as is reasonably practicable, that persons not in his employment who may be affected thereby are not thereby exposed to risks to their health or safety.*'

Octel claimed that Mr Cuthbert's injury was not caused by Octel conducting its undertaking within the meaning of s. 3(1). Cleaning the tank was part of the contractor's undertaking, and Octel had no right to control how its independent contractors worked.

Why is the employer liable?

The House of Lords rejected the company's argument. In Lord Hoffman's words: 'The tank was part of Octel's plant. The work formed part of a maintenance programme planned by Octel. The men who did the work, although employed by an independent contractor, were almost permanently integrated into Octel's larger operations. They worked under the 'permit to work' system (having to obtain authorisation before every job from Octel's engineers). Octel provided their safety equipment and lighting.'

In this case, he found that it was clear that having the tank repaired was part of Associated Octel's undertaking. But he pointed out that in other situations 'there will also be ancillary activities such as obtaining supplies, making deliveries, cleaning, maintenance and repairs which may give rise to more difficulty'.

He gave some useful examples of an employer's duty: 'If he has a repair shop as a part of his plant, that is an ancillary part of his undertaking. Likewise, as in this case, if he has independent contractors to do cleaning or repairs on his own premises, as an activity integrated with the general conduct of his business.' Other activities, though, which are completely separate from an employer's business, such as 'the cleaning of the office curtains at the dry cleaners, the repair of the sales manager's car in the garage [and] maintenance work on machinery returned to the manufacturer's factory' would not form part of its undertaking. Thus: 'The place where the activity takes place will in the normal case be very important; possibly decisive.'

The ACoP supporting the CDM Regulations 2007 contain a useful approach to checking contractor (and other duty holder) competence by developing a checklist based on the Twelve Core Competencies (Table 1.3).

Table 1.3 Twelve Core Competencies

Core Criteria	Requirements
1. H&S policy and organisation (five employees or over)	Have and implement an appropriate policy, regularly reviewed, and signed by the MD or equivalent. The policy should be relevant to the nature and scale of the work and set out responsibilities for health and safety (H&S) management at all levels in the company.
2. Arrangements for ensuring H&S measures	These should set out the arrangements for H&S management and should be relevant to the nature and scale of the work. They should set out how the company will discharge duties under CDM 2007 and other H&S legislation, with a clear indication of how these arrangements are communicated to the workforce.
3. Competent advice: company and construction/sector related	The company and its employees must have ready access to competent H&S advice, preferably from within the company. The advisors must be able to provide general H&S advice and advice relating to H&S issues on site.
4. Training and information	Have in place, and implement training arrangements to ensure employees have sufficient skills and understanding to discharge their various duties. Have a programme of refresher training (e.g. a CPD programme or lifelong learning) that will keep employees updated on legislation and good H&S practice. This applies throughout the company from top management to trainees.
5. Individual qualifications and experience	Employees should have the appropriate qualifications and experience for the assigned tasks, unless they are under controlled and competent supervision.
6. Monitoring, audit and review	Have a system for monitoring procedures, for auditing them at periodic intervals and for reviewing them on an ongoing basis.
7. Workforce involvement	Have, and implement, an established means of consulting with the workforce on H&S.

TABLE 1.3 *(Continued)*

Core Criteria	Requirements
8. Accident reporting and enforcement action; follow-up investigation	Records of all RIDDOR-reportable events for at least the last three years. Have a system for reviewing incidents, and recording the action taken as a result. Record any enforcement action taken against the company in the last five years and action taken to remedy any enforcement-related issues.
9. Subcontracting/consulting procedures (if applicable)	Have arrangements for appointing competent subcontractors/consultants. Be able to demonstrate how to ensure that subcontractors will also have arrangements for appointing competent subcontractors or consultants. Have arrangements for monitoring subcontractor performance.
10. Risk assessment (leading to a safe method of work if need be)	Have procedures in place for carrying out risk assessments and for developing and implementing safe systems of work/method statements. Note: this should include occupational health issues.
11. Cooperating with others and coordinating work with other contractors	Illustrate how cooperation and coordination of the work is achieved in practice, and how others are involved in drawing up method statements/safe systems of work.
12. Welfare provision	Able to show how the appropriate welfare facilities will be in place before people start work on site.

Working at a height

The Working at Height Regulations 2005 are designed to improve safety procedures and practices involving work at any height in any location.

One of the key changes regarding working at height introduced by the Regulations is the removal of the '2.0 m rule'. Under previous legislation, protection from falls from a height were not really specified when the fall distance would be less than 2.0 m. But the 2005 Regulations have removed the 2.0 m rule and any work which involves the potential of falling any distance will need to be reviewed to ensure compliance with the law.

All work at height must be avoided if possible. Where it cannot be eliminated, it must be controlled and managed so that it is done safely, by competent people.

Facilities managers will be responsible for ensuring that their own employees work at height safely and they will have a role to play to ensure that contractors are working safely at all heights.

The 2005 Regulations do not ban the use of ladders and stepladders but they do expect employers and others to consider whether a ladder or stepladder is the safest piece of equipment to use. Consideration must always be given to working on the ground or from mobile elevating work platforms, scaffold towers or similar equipment.

The key points of the Regulations are summarised as follows:

- Work at height is defined as work in any place, including a place at or below ground and obtaining access to or egress from such a place while at work, except by a staircase in a permanent workplace.
- Every employer should ensure that work at height is properly planned, supervised and carried out in a safe manner. The planning should also make reference to emergencies and rescue. Work at height should not be carried out if weather conditions jeopardise safety.
- Work at height should be avoided whenever possible, particularly when it is reasonably practicable to carry out the work safely by other means.
- If work at height is taking place, measures must be taken to prevent a person falling a distance which is likely to cause injury. Protective measures can be utilised to minimise the distance and consequences of the fall.
- Work equipment must be appropriate for the nature of work and loading, allow passage without risk and be the most suitable for the work required. Work equipment must be assembled and used by competent persons who have received appropriate training.
- The Regulations provide specific information on fragile surfaces, ladders, scaffold, working platforms, collective safeguards for arresting falls, personal fall protection system, work positioning systems, rope access and positioning techniques, fall arrest systems and work restraint systems.
- Inspections of equipment should be completed before use, after an event that may affect the equipment (e.g. bad weather) and if scaffold every seven days. The information to be recorded is also detailed within the Regulations.
- Systems should be established to reduce the risk of people on the ground being injured by falling objects and the risk of items or people falling from working platforms such as toe boards and guard rails should be provided.
- Risk assessments should be completed on the task to ensure that the most appropriate equipment and system is being used and appropriate safety measures are in place.

Facilities managers will need to review the working procedures of:

- in-house employees
- contractors: cleaners
- contractors: roof repairs
- contractors: window cleaning
- contractors: maintenance
- contractors: mechanical and electrical works
- contractors: grounds maintenance

Confined spaces

Regulation 1(2) of the Confined Spaces Regulations 1997 defines a confined space as 'any place, including any chamber, tank, vat, silo, pit, trench, pipe, sewer, flue, well or other similar space in which, by virtue of its enclosed nature, there arises a reasonably foreseeable specified risk'. Some areas may, for example, become confined during construction or modification work. Risks can arise from lack of oxygen, chemical residues, dangerous gases, leaks, heat, fire or use of machinery (which may create dust or cause an electric shock).

Under s. 2 of the HSW Act, employers also owe their employees a duty to provide and maintain safe means of access to and egress from places of work. This duty extends to the workforce of other contractors. Regulation 4 of the Confined Spaces Regulations prohibits a person from entering a confined space to carry out work for any purpose where it is reasonably practicable to carry out the work by other means.

In practice, this means that employers should identify another way to do the work, for example, by using a remote camera for inspection, or tools that can be used from outside. If entry is necessary, risk control measures include:

- appointing a supervisor
- ensuring that workers have suitable training and experience, and are medically fit
- isolating hazardous machinery or pipework
- cleaning chemical residues before entering
- providing ventilation
- testing the air
- providing non-sparking or low-voltage tools and protected lighting as appropriate
- providing breathing apparatus if necessary
- preparing emergency arrangements, including how to raise the alarm
- providing lifelines
- having a safe communication system
- having a 'permit to work' system.

Control of Vibration at Work Regulations 2005

These Regulations have been introduced to consolidate existing safety legislation dealing with the ill-health effects of exposure to excessive vibration. Vibration referred to in the regulations is mechanical vibration arising out of or in connection with work. The key points of the Regulations are outlined below.

Where there is a risk of exposure the employer is required to assess the daily exposure to vibration of its employees. The assessment should include control measures required to meet the requirements of the Regulations. When completing the assessment the employer should take into account the magnitude, type and duration of exposure including exposure to intermittent vibration or repeated shocks, effects of vibration on employees, workplace and work equipment, information provided by the manufacturers of the equipment, specific working conditions, alternative equipment and health surveillance information.

The Regulations define daily exposure as the quantity of mechanical vibration a worker is exposed to during a working day, normalised to an eight-hour reference period, which takes into account the magnitude and duration of the vibration.

There are two levels of exposure that require identification and action:

- **exposure action value**: when the level of daily exposure for any worker, if reached or exceeded, requires specified action to be taken to reduce the risk. For hand–arm vibration this is 5 m/s^2 A(8). For whole body vibration it is 1.15 m/s^2 A(8).
- **exposure limit value**: the level of daily exposure for any worker must not be exceeded, except where other levels are defined. For hand–arm vibration this is 2.5 m/s^2 A(8). For whole body vibration it is 0.5 m/s^2 A(8).

Within the regulations there are details on how to calculate the exposure levels. The employer is required to eliminate the vibration at source or reduce it to as low a level as reasonably practicable. The employer is required to ensure that employees are not exposed to vibration above the exposure limit value.

Where the vibration level is usually below the exposure action value but varies markedly, it may occasionally exceed the exposure limit value provided that any exposure averaged over one week is less than the exposure limit value, there is evidence that the risk from the actual pattern of exposure is less than the corresponding risk from constant exposure at the exposure limit value, the risk is reduced and the employees affected have increased health surveillance.

Health surveillance should take place on employees exposed to vibration.

Appropriate information and training should be provided to employees at risk from exposure to vibration.

FIRE SAFETY

The Regulatory Reform (Fire Safety) Order 2005 (RR FSO) came into force in October 2006. This order consolidates a wide range of fire safety legislation into one enforceable order and enables fire authorities to take a consistent approach to fire safety requirements. The RR FSO develops further the requirement for businesses and others to manage fire safety by way of fire risk assessments. The key requirements of the RR FSO are:

- The Fire Safety Order is based on a new general duty of fire safety care, with specific requirements that will need to be met to comply with that duty.
- Fire safety will be based on a risk-based approach. Risk assessment will apply to all workplaces.
- With the abolition of fire certificates, the risk-based approach will apply to non-employees.
- A 'responsible person' will need to be designated to comply with fire safety legislation.
- Empty buildings come under the order and the owner is responsible for fire safety; fire safety is not just required when you have employees.

- People employed to undertake duties that have a bearing on the safety of a building must be competent to do the job. The responsible person must ensure that people, including contractors, are competent.
- Legal action would be taken against a contractor if they have been negligent in their duties or if they have failed to complete their duties or misrepresented themselves in any way. Action could also be taken against the responsible person for failing to appoint a competent contractor.
- Fire safety duties will cover all workplaces and places to which the public has access. Domestic premises are excluded.
- Voluntary workers operating in premises are covered by the order. They will have to be protected for fire safety.
- The responsible person will have a general duty to ensure fire safety of others.
- Preventive and protective fire safety measures must be implemented by the responsible person.
- Fire safety arrangements must be made by the responsible person.
- Dangerous substances will have special arrangements and requirements.
- Provision must be made for firefighting and fire detection.
- Emergency exit routes and exits must be designated and maintained.
- Premises and any firefighting equipment must be maintained by the responsible person.
- The responsible person must appoint a competent person to assist him in his duties of fire safety.
- Emergency procedures must be put in place for serious and imminent danger, and for danger areas.
- Responsible persons shall ensure that adequate information is available to those who will need it.
- Where there are two or more employers within a building there shall be cooperation and coordination to comply with the fire safety regime.
- Information must be given to visiting employees, and those working away or in host employers' premises.
- Employees must be given general fire safety training and must be given specific training on risks, changes in responsibilities, and so on.
- Employees will have general duties to take reasonable care of themselves and others for fire safety.

Fire risk assessment

It is important to carry out a fire risk assessment appropriate to the particular workplace. It is also good practice to involve staff in the process, as they may have identified a potential fire risk of which people higher up the organisation may not be aware.

The two most important questions to ask are:

- How likely is it for a fire to start in my workplace?
- How easy is it for employees, and other people who may be affected, to escape to a place of safety in the event of a fire?

In larger workplaces, it is good policy to carry out a separate inspection for each significantly different section, area or department. The whole of the workplace should be taken into account, including any outdoor areas and any rooms or parts of buildings that are not currently in use.

Even if the workplace has been subject to previous approvals by the various enforcing authorities for other safety, licensing or building legislation, you are still required to carry out an assessment of your fire precautions under the RR FSO. However, if there has been no significant change in the workplace, for example, in the number of employees or the activities which they undertake, it is unlikely that any significant additional fire precautions will have to be provided.

If you do propose to make changes to your fire precautions as a result of carrying out a fire risk assessment, these must not conflict with the controls imposed by other legislation. If in any doubt, you are advised to consult a fire safety officer from your local fire service.

If other employers share your premises, your organisation has a responsibility to ensure that they are made aware of any significant risks and any action you have taken to reduce that risk. In addition, you should take all reasonable steps to coordinate your fire safety measures with those of any other employers who may share your workplace.

The 'five steps' approach

Identify fire hazards
Potential fire hazards in the workplace will include potential sources of ignition, sources of fuel and any hazards associated with the processes carried out in the workplace.

Identify the location of people at significant risk in case of fire
This step needs to take into account not only employees, but other people who may be in the premises, such as customers, members of the public, visitors and contractors. The special needs of any disabled staff and visitors must also be considered. There may be parts of the premises where people are more at risk than others.

Evaluate the risks
This step involves deciding whether existing fire precautions are adequate, or whether improvements are required to remove the hazard or to control the risk. It is necessary to look at any existing fire safety measures provided in terms of:

- the control of ignition and fuel sources
- fire detection and fire warning systems
- means of escape
- means of fighting fire
- maintenance and testing of fire precautions
- fire safety training for employees.

The nature of the risk evaluation will depend very much on the nature of the workplace and the work activities carried out.

Record findings and action taken

The RR FSO requires organisations that employ five or more people to record the significant findings of the assessment and any group of employees identified as being especially at risk. There is a legal requirement to provide employees with 'comprehensive and relevant information'. This means telling employees or their representatives about the risk assessment findings, and perhaps making the formal risk assessment report available to them on request.

Keep assessment under review

It is good practice to carry out an annual review of the workplace to ensure that no new risks have developed as a result of, for example, changes to work processes, machinery, substances or the number of people likely to be present in the workplace. There should also be a reassessment of the workplace if you have carried out alterations or extensions, as they may have affected the fire precautions previously provided.

FIRE RISK ASSESSMENT CHECKLIST

Escape routes

- Are main and emergency stairways protected by self-closing fire doors?
- Is the emergency route clearly sign posted?
- Are there any 'dead end' conditions where escape is possible in one direction only?
- Are all escape routes clear of obstruction?
- Are all exit doors unobstructed externally?
- Are there enough exits?
- Are exit doors free to open at all times (not locked)?
- Are fire doors fitted with 'fire door – keep shut' signs, and is this instruction followed?

Fire defence equipment

- Is the fire alarm system satisfactory for the risk?
- Will it meet current legal requirements?
- Are the fire alarm, hydrants, fire extinguishers/hose reels, sprinklers and emergency lighting maintained by qualified people? Is maintenance recorded in a logbook?
- Is the fire alarm tested weekly?
- Does the fire alarm have automatic fire detectors in corridors, stairways and risk rooms?
- Are routine checks made to ensure that equipment has not been obscured, moved or damaged?

Work equipment and furnishings

- Are all items of portable electrical equipment inspected regularly and fitted with correctly rated fuses?

- Is the wiring of electrical installations inspected periodically by a competent electrical engineer?
- Is the use of extension leads and multipoint adapters kept to a minimum?
- Are flexible electrical leads run in safe places where they will not be easily damaged?
- Is upholstery in good condition?

Cleanliness and tidiness

- Are staff encouraged to tidy their personal workplaces?
- Are the premises kept clear of combustible waste?
- Are metal bins with closely fitting lids available for waste such as floor sweepings?
- Are separate, clearly labelled containers provided for waste and special hazards, such as flammable liquids, paint rags, oily rags?
- Are waste containers removed from the building at the end of each working day or more frequently if necessary?
- Is waste disposal put in a safe place which is not accessible to the public?
- Is the burning of waste on site prohibited?
- Are cupboards, lift shafts, spaces under benches, gratings, conveyor belts and similar places kept free from dust and the accumulation of rubbish?
- Are pipes, beams, trusses, ledges, ducting and electrical fittings regularly cleaned?
- Are areas in and around the building kept free from accumulated packaging materials and pallets?
- Are metal lockers provided for employees' clothing?

Storage

- Are fire doors, exits, fire equipment and fire notices kept unobstructed?
- Are storage areas accessible to firefighters?
- Are stack sizes kept as small as possible?
- Are there adequate gangways between stacks?
- Are stacks stable?
- Are stocks of material arranged so that sprinkler heads and fire detectors are not impeded and are the required clearances beneath this equipment maintained?
- Are excessive quantities of stock avoided?
- Is access to storage areas restricted to those who need to be there?
- Is stock kept well clear of light fixtures and hot service pipes?

Maintenance of buildings

- Is every point of entry to the site and building secure against intruders?
- After closedown of operations are all doors, windows and gates checked and secure?
- Is the building regularly inspected for damage to windows, roof and walls?
- Are the grounds surrounding the premises kept free of combustible vegetation by regular grass cutting and scrub clearance?
- Are all outside contractors supervised while on the premises and their work authorised by 'permit to work' schemes?

Heating and lighting
- Are there restrictions on using unauthorised heaters?
- Are combustible materials at a safe distance from appliances and flues?
- Is care taken that no materials are left on heaters?
- Are portable heaters securely guarded and placed where they cannot be knocked over or ignite combustibles?
- Are goods kept clear of lighting equipment?

Smoking
- Is smoking prohibited in all but designated external 'smoking' areas?
- Where smoking is permitted are there enough ashtrays or other disposal facilities?

Staff training
- Are new staff instructed in fire procedures and shown the fire escape routes on their first day at work?
- Are fire action notices posted throughout the workplace?
- Are there trained fire marshals?
- Is there a designated fire assembly point?
- Have staff had the opportunity to operate a fire extinguisher?
- Do staff know how to deal with the disabled, the public and visitors in the event of an evacuation?

Action plan
- Do you have a prioritised action plan for remedial measures following an assessment?

Means of escape

Once people are aware of a fire, they should be able to proceed safely along a recognisable escape route to a place of safety. In order to achieve this, it may be necessary to protect the route by using fire-resisting construction. The means of escape is likely to be satisfactory if your workplace is fairly modern and has had Building Regulation approval, or if it has been found satisfactory following a recent inspection by the fire authority (and in each case you have not carried out any significant material or structural alterations or made any change to the use of the workplace). However, you should still carry out a risk assessment to ensure that the means of escape remain adequate. If, as a result of the risk assessment, you propose making any changes to the means of escape, you should consult the fire authority (in Scotland you must seek the agreement of the Building Control authority) before making any changes.

When assessing the adequacy of the means of escape, you will need to take into account:

- the findings of the fire risk assessment
- the size of the workplace, its construction, layout, contents, the number and width of the available escape routes and the distances people have to travel to reach them

- the workplace activity, where people may be situated in the workplace and what they may be doing when a fire occurs
- the number of people who may be present, and their familiarity with the workplace
- individuals' ability to escape without assistance.

In some cases, it may be necessary to provide additional means of escape or to improve the fire protection of existing escape routes. If, having carried out your risk assessment, you think this might be the case in your workplace, consult the fire authority and, where necessary, your local Building Control officer before carrying out any alterations.

Fire safety guidance suggests that:

- Other than in small workplaces, or from some rooms of low or normal fire risk, there should normally be alternative means of escape from all parts of the workplace.
- Routes which provide means of escape in one direction only (from a dead end) should be avoided wherever possible as this could mean that people have to move towards a fire in order to escape.
- Each escape route should be independent of any other and arranged so that people can move away from a fire in order to escape.
- Escape routes should always lead to a place of safety; they should also be wide enough for the number of occupants and should not normally reduce in width.
- Routes and exits should be available for use and kept clear of obstruction at all times.

The time for people to reach a place of safety should include the time it takes them to react to a fire warning. This will depend on a number of factors, including:

- what they are likely to be doing when the alarm is raised
- what they may have to do before starting to escape (turn off machinery, help other people)
- their knowledge of the building and the training they have received about the routine to be followed in the event of fire.

Where necessary, these can be checked by carrying out a practice drill.

To ensure that the time available for escape is reasonable, the length of the escape route from any occupied part of the workplace to the exit should not exceed:

- where more than one route is provided, 25 m for a high fire-risk area and 60 m for a low fire-risk area
- where only a single escape route is provided, 12 m for a high fire-risk area and 45 m for a low fire-risk area.

The guidance goes on:

- A doorway of no less than 750 mm in width is suitable for up to 40 people per minute (where doors are likely to be used by wheelchair users the doorway should be at least 800 mm wide).

- A doorway of no less than 1 m in width is suitable for up to 80 people per minute.
- Where more than 80 people per minute are expected to use a door, the minimum doorway width should be increased by 75 mm for each additional group of 15 people.
- For the purposes of calculating whether the existing exit doorways are suitable for the numbers using them, you should assume that the largest exit door from any part of the workplace may be unavailable for use; the remaining doorways should be capable of providing a satisfactory means of escape for everyone present.

Emergency lighting

In general, in premises which have a daytime occupancy only, emergency lighting will only be necessary when there is insufficient natural light for people to make their way out of a building safely if the primary lighting should fail. The need for escape lighting is greater in buildings where visitors are present who are unfamiliar with the building.

Emergency escape and fire exit signs are necessary to indicate any emergency exit doors and routes that are not in common use.

Emergency exits

Emergency exits are vital to ensuring the safety of people using a building. Key rules are:

- Routes to emergency exits and the exits themselves must be kept clear at all times.
- Emergency doors must open in the direction of escape.
- Sliding or revolving doors must not be used for exits specifically intended as emergency exits.
- Emergency doors must not be so locked or fastened that they cannot be easily and immediately opened by any person who may need to use them in an emergency.

Fire notices

Notices giving clear and concise instructions on the action to be followed in case of fire should be prominently displayed throughout the workplace. It is also important to include a designated assembly point in the notices.

Fire equipment

Where necessary, appropriate firefighting equipment, fire detectors and alarms must be provided and maintained. In determining what is appropriate firefighting equipment, account must be taken of:

- the dimensions and use of the buildings at the workplace
- the equipment they contain

- the physical and chemical properties of the substances likely to be present
- the maximum number of people present at any one time.

Fire detection and fire warning systems, emergency lighting (including torches) and firefighting equipment should be checked weekly. There also needs be an annual full check and test carried out by a competent service engineer.

Staff responsibilities

An effective system for ensuring that the fire safety message is spread throughout an organisation is to cascade the responsibility through all staff levels by appointing floor, department or section fire marshals. The basis of this system is that the fire marshals are given a higher level of fire safety training than the average member of staff in order to look after a designated part of the premises.

The role of the fire marshals should be to check for hazards and potential fire risks within their area. They must check that fire safety equipment is working and in place, and be responsible for organising the evacuation of staff, the public and visitors, in the event of a fire alarm. This should include a final sweep of their designated area to ensure that everyone is out and to report accordingly to the person in charge of the assembly point.

It is not necessary for all staff to receive the same high level of training. This would be inappropriate and unmanageable. It is, however, necessary to ensure that all staff know:

- the location and use of escape routes
- the location of their assembly point
- how to use fire equipment provided
- how to summon the fire service.

Competency in fire safety

It is essential that anyone given any responsibilities to manage fire safety or to react in the event of a fire is competent to carry out the tasks allocated. Competency is generally deemed to imply:

- adequate knowledge
- adequate training
- appropriate experience
- adequate knowledge of the building, environment, processes, etc.

Anyone with responsibility for organising fire safety inductions or training for employees, contractors or others must ensure that they have a procedure in place to demonstrate that whatever information has been imparted on the course has been understood by those receiving it. An end of course test would be one appropriate way to record understanding of the information.

DISABILITY DISCRIMINATION

The Disability Discrimination Act 1995 (DDA)

Many businesses will be service providers under the Act and, as such, must ensure that they do not discriminate against customers who have a disability. The provision of business services, food and drink, entertainment and accommodation, to name a few, are 'services', and the DDA applies to services, goods and facilities.

The DDA applies to all parts of the premises accessed by members of the public, including:

- reception areas
- offices
- sanitary accommodation
- car parks
- gardens
- upper floors, etc.
- consulting rooms.

Any part of the premises to which customers have access in order to be provided with services is included in the Act and therefore all parts of the premises need to be accessible to people with disabilities.

The DDA also covers discrimination against employees with disabilities and states that they must not be prevented from obtaining employment because they have a disability. If staff facilities are on upper floor levels and there is no lift, provision may need to be made to give a mobility impaired employee access to their facilities. Alternatively, the employee may be provided with welfare facilities on the ground floor and this would be acceptable under DDA.

Private residential accommodation may need to be accessible to disabled people if it is provided as part of their employment package. If access was not available to the accommodation and the individual was unable to take up the job offer, there may be discrimination.

If private accommodation is suitable for the current tenants it will not need to be adapted to allow the tenants' friends and family to access the premises as this would not be 'providing a service' to disabled persons. However, a residential landlord may need to adapt any letting premises so that disabled tenants could choose to rent the premises.

The first consideration is to make sure that you are not discriminating against disabled people, or that your staff are not discriminating against them. You may believe that you have a non-discriminatory approach to all your customers, but your staff may be adopting a different, and possibly illegal, approach without your full knowledge.

A service provider may not discriminate against any person and may not offer a less favourable service to a disabled person unless it is unreasonable for the business to do otherwise.

Customers with any disability must be equally able to enjoy access to the premises, the environment, facilities, opportunities to buy food and drink and take part in entertainment as any able bodied or non-disabled customer.

The service provider needs to consider whether any customer with a disability will be prejudiced in any way from receiving a friendly, efficient service because of their disability.

If a customer, whether disabled or not, has equal choice as to where they can sit and the facilities they can use then they may not be being discriminated against. If a customer who is disabled can have access to an external smoking area, a food eating area or an entertainment area within certain parts of the premises then it may not be necessary to provide access to, say, mezzanine or first floor bar or restaurant areas.

Areas such as external gardens will need to be made accessible to disabled people unless it contravenes health and safety laws to do so.

It is advisable to facilitate access to the premises for all customers via the main access route, but it may not always be possible for customers with disabilities because the main entrance may have steps and no ramp.

The first consideration should be to provide access to mobility impaired people via the main entrance by providing a ramped access or small stairlift. If this is not practicable, it would be acceptable to provide the disabled customer with an alternative access via a side or rear entrance which was accessible to them. Any alternative entrance must not prejudice the disabled person from gaining access to the premises; they must not be allowed to stand outside the premises waiting to be heard so that doors can be opened. They should not be sent down side alleyways to, for example, rear delivery yards without being accompanied.

If alternative access has to be made available for disabled people there should be clear signs at the main entrance guiding people to the alternative entrance and advising them what actions to take to gain access; for example, press the bell or use the intercom.

Premises should generally be accessible to wheelchairs and door widths need to be of a certain size. If all doors are too narrow in the front entrance, an alternative entrance can be provided.

If physical access for wheelchair users is prevented because door widths are too narrow it would be reasonable, under the reasonable adjustment provisions, for the door width of at least one door to be widened.

However, current wheelchair widths vary and many disabled people use the electric buggy type wheelchairs, which may be wider than the traditional chair. It might not be reasonable to widen existing suitable doors for these wheelchairs, but it would be necessary to have a discussion with the customer to establish whether any alternative solutions are available; for example, they may have use of a more traditional wheelchair in which they could have access to the premises.

Many obstacles faced by disabled people are simply overcome by changing the way a service is given.

Self-service food, for instance, is difficult to access by people using crutches, sticks, the visually impaired and wheelchair users. But self-service food displays are ideal for many people and it would be unreasonable for the business to change their business completely to accommodate disabled people. What would be expected under DDA is for the service provider to offer to provide table service to those customers who need to use it because of a disability. A simple notice could be displayed at the counter advising that table service is available, or more productively, members of staff should be trained to approach, diplomatically, the disabled customer and ask them what, if any, assistance they require.

Customers with hearing impairments or speech difficulties could benefit quite simply from using a piece of paper and a pen to assist them in making their requests known.

Changes in levels are difficult for people with visual impairment and can cause trip and fall hazards. Visually impaired people benefit enormously from good, clear, visible signs which could warn of hazards, good lighting which does not produce glare and colour contrasting in materials. Steps of all types should ideally be colour contrasted so that the edges or nosings are clearly visible, as this helps people to judge distance and depth. Concrete outdoor steps or uneven levels could be painted with white or yellow paint. Internal steps and nosings could be highlighted with suitable paint or rubber finishes.

Customers must not be prevented from accessing special areas because of steps and where this might be the case, a temporary, lightweight ramp could be used to facilitate access.

CHECKLIST FOR GOOD PRACTICE

- Think and plan ahead to meet the needs of disabled customers.
- Do not make assumptions about disabled people based on speculation or stereotypes.
- Think about the wide range of disabilities when planning to make adjustments.
- Ask disabled people themselves what support and assistance they would need when using your services.
- Listen and respond to what people want, not what you think they need.
- Consider how you treat disabled people and let them know how to ask for assistance.
- Take all customer complaints seriously and make sure that any complaints procedure is easy to use.
- Ensure a positive policy to serving disabled people: be inclusive.
- Train all staff in effective communication methods and raise their awareness of the needs of disabled people.
- Introduce positive practices which indicate to all disabled customers that they are welcome and that there is no prejudice in serving them.
- Train staff in the law, disability etiquette, etc. Make sure that disabled people are not patronised: often the worst offence!
- Regularly review practices and procedures.

It is not a legal requirement to keep records, but it is good practice to record what you have done and why. In particular, it may be good practice to record what you have not been able to do and why; for example, any physical alterations which may be very expensive and therefore unreasonable given the turnover and profit of the business.

A simple disability discrimination policy should be written and displayed in staff areas or issued to staff so that you can demonstrate that you have given them information on the importance of not discriminating against customers with disabilities.

Any access audits or premises checks should be recorded as they will demonstrate that you have considered your legal obligations and the steps you were able to take to meet them.

If you provide brochures, for residential accommodation for instance, then include a disability discrimination policy in the document so that customers know how they will be accepted and treated.

Should a dispute arise regarding discrimination under DDA, it would help your defence if you were able to provide evidence that you had considered the requirements of the Act and your duties, and that you had endeavoured to comply as far as was reasonably practicable.

CATERING FACILITIES

Many facilities management contracts include responsibilities for overseeing the contract or in-house catering service. It is important to be aware of the key aspects of food safety and hygiene legislation as a failure to act on say, premises defects, could result in the prosecution of the facilities management contractor.

The Food Safety Act 1990

The Act sets out the standards for the preparation and sale of foods for human consumption. It requires food to be fit for consumption when sold, properly labelled and described, free from foreign objects, free from adulteration and generally non-injurious to health.

The Food Safety Act 1990 falls into the category of an 'enabling act' and paved the way for a raft of food hygiene and food standards regulations. It is enforced by environmental health departments in the main, with some duties falling to trading standards.

The Food Safety Act 1990 applies to 'food business proprietors' and it is important for contractual agreements to state clearly who is running the food business and on behalf of whom. In addition to the food business proprietor, 'any person' can be prosecuted for an offence involving the selling or rendering of food unfit for consumption.

The powers invested in EHOs under the Act can be draconian and enforcement officers have the authority to close food businesses if they believe there is an imminent risk to health.

An authorised officer of the food authority may at all reasonable times inspect food intended for human consumption. It is important to note that there is a presumption under the Act (s. 3) that all food usually intended for human consumption is actually for consumption if it is sold, stored, offered or exposed for sale within premises used for the preparation, storage or sale of foodstuffs. It is for the person accused of having food unfit for consumption to prove that the 'presumption' was invalid and that the food was not for sale.

Most routine food safety inspections are carried out during normal working hours but inspections can occur at other times; for example, during evening functions.

If an authorised officer is of the opinion that food is injurious to health and will cause food poisoning, he or she may do one of two things:

- seize the foods and have them dealt with by a Justice of the Peace
- give notice to the person that until permission is given, the food must not be used for human consumption nor removed from the premises.

If a 'delay' or 'do not move' notice is placed on the foods, a decision on their suitability for consumption must be made within 21 days.

Improvement notices

Should food premises be in such a condition that breaches of the Food Hygiene (England) Regulations 2006 are deemed to have taken place, the authorised officer can serve an improvement notice on the food business proprietor. An improvement notice is a legally binding notice served under s. 10 of the Food Safety Act 1990 and requires the food business proprietor to remedy breaches of law within a given timescale.

Improvement notices must:

- state the officer's grounds for believing that the proprietor is failing to comply with the regulations
- state the precise reasons why the regulations are being contravened
- state the measures necessary to rectify the contraventions and a time within which they must be completed.

The time limit for works to be carried out on an improvement notice must be at least 14 days as this allows the recipient time to lodge an appeal against the notice. Appeals are lodged in the Magistrates' Court. Failure to comply with an improvement notice is an offence which carries a maximum fine of £20,000 if proceedings are taken in the Magistrates' Court.

There is a defence of due diligence in respect of any prosecution under the Food Safety Act 1990. The food business proprietor or other person charged must show that they took all reasonable precautions to avoid committing the offence, or exercised all due diligence in avoiding an offence. It may be a defence for a person accused of food safety offences to cite that some other person was liable for the offence through their act or default.

The citing of 'another person' could be the facilities management company if the contract caterer believes that the offence was caused by the negligence or inaction of the facilities manager in respect of actioning repairs, maintenance programmes or pest proofing, for example.

Prohibition order

If during an inspection of the food premises, the EHO believes that there is a risk of injury to health, they may apply to the Magistrates' Court for a prohibition order which will prohibit the premises operating a food business. Such orders are made where premises are unsanitary, infested or filthy, or where practices are so bad that food is at risk of contamination. The court may grant the application and the premises will close until such time as the defects have been remedied and the order is lifted.

Emergency prohibition notice

In certain circumstances, food premises are so unsanitary, defective or filthy that there is 'imminent risk' of injury to health in the opinion of the officer. In these cases, the officer can serve an emergency prohibition notice, which requires the food business to close immediately. The emergency order must be ratified as soon as possible (and within three days) by the Magistrates' Court. Once ratified, the food business must remain closed until such time as the local authority is satisfied that the 'imminent risk' to health has been removed. While the premises are closed, a statutory notice advising of the closure must be displayed in a prominent position within the premises for public view.

Direct to court

Finally, in respect of enforcement, food authorities can dispense with all service of statutory notices and proceed direct to court with a prosecution for food hygiene offences. Often, when prosecutions are likely, written notice is given prior to the issuing of summonses.

As with all legal issues, if summonses are issued, either the facilities manager or contract caterer or both will need legal advice. Fines per offence can be up to £5000 in the Magistrates' Court.

The Food Premises Registration Regulations 1991 (amended)

These regulations require that all food businesses register with the local authority, supplying key information on contact details and types of food sold, etc. Any significant change in food business ownership for existing food businesses should also be notified. The registration process is to enable the local authority to compile a register of its food businesses so that, in emergencies, such as food scares and product recalls, the Food Safety Team can effectively utilise their resources to safeguard the public's health.

The Food Hygiene (England) Regulations 2006

Consolidating food hygiene regulations came into force on 1 January 2006. The Food Hygiene (England) Regulations 2006 (with similar Regulations for Wales and Northern Ireland) unify food safety standards across Europe and revoke earlier Regulations. The Regulations are enacted alongside Supporting Community Regulations; namely, 852/2004, 853/2004 and 854/2004.

The Regulations can be summarised as follows:

- Food businesses will need to implement a food safety management system based on the seven principles of hazard analysis critical control point (HACCP).
- Written records will need to be kept.
- Staff will need to be trained in the application of HACCP principles: especially those members of staff responsible for the development and implementation of the HACCP system.
- Food premises must be registered with the local authority.
- Temperature control measures must be followed.
- Premises must meet specific standards to ensure food safety.
- Enforcing authorities can serve new hygiene improvement and hygiene prohibition notices.
- Certain premises supplying food of animal origin will need to be approved.
- Enforcement officers will have the powers to sample foodstuffs.
- Businesses will be able to have a due diligence defence.
- Some offences may be attributable to other persons.
- Provisions are included to prevent officers being obstructed when carrying out their duties.

The Regulations cross-reference to the Community Regulations. These Regulations set out the specific requirements for premises design, sanitary conditions, food hygiene practices, etc.; they broadly contain similar requirements to the Food Safety (General Food Hygiene) Regulations 1995.

Facilities managers must ensure that any catering concessions, in-house catering teams, etc., comply with the new legislation and should implement monitoring reviews.

Premises

These must be maintained in good condition so as to ensure that surfaces and structures are easy to clean and free from defects. Surfaces in food rooms must be 'smooth, impervious and easily cleanable'; for example, tiles, continuous non-porous sheeting or proprietary materials, epoxy resin floors and vinyl sheeting. Wood is not recommended as it is porous, harbours bacteria and splinters easily, causing the potential for 'foreign object in food complaints'.

Pest-proofing

All food rooms shall be so maintained so as to prevent the ingress of pests, including rodents, cockroaches and flies. Proofing strips on the gaps to doors, shutters, etc., are essential. Mesh screens against windows and openable doors are required. Fill in voids and inaccessible harbourage places. Pests spread diseases on their bodies, in their faeces and urine, and can contaminate food and work surfaces.

Services

All food rooms in which open food (i.e. food not wrapped or placed in containers) is kept or prepared must be equipped with a supply of hot and cold water (or suitably temperature-controlled supply), adequate drainage, sinks and hand wash basins. Food rooms must also be adequately ventilated and have reasonable ambient temperature controls.

Refuse and waste material

Waste must not be allowed to accumulate in food rooms as it is a source of food poisoning contamination and attracts pests. Adequate refuse facilities must be provided and regular refuse collections are essential.

Drainage facilities (including grease traps)

These must be maintained in hygienic conditions and grease traps cleaned and emptied regularly as they can soon become unsanitary.

Equipment

Equipment must be maintained in a safe and hygienic condition. If statutory inspections are necessary, it will usually be the facilities manager who has to arrange them. Do not forget pressure boilers, steam ovens and combination boilers, etc. All will need statutory inspections, as will any service lifts, hoists and exhaust ventilation systems.

Management systems

A food business proprietor must identify the hazards associated with the food production process and implement suitable controls to reduce the risks. This is commonly known as HACCP, and the facilities manager should be satisfied that the caterer has addressed these issues.

Hazard analysis critical control point

Hazard analysis is a system designed to ensure that food and drink are safe and hygienic. It is now a legal requirement that the proprietor of every food business has in place a documented hazard analysis system.

The Food Hygiene (England) Regulations 2006 require food businesses to implement a food safety management system within the HACCP principles.

Hazard analysis

Hazard analysis involves looking at the operation of your food business step by step, starting from the selection of ingredients and suppliers through to the service of food to the customer. The HACCP system is based on five key steps.

Step 1: Analysis of the potential food hazards in a food business operation

Determine the hazards that could cause harm to the consumer. There are three main types of hazard:

- bacteria or other microorganisms (e.g. viruses) that may cause food poisoning
- chemicals, for example, cleaning chemicals, pesticides
- foreign objects, for example, glass, metal, plastic, flies, hair.

Step 2: Identification of the points at which food safety hazards may occur

After determining the hazards, identify at which step in the preparation or display operation they may occur. The main steps usually are:

- purchase
- delivery/receipt
- storage
- defrosting
- storage after defrosting
- preparation
- cooking
- portioning
- cooling
- storage after cooling
- reheating
- hot-holding and service.

Step 3: Deciding which of the points identified are critical to ensuring food safety

Critical points are the stages at which the hazards must be controlled to ensure food safety. This is done by eliminating or reducing the hazard to a safe, acceptable level.

The stages or steps within the food business operation that need to be controlled are anywhere where the food may become contaminated or bacteria may grow and survive.

Step 4: Identification and implementation of effective control and monitoring procedures

You need to decide what suitable controls can be introduced to eliminate or reduce the hazard to a safe level; for example, if the potential hazard is growth of bacteria in food during chilled storage, suitable controls would be to ensure that the refrigeration equipment is operating at the correct temperature below 8 °C and to check use-by dates to ensure that food is not out of date.

Once the controls have been set, some will be specific and easy to measure, while others may be more difficult to measure and will rely on visual checks; then the procedure for monitoring the control points must be decided upon.

The monitoring of the critical control points is to ensure that the targets set are working. At this stage the frequency of checks should be set for each critical control point and the method for recording the findings should be decided upon.

As well as knowing the hazards involved and knowing and monitoring suitable safety controls, it is also essential to know what corrective action to take if monitoring gives results outside acceptable targets. For example, if chilled storage temperatures are too high, it is important to establish what action has been taken and document it, for example, engineer called to undertake repair or product discarded.

Step 5: Review of the system

The final step is to review the system to ensure that it continues to work and you are still providing safe and hygienic food to your customers. The Regulations require periodic reviews to be carried out to ensure that the assessments are still relevant and up to date. Therefore, it is not satisfactory to go through this process as a 'one off' and then forget all about it. For example, your HACCP may need to be reviewed:

- if the checks reveal that the system is not working
- if new menu items are introduced to the range
- if new equipment is introduced
- periodically to confirm that the system is still relevant and working as planned.

Any corrective actions, improvements and revisions to the HACCP plan should be documented.

The seven principles of HACCP

HACCP plans are generally based on seven principles, as follows:

1. Conduct a hazard analysis.
2. Identify critical control points.
3. Establish critical limits for each critical control point.
4. Establish monitoring procedures.
5. Establish corrective actions.
6. Establish record-keeping procedures.
7. Establish verification procedures.

Food safety: good practice standards

What are the main hazards?

- poor temperature controls
- inadequately reheated foods
- foods left out at ambient temperatures
- inadequate cooking: failing to reach 75 °C
- keeping food past its use-by date
- poor personal hygiene practices
- cross-contamination from raw foods to cooked foods
- poor standards of cleanliness
- pest infestations
- reheating foods more than once.

What can be done to eliminate the hazards?

Food poisoning is not inevitable and by following some simple food safety rules you can be confident that your food operation will be both a gastronomic delight and safe for customers.

In effect, you can think of the 'six Ps':

- preventing contamination
- preventing dirty premises
- proper temperature controls
- personal hygiene
- pest prevention
- proper practices and procedures in cooking.

Preventing contamination

- Keep raw and cooked foods separate.
- Store raw foods below cooked foods in the fridge.
- Disinfect or sanitise work surfaces.
- Use different coloured chopping boards for different food preparation.
- Keep foods covered.
- Look out for foreign objects that could fall into foods.

Preventing dirty premises

- Clean as you go.
- Move equipment and clean underneath it and behind it.
- Wipe down walls.
- Use a disinfectant, degreaser or detergent, or sanitiser.
- Clean down at the end of the day.
- Devise cleaning schedules and stick to them.

Proper temperature controls

- Keep foods cold: below 8 °C, or even below 5 °C.
- Make sure that fridges work at temperatures between 1 and 4 °C.
- Cook foods thoroughly: above 75 °C.
- Keep foods hot: above 63 °C.

Personal hygiene

- Proper hand washing is the key to good food hygiene: use hot water, soap, nailbrush and paper towels.
- Do not cough, sneeze or spit over food.
- Wear protective over clothing.
- Do not wear jewellery.

Pest prevention

- Keep flies out of the kitchen.
- Use a fly screen or an insectocutor.
- Put pest proofing strips on doors.
- Watch out for mice droppings and cockroaches.
- Keep food rooms clean.
- Move stock regularly to clean.
- Keep drains clean and disinfected.

Proper practices and procedures in cooking

- Implement HACCP, that is, identify hazards and control them.
- Cook food thoroughly and serve immediately.
- Do not keep food at ambient temperatures for more than four hours.
- Sanitise work surfaces regularly.
- Cook foods when needed where possible: do not cook too far in advance.
- Do not reheat foods more than once.
- Defrost foods thoroughly.
- Cool foods rapidly, within 90 minutes, and put in the fridge.
- Keep temperature records.
- Do not use food past its use-by date.
- Throw away food when out of date or label 'not to be used'.
- Keep records of what you do for 'due diligence'.

The Food Labelling Regulations 1996

The most important legal requirement in the Regulations relates to the sale of perishable foods and the need for them to display use-by dates or best before dates.

It is an offence to sell food past its use-by date. Foods that are susceptible to the growth of pathogenic bacteria must display a use-by date determined by the manufacturer. Other perishable foods must display a best before date. It is not an offence to sell food past a best before date unless it is unfit for consumption, or not of the 'nature, quality or substance' demanded by the customer.

All foods must display a label with the name of the food, a list of ingredients and details of any special storage conditions. The name of the manufacturer, packer or seller must also be displayed on the label.

The Weights and Measures Act 1985

Foods must generally be sold in specified quantities or weights. Metrification applies to the UK, and loose fruit and vegetables, meats, fish and virtually all products must be sold in grams or kilograms, litres and other metric quantities. Beer is currently still exempt and may be sold in half-pint measures.

Health and safety in catering units

Health and safety in catering units is also a key priority and a significant percentage of worker accidents and injuries occur in kitchens and food service areas. Slips, trips and falls are common in the catering industry and are often caused by poor maintenance, inadequate cleaning or a failure to clear up spillages.

Kitchen temperatures are often a cause of concern and high temperatures increase food contamination risks (as food may be exposed to temperatures which enable bacteria to multiply) and employee accident rates. Facilities managers must ensure that 'suitable and sufficient' temperatures are maintained in kitchens.

The Provision and Use of Work Equipment Regulations 1998 require all work equipment to be maintained in a safe condition. Regular maintenance checks will be needed of all catering equipment as well as associated records. Facilities managers should ensure that equipment is inspected by competent persons and must either arrange this or ensure that the caterer does so. Frequency of inspection is recommended as shown in Table 1.4.

Finally, facilities managers may be responsible for safety signs in catering areas. These should follow the principles of the Health and Safety (Safety Signs and Signals) Regulations 1996 and depict residual hazards, prohibition instructions, mandatory procedures, safe areas, etc. Ensure that pipes carrying flammable substances are appropriately labelled; for example, 'gas pipes'.

The catering facilities provided by any employer, whether directly or in-house or outsourced to a contract catering company, must be regularly inspected and monitored if good practices are to be maintained and safe food is to be prepared and sold. A major food poisoning outbreak for any company could seriously jeopardise their efficiency and the resulting publicity will be an unacceptable business risk. Where facilities management contractors have catering facilities within their premises, they must ensure a high-profile auditing and monitoring regime.

Licensing

Many in-house catering facilities include either a social club or function facility which will be licensed to sell intoxicating liquor and/or provide public entertainment or dancing.

Premises need to be licensed to sell alcohol and it may fall to the facilities manager to make the application to the licensing authority. It is an offence to sell alcoholic beverages without a licence unless deemed to be a private members' club, where there are some exemptions.

Facilities managers may need to be familiar with the Licensing Act 2003. Premises will need to be licensed and so will individuals. Opening hours will be as agreed with the licensing authority. All those applying for premises licenses will need to submit an operating schedule plan setting out how they propose to control key aspects of running the venue. Appeals against refusals will be held in the Magistrates' Court.

Table 1.4 Recommended Frequency of Inspections

As appliances	Every 12 months
Pressure cookers, pressure fryers, steam pans, steam pipes, water boilers, etc.	According to a written scheme of examination set by a competent person
Electrical appliances in kitchens	Combined inspection and test approximately every 6–12 months
Lifting equipment, hoists, etc.	Every 12 months if not carrying people
Fire alarm and firefighting equipment	Minimum of annual maintenance

Facilities managers will need to ensure that they have both premises and individual licenses if they are intending to sell alcoholic beverages for consumption on or off the premises.

Any temporary events or corporate hospitality events will come under the Licensing Act 2003 and facilities managers will need to consult a licensing lawyer to ensure that they have the correct permissions in place to host the event.

ENVIRONMENTAL PROTECTION AND LIABILITY

There is, of course, a legal regime for prosecuting firms that accidentally or intentionally pollute the air, water or land during their industrial processes. But for facilities managers, the important new development in environmental law is the scrutiny under which every aspect of every company's environmental performance is starting to come.

The regime is subtler: if you forget to turn the lights off at the end of the day, you will not end up in a courtroom, but you will feel the financial effects when your tax bill arrives. So whereas energy efficiency has, until recently, been a matter of choice, it is now starting to be regulated. There is also pressure to report on environmental performance, so that energy wasters will find it harder to conceal their poor record.

Climate Change Levy

In 1997, the UK government committed the country to reducing substantially greenhouse gas emissions such as carbon dioxide and methane by 2010.

Non-domestic energy use is one of the government's main targets, and businesses are feeling the effect as the first fuel bills arrive containing the Climate Change Levy (CCL). The levy works by charging businesses and public sector organisations an extra sum on their fuel bills for every kilowatt hour (kWh) of electricity, coal, natural gas and liquid petroleum gas they use. In other words, unlike value added tax (VAT), which is charged on the cash value of goods and services provided, the CCL is based on the energy value of the fuel supplied. This means that getting a cheaper supply will not reduce the amount of tax payable. Load management, that is, moving energy-thirsty activities to periods of lower cost electricity, will be similarly ineffective at reducing the amount to be paid.

Exemptions

The CCL is not to be applied across the board; for example, the following will be exempt:

- small users and charities who pay VAT on fuel at 5 per cent
- fuels used by the transport sector
- fuels used for energy generation
- oils already subject to excise duty
- electricity generated from new renewable sources such as solar and wind power

- fuels used jointly as a feedstock and an energy source within the same process, for example, coke in steel-making
- electricity used in electrolysis processes, for example, the chlor-alkali process, or primary aluminium smelting
- fuel used by good quality combined heat and power (CHP) schemes as certified by the Quality Assurance Programme (CHPQA), a voluntary programme which has developed a universal method to assess, monitor and certify the quality of CHP.

WASTE MANAGEMENT

The UK is subject to numerous European Directives on waste, and a key part of the government's strategy on waste reduction is to place recovery and recycling targets on business for a wide range of waste materials.

Producer Responsibility Obligations (Packaging Waste) Regulations 1997

Any business handling more than 50 tonnes of packaging and with financial turnover of more than £2 million is obligated if it is involved in manufacturing raw materials for packaging, converting materials into packaging, filling packaging, selling packaging to the final consumer; or importing packaging or packaging materials into the UK.

A facilities management company may be asked by a client to assume responsibility for compliance with the regulations.

Any business that is obligated by the regulations must:

- register with the Environment Agency (or other parallel enforcement agencies in the devolved administrations), pay a fee and provide data on packaging handled by the business in the previous year (costs and calculations are always based on the previous year)
- take reasonable steps to recover and recycle packaging waste
- certify that the necessary recovery and recycling has been carried out
- or join a registered compliance scheme in order to have their obligations discharged.

Evidence of recovery and recycling targets being met is demonstrated by the use of packaging waste recovery notes (PRNs) and packaging waste export recovery notes (PERNs). Compliance schemes buy PRNs and PERNs on behalf of all their members and it is their responsibility to show that they have enough PRN/PERNs to cover the recycling and recovery targets. Current targets set by the government are:

- 59 per cent recovery
- 19 per cent recycling.

These targets are due to rise in line with European Directive targets. Each PRN required carries a cost, currently between £22 and £30 per tonne.

Minimising waste

The less packaging a business handles the less tonnage there will be to recycle and recover and the lower the cost of regulatory compliance will be. Even for those businesses not affected by the Packaging Regulations, waste minimisation is a sound business and environmental strategy. Whether the Packaging Regulations apply or not, there are costs associated with waste removal, landfill and special disposal requirements.

Good business practice on waste minimisation suggests an audit of all products, packaging and waste used and produced in the course of the business. Consider what is essential packaging to product safety and what can be eliminated as unnecessary. Review products purchased with suppliers so that they can understand that packaging is superfluous. Set up workshops with key suppliers on waste minimisation. Set targets so that achievements can be quantified.

Facilities managers who have responsibilities for clients' premises which include catering and hospitality facilities may be required to separate food waste from other waste products. Under the EU Landfill Directive, the UK is committed to reducing the amount of biodegradable waste in landfill. Greater emphasis is being placed on composting. Separate containers for refuse may be required at premises and caterers should be encouraged to minimise food and other organic waste.

Duty of care in respect of waste

The duty of care is a law which states that all reasonable steps must be taken to keep waste safe, that it is handled by authorised persons and is disposed of legally. Failure to comply with the duty of care requirements is an offence carrying an unlimited fine, under the Environmental Protection Act 1990 (EPA).

The duty applies to anyone who produces, imports, keeps, stores, transports, treats or disposes of waste. Waste can be anything which the business produces.

Waste transfer notes are required to be completed for commercial waste. The facilities manager will probably be responsible for ensuring compliance.

Waste can only be removed by authorised contractors: this is to prevent, as far as possible, flytipping. Their registration details may be checked with the Environment Agency or from their registration certificate.

When waste is passed from one person to another, the person taking the waste must have a written description of it. The transfer note must be completed and signed by both parties involved in the transaction. One transfer note can cover the same type of waste for up to 12 months; for example, weekly trade waste collections.

The transfer note must include:

- what the waste is and how much there is
- what sort of containers it is in
- the time and date the waste was transferred
- where the transfer took place
- the names and addresses of both persons involved in the transfer
- whether the person transferring the waste is an importer or producer of the waste

- details of the authorised persons category for handling waste, for example, for transport or disposal
- the certificate number of the registered waste carrier and the local Environment Agency office which issued it
- the licence number of the waste management licence and the Environment Agency details
- any reasons for any exemption
- the name and address of any broker if appropriate.

Copies of transfer notes must be kept by both parties for two years. They may need to produce these as evidence if, for instance, they are involved in a prosecution for illegally disposing of waste.

Waste Electrical and Electronic Equipment (WEEE) Regulations

The WEEE Regulations apply to electrical and electronic equipment (EEE) in the categories listed below with a voltage of up to 1000 V for alternating current or up to 1500 V for direct current.

You will need to comply with the WEEE Regulations if you generate, handle or dispose of waste that falls under one of 10 categories of WEEE:

- large household appliances
- small household appliances
- information technology (IT) and telecommunications equipment
- consumer equipment
- lighting equipment
- electrical and electronic tools
- toys, leisure and sports equipment
- medical devices
- monitoring and control equipment
- automatic dispensers.

Schedule 2 of the WEEE Regulations provides examples of products falling within these categories, which include equipment such as:

- large cooling appliances
- fridges and freezers
- washing machines
- dishwashers
- cookers
- electric heaters
- air conditioning appliances
- exhaust ventilation
- IT and communications equipment.

All of which could come into the possession of a facilities management company for disposal. Such electrical equipment can no longer be disposed of without proper procedures in place to ensure that it is dealt with under WEEE. In particular, waste carriers must be licensed with the Environment Agency.

The WEEE Regulations apply to importers, producers, retailers and users of EEE, and to businesses that treat or recover WEEE.

The Regulations aim to:

- reduce waste from electrical and electronic equipment
- encourage the separate collection of WEEE
- encourage treatment, reuse, recovery, recycling and sound environmental disposal of WEEE
- make producers of EEE responsible for the environmental impact of their products
- improve the environmental performance of all those involved during the lifecycle of EEE

The environmental regulator will enforce the producer responsibility aspects of the WEEE Regulations regarding collection, disposal and processing of WEEE.

Site Waste Management Plans Regulations 2008

All construction projects which started after April 2008 and which have a construction value of more than £300,000 must have a site waste management plan. This provides a structure for waste delivery and disposal at all stages of the construction project. Typically, the plan will identify:

- who will be responsible for resource management
- what types of waste will be generated
- how the waste will be managed: whether it will be reduced, reused or recycled
- which contractors will be used to ensure that the waste is correctly recycled or disposed of responsibly and legally
- how the quantity of waste generated from the project will be measured

This is no set format for a site waste management plan, although various templates are available via government departments and online advice providers (e.g. www. netregs.gov.uk).

The Regulations will be enforced by the Environment Agency and potentially, both the HSE and local authorities. Fixed penalty fines for not having a site waste management plan are fixed at £300, or court fines of up to £50,000 could be levied.

POLLUTION

Environmental protection addresses pollution to air, land and water, under the Environmental Protection Act 1990 and the Environment Act 1995, plus numerous regulations set down the controls expected to safeguard the environment.

Businesses can pollute the environment by:

- emissions to atmosphere from industrial processes
- effluent discharges to water courses
- contamination of land
- noise.

Environmental compliance is monitored by both the Environment Agency and local authorities and statutory inspections for compliance are undertaken by all enforcing authorities.

A business may be creating a 'statutory nuisance' under the EPA by emitting noise that is detrimental to the neighbourhood (e.g. noisy air conditioning plant keeping residents awake at night), by creating accumulations that are 'noxious', by creating smell nuisances or by emitting atmospheric pollutants.

The enforcing authorities have powers under EPA to:

- serve statutory notices requiring remedial works to abate the nuisance
- prohibit the continuation of the nuisance
- undertake remedial measures to abate the nuisance
- prosecute the offender.

Failure to comply with statutory notices can carry an unlimited fine and/or imprisonment. Daily fines can also be levied for each day that the nuisance continues.

Statutory nuisances

There is no simple definition of a statutory nuisance as a wide range of situations is described in various Acts of Parliament. In general, a statutory nuisance poses a threat to someone's health or well-being or interferes with the person's use and enjoyment of land, and it must be more that just nuisance or annoyance.

Statutory nuisances usually need to occur for some length of time, and to be regular and consistent and frequent. One-off instances are less likely to be referred to as statutory. Judgement as to whether a nuisance exists must have regard to what most people would regard as reasonable.

Statutory nuisances usually involve noise, smells, refuse accumulations, insect and pest infestations, artificial light, fumes, gases, animals, dust and effluvia.

EHOs are authorised to investigate statutory nuisances and they use their training and experience to determine whether they believe a statutory nuisance exists. It may be worth challenging their opinion: one officer can class something as a statutory nuisance and another one may not. Noise nuisance is probably the easiest nuisance for an EHO to class as a statutory nuisance because there are various guidelines and documents which set down acceptable noise limits, such as British Standards, WHO and Environment Agency.

Investigations for statutory nuisances usually follow a complaint from a nearby resident or local business. EHOs have a duty to investigate all complaints alleging statutory nuisance.

The EHOs will want to satisfy themselves that a nuisance exists, so they will often visit several times and take noise readings or samples of air/dust/effluvia, etc. Photographs form a key part of the evidence. Diaries of when residents are affected by the nuisance are often encouraged, especially in relation to noise. Such diary entries can be used as evidence, although usually EHOs will want to have experienced the nuisance for themselves.

Once a statutory nuisance has been identified the EHO can serve an abatement notice under the EPA or other legislation. The abatement notice requires steps to be taken to abate the nuisance within a certain timescale: always no less than 21 days and sometimes over several weeks. Time limits have to be reasonable to allow the proprietor to evaluate remedial measures, etc.

If the nuisance is not abated within the time stipulated and no extension of time has been requested, the person on whom the notice is served commits an offence and can be prosecuted. Fines can be up to £20,000 and the court can order the council to take the necessary remedial steps and recharge the business or person.

Noise

EHOs have a wide range of powers to deal with noise nuisance from business premises (and also domestic dwellings) and can serve noise abatement notices or seize any equipment causing the noise, such as disco equipment.

Under the Antisocial Behaviour Act 2003, EHOs (or other authorised officers) can close noisy premises where they cause a public noise nuisance. A closure order is served for a maximum period of 24 hours and the premises (usually a licensed premises or a venue holding a temporary event) will have to stay closed for the period specified. Failure to comply with the closure order can lead to a fine of up to £20,000.

Local authorities also have additional powers under the Clean Neighbourhoods and Environment Act 2005 to serve fixed penalty notices on licensees for causing a noise nuisance at night. Local authorities must give warnings to premises owners or operators about the noise nuisance before serving a fixed penalty notice. Fines for fixed penalties are £500 per notice.

Noise from outside drinking areas used by people who want to smoke is becoming an increasing problem and some premises operators could find themselves served with a noise abatement notice to stop customers from causing a noise nuisance from talking or otherwise using the external area. Noise does not always need to be from amplified sound.

Air conditioning plant, chiller units, ventilation fans and other equipment can cause noise nuisance and the owner of the premises or business operator may have to take steps to enclose plant and equipment acoustically or change fan speeds or types.

Noise nuisance is a complicated subject and there are defences available to businesses, so it may be worthwhile consulting your lawyers as soon as you have been advised that you are causing a noise nuisance.

Smells, dust and fumes causing a nuisance

Cooking odours emanating from business premises are a common cause of complaint to councils, and often statutory nuisance is easily proved. Increasingly, some councils are receiving complaints about the smell from tobacco smoke and even though the premises owner has no choice about asking customers to smoke outdoors, they could still be charged with a statutory smell nuisance.

Complaints are investigated in a similar way to noise, that is, the EHO has to establish a statutory nuisance which affects the health or well-being of an individual or affects the 'quiet enjoyment' of one's home.

Nuisance abatement notices are served and a failure to abate the nuisance could cost up to £20,000 in fines. A number of defence pleas could be made but, again, it is best to seek legal advice.

Artificial light

Artificial light is a newcomer to the list of statutory nuisances and, broadly, any artificial light from commercial premises which is so intrusive as to interfere with someone's use of their premises could be a statutory nuisance and the council could take appropriate statutory nuisance action.

Abating statutory nuisances

There is a wide range of actions that all businesses can take to reduce or eliminate the causes of statutory nuisances. Noise can be turned down, windows and doors kept closed, speakers moved, and plant and equipment properly maintained (poorly maintained equipment is a major cause of noise nuisance). Smells can be reduced by cleaning and changing filters and fans or by introducing more pleasant counter-odours.

Remedies are many, but the experienced licensee would be wise to seek proper legal and professional advice before embarking on any actions. Very often, lots of money is spent on the problem but the nuisance is not abated. So, a little bit of planning and preparation, and proper consultation, will pay great dividends.

Conditions may be attached to any premises licence about noise and smell nuisances. If you breach a condition you could lose your licence and see your premises closed.

CORPORATE SOCIAL RESPONSIBILITY

Corporate social responsibility (CSR) has come to the fore over recent years and is driven, to a large extent, by the many independent reports on corporate governance. 'Stakeholder value' has become a common maxim over the years and CSR requires companies to set out their social policies, corporate governance procedures and stakeholder value initiatives. In particular, Turnbull has set the agenda on business risk management and companies know-how to address how they effectively manage all risks to the business and how they intend to maximise and protect shareholder/stakeholder value.

SUPPLY CHAIN MANAGEMENT

The increasing acceptability of shared responsibility for health and safety, fire safety and environmental protection has led to a new partnership approach in supply chain management.

When contractors, suppliers, consultants and others are appointed it is important to understand the approach they have to their business and compliance issues, and to determine whether they fit in with your own culture and business ethics. CSR requires a more transparent relationship with suppliers and the current expectation is that those companies who have better and more robust policies, etc., should be working with their suppliers to help move them towards continuous improvements.

Facilities managers should play a leading role in developing supply chain management policies and practices.

2 Complying with the Law on Staff, Casual and Contract Workers

Pat Perry, Jackie Le Poidevin and Louis Wustemann

FLEXIBLE WORKING FOR EMPLOYEES

Reforms

These reforms are an attempt to enshrine the right to flexible working in law and not just for traditional 'employees'. Section 230 of the Employment Rights Act 1996 defines a worker as an individual who has entered into a contract of employment or any other contract, whether express or implied, which undertakes to 'perform personally any work or services for another party to the contract whose status is not by virtue of the contract that of a client or customer'. This includes individuals such as information technology (IT) contractors who are not genuinely self-employed, but who work under a personal service contract. It will also cover some casual or temporary agency workers, and other 'atypical' workers. Facilities managers will need to clarify the employment status of any contract or agency workers to whom they have outsourced work in the light of the ongoing changes in legislation.

Flexible working right of request

The Employment Act 2002 introduced a right for employees with caring responsibilities for young children to ask to change their working pattern or place of work. The legislation does not give employees the right to work flexibly, but it places the obligation on employers to consider requests seriously.

Parents of children under the age of six (or under the age of 18 where the child is registered disabled) are able to apply to change their work pattern to care for the child. A parent is defined as anyone with responsibility for raising the child, including guardians, adoptive and foster parents, or anyone who is the partner of one of the above and lives with the child.

In April 2007 the right to request flexible working was extended to include those caring for relatives.

Eligible employees can ask to change:

- their hours of work (e.g. part-time working, jobsharing, term-time working)
- their working times (e.g. variable hours, staggered hours)
- their workplace (e.g. homeworking or, if available, an office nearer home).

114

The employee's application has to be made in writing (email and fax are acceptable) and must state the working pattern applied for and when the applicant wants to make the change, the effect, if any, that the applicant believes the new working pattern would have on the employer and how this effect could be offset.

Within 28 days of receiving the application (or of its being sent electronically), the employer must either accept the request and write to the employee notifying them of the variation to their contract and when the change will be effective, or hold a meeting with the employee to discuss the request at a convenient place and time for both the employer and the employee. At the meeting (and at any appeal meeting afterwards) the employee has the right to be accompanied by a colleague or trade union representative. Breaching this provision results in a compensation payment of up to two weeks' pay (limited to £280 per week at present).

Following the meeting, if the employer wishes to refuse the application, they must inform the employee within 14 days in writing, explaining the reasons for the refusal and the procedure for the employee to appeal against the decision.

Reasons for refusal

Employers can refuse flexible working requests for any of the following reasons:

- the burden of additional costs
- detrimental effect on ability to meet customer demand
- inability to reorganise work among other staff
- inability to recruit extra staff
- detrimental effect on quality
- detrimental effect on performance
- insufficient work during the periods the applicant has requested to work
- planned changes to the business.

The explanation in writing of these reasons to the employee after the first meeting must be 'sufficient'.

Right of appeal

If the employee decides not to accept a refusal of their request, they have the right to appeal the decision within 14 days of receiving the refusal.

The employee must notify the employer in writing giving details of the reasons for the appeal. Within 14 days of receiving notice of the appeal, the employer must either accept the original request and inform the employee in writing, or hold another meeting to discuss the appeal with the employee.

Once a request has been refused or the appeal process is exhausted, an employee must wait for 12 months from the date of the first request before making another application.

If an employee believes the employer did not follow the regulations in dealing with their request or rejected the application on the basis of false information, they can bring a complaint to an employment tribunal, within three months of the rejection of the request.

The legislation also protects any employee dismissed because they made an application, appealed against a refusal or brought a complaint because the employer did not follow the correct procedure. Employees in this position may bring complaints for unfair dismissal.

The 48-hour week

The European Working Time Directive was officially introduced as a piece of health and safety legislation, to reduce the risk of ill-health caused by overly long working hours. Regulatory working hours was also thought to be beneficial to all workers as it would introduce an element of work–life balance.

Organisations that allow employees to work flexible hours or from home will need to keep track of their hours to ensure that they comply with the working time legislation. Alternatively, they can ask employees to opt out of the minimum requirements, but this could be bad for homeworkers' stress levels and ultimate productivity, since it can be difficult for them to know when to draw a line under their work each day.

The Working Time Regulations 1998 (as amended) provide that:

- Workers' maximum working week must average 48 hours, normally calculated over a 'reference period' of 17 weeks.
- Workers can choose to sign a written agreement that they will work more than the 48-hour limit: this is an 'opt-out'. Opt-outs must be agreed with individual workers; a reference in a contract to a collective agreement that allows working beyond 48 hours is not sufficient.
- Workers who opt out can cancel the agreement provided they give the employer at least seven days' notice, or longer (up to three months) if this has been agreed.
- The regulations define working time as time when an individual is 'working, at their employer's disposal and carrying out his activity or duties': this includes travel which is part of the job, working lunches, job-related training and 'on-call' time when a worker is required to be at their place of work.
- Working time does not include commuting time, lunch breaks, or non-job-related evening classes or day-release courses.

Rest periods

Workers required to work more than six hours continuously are entitled to a 20-minute rest break. The employer can decide the exact timing, but the break should be during the six-hour period, not at the beginning or end of it. Employers must make sure that workers can take their rest breaks.

Young workers are entitled to a rest break of 30 minutes if required to work for any continuous period of more than four and a half hours. This entitlement can be changed in exceptional circumstances, in which case the young worker should receive 'compensatory rest' within three weeks (see later in this chapter, Exceptions to the regulations: Agreements).

Annual leave

The Working Time Regulations entitle workers to 4.8 weeks' paid annual leave each year, that is, 24 days if you work a five-day week. There is no qualifying period for this entitlement. From 1 April 2009, the leave entitlement will rise from 4.8 weeks to 5.6 weeks or 28 days if you work a five-day week.

Workers accrue annual leave during their first year of employment at the rate of one-twelfth of their entitlement per month worked, rounded up to the nearest half day. So if a worker requests leave after three months' service, their entitlement will be three-twelfths (one-quarter) of 24 days; that is, six days.

The current 4.8-week entitlement includes bank holidays; there is no statutory right for bank holidays to be granted as leave. Employers can specify the times that workers take their leave, for example, over Christmas. Workers must give the employer notice that they want to take leave. When their employment terminates, they have the right to be paid for any leave not taken.

Nightworkers

Facilities managers who employ in-house or external security personnel may need to be aware of the provisions on nightworking (but see below, Exceptions to the regulations; Special circumstances). The regulations define a nightworker as someone who works at least three hours a night 'as a normal course'. Such workers should not work more than eight hours daily on average. There has been a court ruling that an individual who worked nights for one-third of their working time was a nightworker.

Night-time is normally between 11 pm and 6 am, although workers and employers may agree to vary this. If they do, night must be at least seven hours long and include midnight to 5 am.

Where a nightworker's work involves special hazards or heavy physical or mental strain, there is an absolute limit of eight hours on their working time each day; this is not an average.

Adult workers assigned or transferred to nightwork qualify for free health assessments. Nightworkers under the age of 18 are entitled to free 'health and capacities' assessments. Any worker experiencing health problems associated with nightwork must, if a doctor advises, be transferred to suitable daywork where possible.

Exceptions to the regulations

According to guidance on the Regulations issued by the Department of Business Enterprise and Regulatory Reform, there are four types of exceptions where parts of the Working Time Regulations may not apply: agreements, special circumstances, and unmeasured and partly unmeasured working time.

Agreements

Workers can agree with their employer to vary nightwork limits and the right to rest periods and rest breaks, in return for 'compensatory rest'. They may also agree to extend the reference period for calculating hours worked up to 52 weeks.

The compensatory rest provision allows workers to take their total weekly rest of 90 hours in a different pattern from that set out in the regulations. The principle is that everyone gets their entitlement in the end, although some rest may come slightly later than normal.

These agreements can be made by collective agreement (between the employer and a trade union) or by a workforce agreement, usually made with elected representatives of the workforce. If a worker has any part of their conditions determined by a collective agreement they cannot be subject to a workforce agreement. A workforce agreement can apply to the whole workforce or to a group of workers.

Special circumstances

The nightwork limits and the right to rest periods and rest breaks do not apply where:

- workers work far away from where they live and want to work longer hours over fewer days to complete a task more quickly
- workers constantly have to work in different places, making it difficult to work to a set pattern
- the work involves security or surveillance to protect property or individuals
- the job requires round-the-clock staffing or there are busy seasonal peak periods
- an emergency occurs.

In these cases, the reference period for the weekly working time limit is extended from 17 to 26 weeks, and workers are entitled to compensatory rest.

Unmeasured working time

Apart from the entitlement to paid annual leave, the regulations do not apply if a worker can decide how long they work, for example, a senior manager or director. The regulations state that a worker falls into this category if 'the duration of his working time is not measured or predetermined, or can be determined by the worker himself'.

Partly unmeasured working time

This exception refers to workers who have an element of their working time predetermined, but otherwise decide how long they actually work. The guidance states: 'Additional hours which the worker chooses to do without being required to by his employer do not count as working time; therefore, this exception is restricted to those that have the capacity to choose how long they work. The key factor for this exception is worker choice without detriment'.

This exception does not apply to:

- working time which is hourly paid
- prescribed hours of work
- situations where the worker works under close supervision
- any time where a worker is expressly required to work, for example, to attend meetings
- any time which a worker is implicitly required to work, for example, because of possible detriment if the worker refuses.

So, if a facilities manager willingly chooses to work longer hours without recompense on their own initiative, they may. But if they feel pressured to work unpaid overtime because the company culture demands it, the company is in breach of the regulations.

Road transport workers

Separate regulations covering the working time of road transport workers came into effect in April 2005. The Road Transport (Working Time) Regulations 2005 typically apply to drivers of vehicles fitted with tachographs such as goods vehicles heavier than 3.5 tonnes and coaches. Drivers and crew workers may not work more than an average of 48 hours during the applicable reference period – in the absence of any local agreement a default arrangement specifies set 17 or 18-week reference periods each year – and not work more than 60 hours in any single week. Transport workers are not allowed to opt out of the average 48-hour week.

Enforcement

Enforcement of the Working Time Regulations is split between different authorities. The Health and Safety Executive (HSE) and local authority environmental health departments enforce the weekly limits on working hours. The employment tribunals enforce the entitlement to rest and leave.

PART-TIME WORKERS

The Part-time Workers (Prevention of Less Favourable Treatment) Regulations 2000 implement the European framework directive on part-time working. The regulations allow part-time workers to bring a compensation claim if they have been treated less favourably, on a pro rata basis, than a comparable full-time worker. This right applies both to the terms of the part-timer's contract and 'to any other detriment by any act, or deliberate failure to act' by the employer. The right applies to workers, not just employees.

Previously, part-timers, 80 per cent of whom in the UK are women, had to prove indirect discrimination under the Sex Discrimination Act 1975 (SDA) in order to win damages for unfair treatment.

Comparable full-timers

Part-timers can only prove less favourable treatment if they can compare themselves with another actual worker (a 'comparator'). Under the SDA and the Race Relations Act 1976 (RRA), the comparator can be hypothetical. A comparable full-time worker must:

- do broadly similar work
- have similar qualifications, skills and experience
- work for the same employer at the same establishment, or if there is no available comparator there, work for the same employer at a different establishment.

The regulations list six different types of contract. The effect is that, for example, a part-time worker on a fixed term contract may only compare their situation to that of a full-time worker on a fixed term contract. They may not compare themselves to a full-time employee on a fixed term contract, or to a full-time worker on a different kind of contract.

A part-timer who switches to full-time work or returns part time to the same level of job within 12 months (e.g. a maternity returnee) can compare their new position to their old one. They can, in other words, be their own comparator.

Pro rata principle

Part-timers are entitled to the same pay, sickness and maternity pay, access to pensions, training, leave and redundancy selection criteria as full-time workers, calculated on a pro rata basis. Employers should ignore part-time status when they make promotion decisions or give bonuses, shift allowances or unsociable hours payments. They should ensure that training, assessments, and so on are arranged so that part-timers can attend. Excluding part-timers from profit-sharing or share option schemes will normally be unlawful. It will not be enough for employers to argue they could not provide the benefits pro rata. The decision not to provide them must be objectively justified.

The regulations do permit one area of difference between part-timers and full-timers. Part-time workers are not entitled to full-timers' overtime rates until they have exceeded the normal full-time hours.

Justification

The employer can defend less favourable treatment of part-timers by arguing that it was objectively justified. Drawing from sex, race and disability discrimination case law, this is likely to mean that the employer must have a legitimate business objective for the less favourable treatment, and must have chosen reasonably necessary means of achieving this objective.

Remedies

Part-time staff who believe that their rights have been breached may ask their employer, in writing, why they have been treated less favourably. The employer must respond, again in writing, within 21 days. The aim is to give both parties the chance to resolve their disagreement without having to go to a tribunal.

However, if the employee does bring tribunal proceedings, the employer's written statement may be used as evidence. If the employer fails to provide a written statement, the tribunal is entitled to infer that the individual's rights were infringed.

The worker must bring the complaint within three months of their rights being breached. The tribunal may:

- award unlimited compensation
- make a declaration of each party's rights
- recommend that the employer takes action to remedy the fault within a specified period.

In practice, employees may prefer to bring an indirect sex discrimination claim, because they may receive an award for injury to feelings. This is specifically excluded from the Part-time Workers Regulations.

The regulations also protect part-timers from being either dismissed or subjected to a detriment because they sought to use their rights.

Regulatory guidance

The regulations offer protection to individuals after they start to work part time. They do not require employers to give part-time work to those who ask for it. The Guidance to the Part-time Workers Regulations also encourages employers to offer part-time work.[1]

GUIDANCE TO THE PART-TIME WORKERS REGULATIONS

To facilitate requests to work part-time, employers should:

- review whether any vacant posts could be done part-time
- consider, when asked by a potential part-timer, whether part-time arrangements can fulfil that position's requirements
- maximise the range of posts at all levels designated as suitable for part-timers or jobsharers
- take requests to jobshare seriously and, in larger organisations, maintain a database of those interested in jobsharing
- take requests to change to part-time work seriously, and explore with workers, if possible, how to bring about this change
- consider having a procedure to discuss with full-time workers whether they wish to change to part-time work
- review how they advertise their vacancies
- communicate with staff representatives on part-time issues.

LEAVE FOR PARENTS AND CARERS

The legal minima for paid leave for working parents are set out below.

Maternity leave

All female employees are entitled to a maximum of 26 weeks ordinary maternity leave (OML) regardless of their length of service, whether they work part or full time and whether they are on temporary or permanent contracts. They may not qualify for statutory maternity pay or maternity allowance but will still be eligible for OML.

---

[1] See www.berr.gov.uk.

An employee's OML cannot begin before the 11th week before their expected week of confinement (EWC) and starts on the earlier of:

- the date she notifies to her employer as the start of her OML
- the first day before the beginning of the fourth week before the EWC, where she is absent for a pregnancy-related reason or
- the day of childbirth.

If an employee has been employed by her employer for at least 26 weeks by the 15th week before the EWC, she qualifies for statutory maternity pay (SMP) during the period of OML at the flat rate set by the government each year. SMP will be paid for 39 weeks.

All pregnant employees are entitled to OML of 26 weeks plus additional maternity leave of 26 weeks, making the total of maternity absence equivalent to 52 weeks.

Employees have the right to return to work on the same terms and conditions as those before they went on maternity leave.

Paternity leave

Employees with 26 weeks' continuous service by the 15th week before their partner's EWC may take up to two weeks' paid paternity leave during the two months after the birth (or after the adoption date for adoptive fathers). Statutory paternity pay (SPP) is set at the same rate as SMP.

The government plans to introduce a right to 26 weeks' paid additional paternity leave for fathers whose partners return to work after the first 26 weeks' of their maternity leave, but the new right will only apply after the entitlement to paid maternity leave is extended to 12 months (see Maternity leave, above).

Adoption leave

Employees who have 26 weeks' continuous service before they receive notification of matching with a child for adoption can take up to 26 weeks ordinary adoption leave (OAL), starting no earlier than 14 days before the expected date of placement of the child. Either adoptive parent can take OAL and the other will then be eligible for two weeks' paid leave at the time of placement on similar terms to paternity leave (see above). During OAL employees receive statutory adoption pay (SAP) paid at the same rate as SMP.

Employees are also entitled to a further period of unpaid additional adoption leave (AAL), lasting for 26 weeks (bringing their total adoption leave entitlement up to one year).

Parental leave

Employees qualify for parental leave if they:

- have one year's continuous employment with their existing employer and
- have legal parental responsibility for a child who is under five years of age or 18 if disabled

or

- have one year's continuous employment with their current or previous employer and
- have legal parental responsibility for a child under five years of age or 18 if disabled.

Provided they have formal responsibility, all mothers qualify for parental leave, as do fathers who were married to the child's mother at the time of the birth or appear as the father on the birth certificate. Adoptive parents, foster parents, step-parents and guardians are also eligible.

Employees may take up to 13 weeks' unpaid leave (18 weeks where the child is disabled) for each child before the child's fifth birthday (18th birthday in the case of a disabled child and fifth anniversary of placement for adopted children).

Employers can change these conditions by collective agreement with employees, as long as the new terms are more generous than the legal minima.

Time off for dependants

Employees have a right to unpaid leave for family emergencies. The entitlement applies in the following situations:

- to provide assistance when a dependant falls ill, gives birth, is injured or is assaulted
- to make care arrangements for a dependant who is ill or injured
- to deal with the death of a dependant
- to deal with unexpected disruption to or the termination of arrangements for a dependant's care
- to deal with an unexpected incident during a son or daughter's education.

Dependants are defined as a spouse, child, parent or person living in the same household (tenants, friends sharing a house, or same-sex couples). Employees are allowed a reasonable time off work to deal with their dependants; no limits are set in the legislation. There is no limit to the number of times an employee can take time off. All periods of leave are unpaid.

CONTRACT AND CASUAL WORKERS

Employment patterns have changed over recent years and many people rely on fixed contract or casual employment opportunities, especially in the service industries. Both parties can benefit from the arrangement.

Fixed term contracts

Around 6 per cent of the UK workforce is employed on fixed term contracts. The Fixed Term Employees (Prevention of Less Favourable Treatment) Regulations came into effect in October 2002 to protect the interest of these workers.

The regulations mirror the similarly named regulations protecting part-timers. Unlike the Part-time Workers Regulations, they apply to the narrower category of 'employees', rather than the more inclusive 'workers'. This means that self-employed people, apprentices, students and trainees and agency workers are not included.

The regulations define a fixed term contract as being one of the following:

- It is made for a specific term that is fixed in advance.
- It ends automatically when a particular task is completed.
- It ends with a specified event (which is not reaching the retirement age).

The regulations create a right for employees on fixed term contracts:

- not to be treated less favourably than a comparable permanent employee
- not to be subjected to detrimental treatment as a result of any act or deliberate failure to act by the employer.

Employees on fixed term contracts are entitled to the same terms in their employment contract as comparable permanent employees, apart from pay and occupational pension terms. The employer can, however, defend its actions if they were objectively justified. The definition of a comparable employee mirrors that used in the Part-time Workers Regulations (see earlier in this chapter, Part-time workers: Comparable full-timers). Fixed term contract conditions can be different from those of permanent employees as long as the conditions are just as good or better, for example, better pay instead of pension rights.

The same problems will arise in giving certain benefits to fixed term employees as occur under the pro rata principle for apportioning part-timers' benefits. Private healthcare and car leases, for example, may only be renewable annually.

The regulations introduced a right for fixed term employees to receive information about any suitable permanent vacancies at their place of work. They also provide that individuals employed for four or more years on one fixed term contract or an unbroken series of contracts automatically become permanent employees if they are re-engaged on a further fixed term contract unless another fixed term can be objectively justified.

Agency workers

The Conduct of Employment Agencies and Employment Businesses Regulations 2003 became law on 6 April 2004.

The Regulations strengthen the rights of agency workers and their hirers. Under the Employment Agencies Act 1973, an employment business which hires out temporary staff is the employer of those staff. But an employment agency, in contrast, simply finds work for individuals, who go on to be employed by the hirer. Most of the provisions of the Regulations apply to both employment businesses and employment agencies.

The general obligations placed on agencies include:

- notifying their terms in writing and in one document to both the hirer and the work seeker; these terms cannot be changed without the hirer's or work-seeker's consent

- not charging the hirer a fee for taking a temporary worker into permanent employment – a 'temp-to-perm' fee – unless there is provision to do so in the contract and the contract gives the hirer the option of a longer hire period instead
- taking steps to ensure both that the work is suitable for the work seeker and that the work seeker is suitable for the hirer.

A key issue surrounding agency workers has been under what circumstances they might become the employee of the hirer. In the past, temporary workers have been supplied on an agency basis, working for a succession of different employers, rather than for a single employment business, so that they fail to accrue the year's continuous service upon which many employment rights depend. There has also been confusion about who the employer is in various situations, and whether agency staff are employees or workers for the purposes of employment legislation.

The problems were highlighted in a 2001 case, Montgomery v Johnson Underwood.[2] Mrs Montgomery tried to claim compensation for unfair dismissal when her contract was terminated, only to be told by the Court of Appeal that she was employed by no one, neither her agency nor her hirer, and therefore had no one she could sue. Various other cases have refined the law on when a temporary worker becomes the employee of their hirer but without a clear definition.

Agency workers do have some employment rights. The Employment Relations Act 1999 gives them the right to be accompanied at a disciplinary or grievance hearing. They are protected from discrimination on the basis of race, sex or disability. They are also protected under the National Minimum Wage Act 1998 and the Working Time Regulations 1998.

The 2003 Regulations aim to help clarify the status of agency workers. Agencies have to make clear whether they are an employment business or an employment agency, and have to have an agreed procedure for dealing with unsatisfactory workers, and this is one of the factors that is taken into account in deciding who, if anyone, employs a temporary worker.

Agency workers' main rights include:

- paid holiday, rest breaks and limits on working time
- no unlawful deductions from wages
- the national minimum wage
- not to be discriminated against under any of the equality legislation.

TRANSFER OF UNDERTAKINGS (PROTECTION OF EMPLOYMENT) REGULATIONS (TUPE)

On 6 April 2006, the revised Transfer of Undertakings (Protection of Employment) Regulations (the TUPE Regulations) came into force. The box at the end of this section summarises the main changes to the previous 1981 TUPE Regulations which the revised 2006 Regulations introduced.

---

[2] Montgomery v Johnson Underwood Ltd [2001] IRLR 269.

These Regulations provide employment rights to employees when their employer changes as a result of a transfer of an undertaking. They implement the European Community Acquired Rights Directive (77/187/EEC, as amended by Directive 98/50 EC and consolidated in 2001/23/EC).

Overview of the TUPE Regulations

Subject to certain qualifying conditions, the Regulations apply:

- when a business or undertaking, or part of one, is transferred to a new employer
- when a 'service provision change' takes place (e.g. where a contractor takes on a contract to provide a service for a client from another contractor).

These two circumstances are jointly categorised as 'relevant transfers'.

Broadly speaking, the effect of the Regulations is to preserve the continuity of employment and terms and conditions of those employees who are transferred to a new employer when a relevant transfer takes place. This means that employees employed by the previous employer (the 'transferor') when the transfer takes effect automatically become employees of the new employer (the 'transferee') on the same terms and conditions (except for certain occupational pensions rights). It is as if their contracts of employment had originally been made with the transferee employer. However, the Regulations provide some limited opportunity for the transferee or transferor to vary, with the agreement of the employees concerned, the terms and conditions of employment contracts for a range of stipulated reasons connected with the transfer.

The Regulations contain specific provisions to protect employees from dismissal before or after a relevant transfer.

Representatives of affected employees have a right to be informed about a prospective transfer. They must also be consulted about any measures which the transferor or transferee employer envisages taking concerning the affected employees.

The Regulations also place a duty on the transferor employer to provide information about the transferring workforce to the new employer before the transfer occurs.

The Regulations make specific provision for cases where the transferor employer is insolvent by increasing, for example, the ability of the parties in such a difficult situation to vary contracts of employment, thereby ensuring that jobs can be preserved because a relevant transfer can go ahead.

The Regulations can apply regardless of the size of the transferred business, so the Regulations equally apply to the transfer of a large business with thousands of employees and to the transfer of a very small one (such as a shop, pub or garage). The Regulations also apply equally to public or private sector undertakings, and whether or not the business operates for gain, such as a charity.

Relevant transfers

The Regulations apply to 'relevant transfers'. A relevant transfer can occur when:

- a business, undertaking or part of one is transferred from one employer to another as a going concern (a circumstance defined for the purposes of this guidance as a 'business transfer'); this can include cases where two companies cease to exist and combine to form a third

- a client engages a contractor to do work on its behalf, or reassigns such a contract, including bringing the work 'in-house' (a circumstance defined as a service provision change).

These two categories are not mutually exclusive. It is possible, indeed likely, that some transfers will qualify both as a business transfer and a service provision change. For example, outsourcing of a service will often meet both definitions.

Business transfers

To qualify as a business transfer, the identity of the employer must change. The Regulations do not therefore apply to transfers by share takeover because, when a company's shares are sold to new shareholders, there is no transfer of a business or undertaking: the same company continues to be the employer. Also, the Regulations do not ordinarily apply where only the transfer of assets, but not employees, is involved. So, the sale of equipment alone would not be covered. However, the fact that employees are not taken on does not prevent TUPE applying in certain circumstances.

To be covered by the Regulations and for affected employees to enjoy the rights under them, a business transfer must involve the transfer of an 'economic entity which retains its identity'. In turn, an economic entity means 'an organised grouping of resources which has the objective of pursuing an economic activity, whether or not that activity is central or ancillary'.

Service provision changes

Service provision changes concern relationships between contractors and the clients who hire their services. Examples include contracts to provide such labour-intensive services as office cleaning, workplace catering, security guarding, refuse collection and machinery maintenance.

The changes to these contracts can take three principal forms:

- where a service previously undertaken by the client is awarded to a contractor (a process known as contracting out or outsourcing)
- where a contract is assigned to a new contractor on subsequent retendering
- where a contract ends with the service being performed in-house by the former client (contracting in or insourcing).

The Regulations apply only to those changes in service provision which involve 'an organised grouping of employees ... which has as its principal purpose the carrying out of the activities concerned on behalf of the client'. This is intended to confine the coverage of the Regulations of cases where the old service provider (i.e. the transferor) has in place a team of employees to carry out the service activities, and that team is essentially dedicated to carrying out the activities that are to transfer (although they do not need to work exclusively on those activities). It would therefore exclude cases where there was no identifiable grouping of employees. This is because, if there was no such grouping, it would be unclear which employees should transfer in the event of a change of contractor. So, if a contractor was engaged by a client to provide, say, a courier service, but the collections and deliveries were carried out each day by various different couriers on an ad hoc basis, rather than by an identifiable team of employees, there would be no service provision change and the Regulations would not apply.

A service provision change will often capture situations where an existing service contract is retendered by the client and awarded to a new contractor. It would also potentially cover situations where just some of those activities in the original service contract are retendered and awarded to a new contractor, or where the original service contract is split up into two or more components, each of which is assigned to a different contractor. In each of these cases, the key test is whether an organised grouping has as its principal purpose the carrying out of the activities that are transferred.

It should be noted that a 'grouping of employees' can constitute just one person, as may happen, say, when the cleaning of a small business premises is undertaken by a single person employed by a contractor.

Exceptions

The Regulations do not apply in the following circumstances:

- where a client buys in services from a contractor on a one-off basis, rather than the two parties entering into an ongoing relationship for the provision of the service.

So, the Regulations should not be expected to apply where a client engaged a contractor to organise a single conference on its behalf, even though the contractor had established an organised grouping of staff, such as a project team, to carry out the activities involved in fulfilling that task. Thus, were the client subsequently to hold a second conference using a different contractor, the members of the first project team would not be required to transfer to the second contractor.

To qualify under this exemption, the one-off service must also be 'of short-term duration'. To illustrate this point, take the example of two hypothetical contracts concerning the security of the Olympic Games or some other major sporting event. The first contract concerns the provision of security advice to the event organisers and covers a period of several years running up to the event; the other concerns the hiring of security staff to protect athletes during the period of the event itself. Both contracts have a one-off character in the sense that they both concern the holding of a specific event. However, the first contract runs for a significantly longer period than the second; therefore, the first would be covered by the TUPE Regulations (if the other qualifying conditions are satisfied) but the second would not.

A second exception to the Regulations occurs:

- where the arrangement between client and contractor is wholly or mainly for the supply of goods for the client's use.

So, the Regulations are not expected to apply where a client engages a contractor to supply, for example, sandwiches and drinks to its canteen every day, for the client to sell on to its own staff. If, however, the contract was for the contractor to run the client's staff canteen, then his exclusion would not come into play and the Regulations might therefore apply.

Transfers within public administrations

Both the Acquired Right Directive and the TUPE Regulations make it clear that a reorganisation of a public administration, or the transfer of administrative functions between public administrations, is not a relevant transfer within the meaning of the legislation. Thus, most transfers within central or local government are not covered by the Regulations. However, such intragovernmental transfers are covered by the Cabinet Office's Statement of Practice 'Staff Transfers in the Public Sector', which in effect guarantees TUPE equivalent treatment for the employees so transferred.

MAIN CHANGES MADE IN THE 2006 TUPE REGULATIONS

The 2006 Regulations introduced:

- a widening of the scope of the Regulations to cover cases where services are outsourced, insourced or assigned by a client to a new contractor (described as 'service provision changes')
- a new duty on the old transferor employer to supply information about the transferring employees to the new transferee employer (by providing what is described as 'employee liability information')
- special provisions making it easier for insolvent businesses to be transferred to new employers
- provisions which clarify the ability of employers and employees to agree to vary contracts of employment in circumstances where a relevant transfer occurs
- provisions which clarify the circumstances under which it is unfair for employers to dismiss employees for reasons connected with a relevant transfer.

The rights and obligations in the 1981 Regulations remain in place, although the 2006 Regulations contain revised wording at some points to make their meaning clearer, as well as reflecting developments in case law since 1981.[3]

The Business Enterprise and Regulatory Reform Department (BERR) has produced a comprehensive guide to TUPE 2006 which is available for download on the government web site. Extracts from the guidance have been used in this section.[4]

EMPLOYEE CONSULTATION

On 6 April 2005 the Information and Consultation of Employees (ICE) Regulations 2004 came into effect to implement the European Union (EU) Information and Consultation Directive. They add a general right for employees to be informed and

---

[3] Refer to: ECM (Vehicle Delivery Service) v Cox [1999] IRLR 559.

[4] See the 'employment relations' section of the BERR website at www.berr.gov.uk.

consulted about workplace issues to existing specific consultation rights on issues such as TUPE transfers and redundancies (see Other consultation rights).

The Regulations introduce a right for employees to be:

- informed about the economic situation of the undertaking for which they work
- informed and consulted about any planned measures which could pose a threat to employment
- informed and consulted with a view to reaching agreement about decisions likely to lead to substantial changes in work organisation or contractual relations, including redundancies and transfers.

Organisations with 50 or more employees have to comply with the Regulations.

If an employer does not set up a consultation structure, a request supported by 10 per cent or more of the workforce in an undertaking triggers the procedure set out under the Regulations. Where employers already have a pre-existing consultation structure, through a recognised trade union or works council, the workforce can be balloted on whether or not it supports the employee request. If there is insufficient support for change, the previous arrangement continues.

If 40 per cent of all employees and a majority of all those who vote are in favour of a new arrangement, an information and consultation arrangement must be negotiated with employees. Where negotiations fail, or the employer takes no action, a prescribed default arrangement applies covering election of representatives, frequency of meetings, and so on.

Employers who do not fulfil their obligations under the ICE Regulations can be challenged at the Central Arbitration Committee (CAC). Appeals can be taken to the Employment Arbitration Tribunal (EAT), which can award penalties of up to £75,000 against employers.

Other consultation rights

The ICE Regulations stand alongside existing consultation rights for employees which apply when:

- collective redundancies or a TUPE transfer are planned
- any measure or new technology is introduced which substantially affects employees' health and safety (see Complying with health and safety law)
- there is a European Works Council.

EUROPEAN WORKS COUNCIL

The European Works Council Directive was implemented in the UK by the Transnational Information and Consultation of Employees Regulations 1999. They require undertakings or groups with at least 1000 employees across the member states and at least 150 employees in each of two or more of those member states to set up European-level information and consultation procedures (usually through a European Works Council agreement).

Redundancies and transfers

The requirement to consult employees before making them redundant is contained in s. 188 of the Trade Union and Labour Relations (Consolidation) Act 1992. The requirement to consult employees before the transfer of a business is contained in the TUPE Regulations 2006. It is easy to confuse the two sets of requirements, so it is useful to summarise both here, even though TUPE is the bigger concern for facilities managers.

Amendments to the requirements came into effect on 28 July 1999. The main changes, according to BERR guidance, are:

- If employees who may be affected are represented by a trade union recognised for collective bargaining purposes, that union now has an automatic right to be informed and consulted over collective redundancies and transfers of undertakings; it may no longer be bypassed by the employer in favour of other employee representatives (although these may also be consulted if the employer so chooses).
- Explicit rules have been introduced for electing employee representatives where there is no recognised union (existing employee representatives may be consulted only if their remit and method of election is suitable).
- Where there is no recognised union and affected employees fail to elect representatives, having had a genuine opportunity to do so, the employer may fulfil its obligations by providing information direct to employees.
- Both union officials and non-union representatives have a right to reasonable paid time off for relevant training.
- Employees are protected against being unfairly dismissed or detrimentally treated for participating in an election of employee representatives.
- The amounts of compensation that employers may be required to pay if they fail to inform and consult have been increased and rationalised.

A collective redundancy is one where 20 or more employees are to be made redundant within a 90-day period. Employers are under no legal obligation to inform and consult representatives if fewer employees are to be dismissed. But the Department of Trade and Industry (DTI) warns that they may face unfair dismissal claims if they fail to inform and consult individual employees who are to be dismissed.

'Affected' employees are not necessarily only those who face redundancy or a transfer to a new employer. If there is a dispute, it is for an employment tribunal to decide whether any particular employee or group of employees was affected, in the light of all the facts.

Consultation about collective redundancies must begin in good time and at least:

- 30 days before the first dismissal if 20–99 employees are to be made redundant
- 90 days before the first dismissal if 100 or more employees are to be made redundant.

Consultation about the transfer of a business must begin 'in sufficient time', but TUPE sets out no specific time period.

On collective redundancy, the employer must consult, with a view to reaching agreement, on how to avoid the dismissals, reduce the number of employees affected and mitigate the consequences of the dismissals. It should disclose:

- the reasons for the proposed dismissals
- the numbers and classes of employees whom it proposes dismissing
- the total number of employees of such description employed
- the proposed criteria for selecting employees for redundancy
- the proposed procedure and timescale for dismissing employees
- the proposed method for calculating redundancy payments.

On the transfer of a business, the employer must inform the union or employee representatives:

- that the transfer is to take place and when
- the reasons for the transfer
- the legal, economic and social implications of the transfer
- the measures that the employer proposes taking in relation to the employees affected.

Selection for redundancy

The main cause of employees complaining to an employment tribunal is that they believe they were unfairly chosen for redundancy. The most easily understood way of choosing employees for redundancy is to use the 'last in, first out' method. But most employers prefer to use a range of criteria against which to 'mark' employees, which may include their attendance record, timekeeping, productivity (quality and quantity of work) and adaptability, and the employer's future needs.

If challenged in a tribunal, the employer must be able to show it used a fair, consistent and objective selection procedure. Employers should avoid vague criteria such as 'attitude', which allow room for prejudice, and discriminatory criteria such as part-time working, pregnancy or trade union membership. They should also take care not to discriminate against disabled employees, whose attendance record, for example, might otherwise mark them out for redundancy.

A redundancy dismissal will also be unfair if the employer has failed to consider suitable alternative employment.

DISABILITY DISCRIMINATION

The main recent change to the laws on discrimination, and a key concern for facilities managers, was the introduction of the Disability Discrimination Act 1995 (DDA). It works differently from the SDA and the RRA, because it does not include the concept of indirect discrimination or require claimants to compare their treatment to another actual or hypothetical employee (a comparator) in order to prove their case.

The DDA imposes duties on all employers (with the exception of the armed forces), suppliers of goods and services and landlords. The stages of the Act so far in force require employers to make adjustments to accommodate disabled employees and jobseekers and to adjust the way they provide services where their current form makes it difficult for disabled people to access them. Part III of the DDA was implemented in October 2004 and any business providing a service to members of the public, customers, clients or tenants has to, where reasonable, remove, alter or avoid any physical feature of their premises that makes it unreasonably difficult for a disabled person to use a service.

The 1995 Act was amended by the Disability Discrimination Act 2005 and introduced new requirements for transport, public authorities and private members' clubs.

The DDA defines 'disability' as a physical or mental impairment which has a substantial and long-term adverse effect on a person's ability to carry out normal day-to-day activities. 'Long-term' means the disability must last or be expected to last for at least 12 months.

Normal day-to-day activities are those involving:

- mobility, dexterity or coordination
- continence
- ability to lift, carry or move everyday objects
- speech, hearing or eyesight
- memory or ability to concentrate, learn or understand
- perception of risk.

The Disability Discrimination (Blind and Partially Sighted Persons) Regulations 2003 which came into force on 14 April 2003 mean that the blind or partially sighted are deemed disabled for the purposes of the DDA. In December 2005 the definition of disability was extended to include individuals with mental illnesses that are not clinically recognised to be mental impairments, and people with cancer, HIV infection or multiple sclerosis, whether or not the conditions have an effect on their ability to carry out normal day-to-day activities.

An employer discriminates against a disabled person if, for a reason related to their disability, it treats them less favourably than other people, and it cannot show that this treatment is justified. Discrimination also occurs if the employer could make a 'reasonable adjustment' to work arrangements or premises to accommodate a disabled person, and its failure to do so cannot be justified.

Discrimination may be justified in limited circumstances. For example, employers and service providers must not do anything that would endanger health and safety, although they can no longer use 'fire risk' as a blanket excuse for excluding disabled people from buildings. Nor does the DDA require them to do anything that would prevent them from complying with other legislation such as listed building regulations.

Reasonable adjustments

Facilities managers will be involved in making reasonable adjustments in so far as they involve changes to the premises and acquiring or modifying equipment. See the box later in this chapter, How accessible are your premises?, for examples of changes that may be necessary to accommodate either employees or members of the public who have a disability.

Facilities managers need to be aware that the responsibility for making adaptations does not rest solely with them. There are also many other adjustments that can be made to accommodate a disabled employee, which involve changing the way the workplace is organised. These include:

- allocating some of the disabled person's duties to someone else
- transferring the person to fill an existing vacancy
- altering the person's working hours
- assigning the person to a different place of work, such as their home
- allowing absences for rehabilitation, assessment or treatment
- giving training
- modifying instructions or reference manuals
- modifying assessment procedures
- providing a reader or interpreter
- providing supervision.

In considering what is 'reasonable', employers are entitled to take into account:

- how far making the adjustment would prevent the effect of the disability
- how practicable it is to make the adjustment
- the financial costs of the adjustment and the scale of the disruption
- their resources
- the availability of financial or other assistance.

The duty to make reasonable adjustments to physical features of premises applies only to your own premises. There is no duty to make adjustments to a disabled worker's own home, even if they work from there, or to other premises that the person may visit during their work.

Nevertheless, facilities managers are likely to be involved in setting up employees to work at home if this is the adjustment needed to allow a disabled employee to continue working. Test cases in the USA relating to the Americans with Disabilities Act (which formed the pattern for the DDA) suggest that denying an employee the opportunity to telework would be failure to make a reasonable adjustment in many cases. Where employees work from home, their employers have the same duty of care to maintain a healthy and safe working environment there as in their main sites.

There is no qualifying period that a disabled person must have worked to be eligible for rights under the DDA. Indeed, a number of compensation claims have involved candidates applying for jobs who were not provided with suitable conditions during their interview.

Public access

Those providing a service to members of the public have to comply fully with the DDA. Specifically:

- Since December 1996, it has been unlawful for service providers to treat disabled people less favourably.

- Since 1 October 1999, service providers have had to make reasonable adjustments for disabled people, such as providing extra help or making changes to the way they provide their services.
- Since October 2004, service providers have had to consider making reasonable adjustments to the physical features of their premises to overcome physical barriers to access.

For the purposes of the DDA, service providers include public authorities, private companies, charitable organisations and individual suppliers. Examples of services to the public are:

- use of any place where members of the public are allowed to enter
- communication and information services, for example, telecoms services
- accommodation, for example, hotels and boarding houses
- banks, financial and insurance services
- places of entertainment, recreation or refreshment
- services of any profession or trade or public or other authority, including schools, hospitals and surgeries
- commercial businesses: shops, broadcasting companies and transport hubs.

Building alterations

Part III of the DDA makes it unlawful for those providing goods, facilities or services to the public and those selling, letting or managing premises to discriminate against disabled people in certain circumstances. There is no exemption for small businesses from this provision.

Service providers must identify any physical barriers that would prevent or seriously hamper disabled people from using the service. These barriers include any temporary or permanent feature of the design or construction of premises, fixtures and fittings or any other physical element of the premises or their approaches and exits.

Once a barrier is identified the organisation has a choice of removing it, altering it so that it no longer forms a barrier, or providing a reasonable means of avoiding the feature. The only circumstances in which failure to make reasonable adjustments is justified are where:

- the adjustment would endanger the health and safety of others
- a disabled person is incapable of entering into a contract or giving informed consent to a legally enforceable agreement
- the adjustment would prevent you from providing the service to the public.

Research in 2001 for the Department for Work and Pensions found that the costs of implementing the provisions of Part III of the DDA included an average £1300 for wheelchair access to over £3000 for accessible toilets and over £12,000 for hoists or lifts. However, for most kinds of adjustment the average initial cost was between £100 and £1000. Grants may be available to make adaptations to the workplace or provide special equipment for employees.

Leasing premises

If your organisation rents its premises, the lease may impose conditions that prevent certain alterations. In this case, the DDA overrides the terms of the lease. The organisation should write to the landlord asking for permission to make the alteration in order to comply with the duty of reasonable adjustment. You do not have to make the alteration until the landlord has given permission. The landlord must reply within 21 days or within a longer time if reasonable. The landlord cannot unreasonably withhold consent for an alteration, but may attach reasonable conditions. The DDA does not state whether landlords can reclaim the cost of any adjustments through increased service charges. This would need to be provided for expressly in the lease. Those selling or renting property do not have to make adjustments to the property to make it accessible; that is the job of the tenant or buyer.

There is no clear ruling in the legislation as to whether a landlord or tenant is responsible for adjustments to the common parts of a building. A tenant in a single let would probably be responsible, but in multilet premises the landlord would probably have responsibility since they are providing services to the public.

If a disabled person brings a discrimination claim, the employer can ask the tribunal to call the landlord as a party to the proceedings. The tribunal will then decide whether the landlord has unreasonably withheld consent for or imposed unreasonable conditions on making alterations. If the tribunal orders the landlord to pay compensation, it cannot require the employer to do so as well, but it can authorise the employer to carry out a specified alteration.

Complaints and compensation

Employees with disabilities who believe they have been subjected to discrimination may complain to an employment tribunal within three months of the act complained of. The tribunal can:

- make a declaration of each party's rights
- order the employer to pay compensation
- recommend that the employer take action to remedy the problem (e.g. by making a reasonable adjustment) within a specified period.

If the employer fails to follow a tribunal recommendation without reasonable justification, the tribunal can award compensation, or increase an existing award. Compensation is unlimited, and ran into six figures even in one of the earliest disability discrimination cases, British Sugar v Kirker.[5] The damages can include an award for injury to feelings.

In line with the SDA and the RRA, the DDA makes employers vicariously liable for discriminatory acts by employees carried out in the course of their employment. This is unless the employer can prove it took reasonably practicable steps to prevent the employee committing such acts.

---

[5] British Sugar Plc v Kirker [1998] IRLR 624.

Contract workers

Again like the SDA and the RRA, the DDA outlaws discrimination by those who hire contract workers. Facilities managers who contract work out to a disabled person or someone who becomes disabled must not discriminate:

- in the terms on which they allow the worker to do the work
- by not allowing the worker to do or continue to do the work
- in the way they give the worker access to benefits or by refusing to give access to benefits
- by subjecting the worker to any other detriment.

The hiring organisation must include contract workers in the count of their employees for the purposes of determining whether the small firms' exemption applies under the DDA. See Chapter 1 for information on DDA for service providers.

Disability Rights Commission (now Equality and Human Rights Commission)

The Equality and Human Rights Commission was formed in 2007 and brings together the Disability Rights Commission, Equal Opportunities Commission and the Commission for Racial Equality. The new Commission is responsible for setting policy in all areas of equality and human rights.

The Disability Rights Commission now forms a subsection of the main commission and is still the major policy maker for disability issues. The Commission will still arbitrate in disability disputes and will act to assist individuals in taking service providers to court for non-compliance with DDA 1995, and so on.

HOW ACCESSIBLE ARE YOUR PREMISES?

- Are there disabled parking spaces with space for a wheelchair alongside a car?
- Are signs clearly written?
- Is there a ramp to the entrance?
- Are steps provided with a handrail, good lighting and clearly marked edges?
- Are glass doors marked for visibility, and are there handles at the correct height?
- Is there an induction loop in the reception area, and is the reception desk the right height?
- Are walkways free of obstructions?
- Are entrance doors wide enough?
- Is seating height adjustable?
- If telephones are fixed on the wall, can they be reached? Are minicoms provided?
- Are flashing and audible alarms provided? Are personal vibrating alarms provided for disabled employees?

UNFAIR DISMISSAL

The Employment Relations Act 1999 reduced the service requirement for continuous employment to one year and raised the maximum compensatory award to £50,000. It also index linked the award, which now stands at £63,000. The changes mean not only that more people are entitled to bring a claim, but also that it is more worthwhile to bring a claim, especially for higher earners.

The Act also abolished the practice of those employed on fixed term contracts of a year or more waiving their right to claim unfair dismissal. The waivers had been intended to prevent workers suing their employer for ending their contract when it expired. But the government was concerned that unscrupulous employers were denying individuals their employment rights by forcing them into a series of short-term contracts.

Automatic unfair dismissal

Employment tribunals will rule that a dismissal was automatically unfair if the employee was sacked:

- for exposing poor health and safety (protected under s. 100 of the Employment Rights Act 1996)
- for maternity-related reasons
- for asserting a statutory right
- as part of a redundancy exercise for any of the above reasons
- for being a member of a trade union, or participating in union activities
- for having a spent conviction or failing to disclose it
- as part of a TUPE transfer
- for refusing to work on Sundays (shopworkers only)
- for blowing the whistle on fraud, malpractice or dangerous practices, in breach of the Public Interest Disclosure Act 1998
- for acting as a companion during a disciplinary or grievance hearing.

The Employment Relations Act additionally made it automatically unfair to dismiss an employee for a reason connected with:

- parental leave
- pregnancy and childbirth
- ordinary, compulsory or additional maternity leave
- time off for domestic reasons
- participation in the first eight weeks of industrial action
- acting as a companion during a disciplinary or grievance hearing.

In most of these situations, the employee does not need to have been employed for a year before bringing a claim.

A new statutory dismissal and disciplinary regime came into force in November 2004, bringing with it new rules that all employers must follow. The rules provide a standard three-step dismissal and disciplinary procedure which applies when an employer is considering dismissing an employee. This includes dismissal on grounds of capability or conduct, redundancy, non-renewal of a fixed term contract and compulsory retirement. The three steps are:

- written notification of the employee of the conduct that has triggered the disciplinary or dismissal procedure
- a meeting to discuss any allegations, followed by notification of the decision and the employee's right to appeal if it is unfavourable
- a further meeting if the employee appeals, followed by notification to the employee of the employer's final decision.

There is also a modified two-step procedure for cases of alleged gross misconduct. Full details are available on the DTI web site. In most cases tribunals will automatically rule that dismissal was unfair if the employer did not follow the relevant statutory procedure.

DISCIPLINARY AND GRIEVANCE HEARINGS

A disciplinary hearing is one that could lead the employer to:

- administer a formal warning
- take some other action in respect of the worker
- confirm a warning issued or some other action taken.

A grievance hearing is one which 'concerns the performance of a duty by an employer in relation to a worker'.

The Employment Rights Act 1996, s. 3(1), imposes a duty on employers to include in the written statement of terms and conditions of employment issued to employees a note specifying the name or job title of a person in the organisation to whom an employee can apply for the purpose of seeking redress of any grievance relating to his or her employment, and how such an application should be made.

In October 2004, new minimum legal standards contained in the Employment Act 2002 came into force. They require employers as a minimum to follow a three-stage process:

- First, the employee must set out the grievance in writing to the employer.
- Second, the employer must invite the employee to a meeting to discuss the grievance.
- Third, if the employee wishes to appeal, the employer must invite them to a further meeting, preferably with a more senior manager.

Employees also have the right to be accompanied at disciplinary and grievance hearings. The right is set out in s. 10 of the Employment Relations Act 1999. Employers must, if the worker in question makes a 'reasonable' request, permit them to be accompanied by a single companion of their choice.

The companion is allowed to address the disciplinary or grievance hearing, but not to answer questions on the worker's behalf. They may, however, confer with the worker during the hearing.

A person is entitled to accompany the worker if they are:

- a trade union official as defined in ss. 1 and 119 of the Trade Union and Labour Relations (Consolidation) Act 1992
- a trade union official whom the union has reasonably certified in writing has experience of or training in acting as a worker's companion
- another of the employer's workers.

There is no qualifying period for enjoying the right to be accompanied. There is no right to be accompanied during an informal interview, although such an interview should be terminated if it becomes clear part-way through that formal action may be needed. The employer must reschedule the hearing if:

- the chosen companion is not available at the employer's proposed time, and
- the worker proposes a reasonable alternative time within five days of the original date.

A worker may complain to a tribunal if they were not allowed a chosen companion or to reschedule a hearing. The maximum compensation is two weeks' pay, currently capped at £280 a week. More seriously, the worker will almost certainly also claim that any dismissal following the disciplinary hearing was unfair.

Workers who act as companions are also protected from any detriment while seeking to exercise their right to accompany a fellow worker. If they are dismissed, the dismissal will automatically be unfair, regardless of their length of service.

WORKPLACE SURVEILLANCE

Used properly, email and the Internet are a real asset to businesses, allowing employees to communicate rapidly with clients and carry out research online. Used improperly (e.g. to forward gossip and indecent photographs), they can not only affect productivity, but also expose the employer to legal action. These problems explain the growing popularity of monitoring employees' email and Internet use, and the introduction of laws aimed at ensuring that such surveillance does not infringe employees' right to privacy.

Surveys suggest that more than two-thirds of pornographic sites are viewed during working hours, more than 50 per cent of employees shop online at work, and 90 per cent of economic computer crime is committed by employees.

Nevertheless, employees have a right to privacy when they use telecommunications systems, as the Halford case[6] showed in 1997. Ms Halford, a senior police officer, alleged that her private work telephone was tapped without her knowledge. The European Court of Human Rights found this breached her right under Article 8 of the European Convention on Human Rights to respect for private and family life, home and correspondence.

---

[6] Halford v United Kingdom [1997] IRLR 471.

At the end of 2000, the government and regulators introduced various controls that seek to resolve the conflict between employers' need to protect their interests through electronic surveillance and employees' right to privacy. However, much of the legislation has been introduced piecemeal and there appear to be inherent legal conflicts.

Why monitor?

There are several reasons why employers are keen to monitor email and Internet use. These include the following legal issues and concerns:

- Employees may unintentionally form binding contracts through email; the tone may be chatty, but an email's legal effect is the same as a formal letter (see Negotiating by email, in Chapter 9).
- Carelessly worded emails could be defamatory or libellous; in a well-publicised case,[7] Norwich Union was sued after two of its employees exchanged defamatory emails about Western Provident.
- Downloading or disseminating copyright material (e.g. as an email attachment) could lead to intellectual property disputes.
- Employers could face criminal charges if employees download pornography or hack into external computer systems.
- Employers might be vicariously liable for emails or downloads that harass fellow employees. In Morse v Future Reality,[8] a woman employee who shared a room with men who downloaded explicit material from the Internet resigned and claimed sex discrimination. Her employer was found liable because it was aware of the men's behaviour and failed to prevent it.
- Employees could hack into their employer's confidential database, putting it in breach of the seventh data protection principle (ensuring data remains secure).
- Employers may commit a breach of confidence if an employee forwards confidential material emailed by a client; or employers may face their own trade secrets or confidential information being disclosed.
- Employers can be liable for unlicensed software on their system, downloaded by employees.
- Monitoring how long employees spend typing may be useful for health and safety purposes, to prevent repetitive strain injury.

RIP to privacy?

In October 2000, the government introduced the Regulation of Investigatory Powers Act (RIPA). This restricts employers from intercepting employees' emails and telephone calls unless they believe both employees and the other party to the communication have agreed. What constitutes reasonable grounds for believing that consent has been given is not, however, entirely clear.

---

[7] Western Provident v Norwich Union (2000) unreported.
[8] Morse v Future Reality Ltd (1996) unreported.

On 24 October 2000, the government tried to clear up the confusion by bringing in extra regulations under the Act. These are the Telecommunications (Lawful Business Practice) (Interception of Communications) Regulations 2000. These allow employers more scope to monitor employees' communications, provided they meet each of the following three criteria:

- The communication being monitored must be made in the course of business.
- The employer must have taken all reasonable efforts to inform employees about the monitoring (there is no requirement to inform the party with whom they are communicating).
- The monitoring must be for one of the lawful purposes set out in the regulations, including:

  - to establish facts relevant to the business
  - to ascertain employees' compliance with the law and self-regulatory policies or procedures
  - to ascertain whether employees are meeting the organisation's standards
  - to prevent or detect crime
  - to investigate or detect unauthorised use of the telecommunications system
  - to ensure the system's effective operation.

Organisations may also monitor (but not record) communications without consent in order to:

- determine whether or not communications are business related
- monitor communications to a confidential support helpline.

An employer who routinely monitors email traffic to ensure that the system is used for business purposes only will act within the regulations. Employers also do not need to obtain the prior consent of both senders and recipients. They are simply required to make all reasonable efforts to inform them that the messages may be monitored.

Laws on privacy

There is no privacy law as such in the UK. Employers who believe that the regulations give them a green light to snoop on staff may, however, be in for a shock. There is an implied term of mutual trust and confidence in all employment contracts, which controls, to some extent, the behaviour that employees should reasonably have to endure.

In addition, there is the Human Rights Act 1998, which took effect on 2 October 2000. This enshrines Article 8 of the European Convention on Human Rights (see above, Workplace surveillance) in UK law, and blanket monitoring of telephone calls, emails and Internet use could violate this right. The Act is directly enforceable only against public bodies, but private companies are also likely to be affected. This is because the courts are public authorities, and have to interpret any ambiguities in the law in accordance with the Act. So they could declare the RIPA incompatible with the European Convention on Human Rights, which would put pressure on (although not oblige) the government to change the legislation.

Private companies might feel the effects of the Human Rights Act more directly if they dismiss an employee and, for example, want to use an abusive email to defend their decision before an employment tribunal. If the employee can show that the email was obtained in breach of their human rights, the tribunal may not allow the evidence, destroying the employer's justification for the dismissal.

Employers can, however, justify intrusion into private and family life if their reasons were to protect the rights and freedoms of others. So, for example, they can defend surveillance by citing their right not to have employees wasting their time at work, or the right of other employees not to receive offensive emails.

Personal data code

Finally, the Information Commissioner (formerly known as the Data Protection Commissioner) has issued a Code of Practice setting down standards for telephone and email monitoring.

The Code does not add any new legal obligations, but compliance with its provisions could be taken into account by tribunals in cases where an employer is alleged to have breached the Data Protection Act.

The Code requires employers to establish a specific business purpose for monitoring, and make all affected staff aware of its operation.

The Code also requires the employer to assess the impact on the privacy, autonomy and legitimate rights of staff. Any adverse impact on the employee must be justified by the benefit to the business and others, and employers should carry out an impact assessment for this purpose. Where the assessment shows that monitoring is justified the Code does not require employers to gain consent from the employees. The Code can be found at the data protection web site.

It is also important to note that the Data Protection Act 1998 itself requires employers to:

- process data fairly and lawfully (first data protection principle)
- obtain personal data for a lawful purpose and not to process it in a manner incompatible with that purpose (second data protection principle).

These principles apply regardless of the Telecommunications (Lawful Business Practice) (Interception of Communications) Regulations 2000.

Practical considerations

Whatever the legislation permits employers to do, there are sensible business reasons for avoiding blanket monitoring. Not only will Orwellian surveillance tactics be bad for industrial relations, but also the logistics of tracking every employee's every communication are mind-boggling.

Similarly, the simplest policy from a legal viewpoint might be an outright ban on personal calls, emails and Internet use. This may cut down on the employer's areas of vulnerability, but will disillusion existing staff and deter prospective employees. The problem is not, however, entirely new. Employers are used to dealing with personal use of the telephone, and a policy that expands on this, allowing limited and reasonable use of the system, including email and the Internet, is a sensible way forward.

Email and Internet use policies

Rather than waiting for the authorities to sort themselves out, facilities managers should ensure that they draw up clear guidelines for employees on using email and the Internet. In the event of abuse, these will allow the employer to dismiss the employee in accordance with the disciplinary procedure as long as the guidelines have been communicated to employees.

Employers cannot, however, change employees' existing contracts without their agreement. It is therefore important to make changes in a reasonable manner, and to explain why they are necessary. As well as inserting clauses into employment contracts, it is advisable to use on-screen warning messages when staff log on.

Suggested contents of an email and Internet use policy are:

- **Access**: covering who within the organisation is authorised to access the Internet and write and receive email and for what broad purposes.
- **Acceptable use of the Internet**: defining unacceptable downloads (such as unauthorised software), banning inappropriate access to web sites or downloading inappropriate images or other material.
- **Acceptable use of email**, including etiquette for normal use, a ban on transmission of pornography, sexually and racially offensive messages, defamatory material, restrictions on forwarding mail inside and outside the organisation.
- **Guidance** on use of email and the Internet for personal use, for example, a complete ban or limited use during breaks.
- **Circumstances** where email and Internet use may be monitored.
- **Penalties** for breach of the policy.

MINIMUM WAGE

The National Minimum Wage Act was passed in 1998, and has been amended by subsequent secondary legislation. As of 1 October 2007, workers aged 22 and over must be paid at least £5.52 an hour (£5.73 per hour from 1 October 2008), and workers aged 18–21 must receive a 'development' rate of at least £4.60 an hour (£4.77 from October 2008). The development rate also applies to workers aged 22 and over who are receiving accredited training in the first six months of a job with a new employer. Young people, that is, those between 16 and 18 years, must receive £3.40 per hour (£3.52 from October 2008). Apprentices are exempt from the minimum rate.

It is a criminal offence to refuse to pay the national minimum wage, to obstruct compliance officers or not to keep proper records; fines for these offences can be up to £5000. The National Minimum Wage Helpline is: 0845 6000 678.

NEW EMPLOYMENT/CONTRACT LEGISLATION

A timetable for new employment legislation is summarised in Table 2.1.

Table 2.1 Timetable for new employment legislation

Area of reform	Date of change	Summary
General employment law	2008/2009	Proposals for a new Employment Act 2008 to introduce more flexible working, clarify procedures for the national minimum wage, and introduce more informal dispute resolution practices
Maternity and paternity leave	October 2008	Amendments to remove the distinction between paid ordinary maternity leave and unpaid additional maternity leave
Work and families	April 2009 and April 2010	Increases statutory holiday entitlement, extends maternity and adoption leave, allows mothers to transfer part of maternity leave to fathers
Fixed-term employees	To be confirmed	Amendments to existing regulations to allow agency workers to have statutory sick pay irrespective of duration of contracts
Equality	Possibly 2009–2010	A single Equality Bill is proposed to streamline all existing legislation
Flexible working	To be confirmed	Proposals to extend the right to flexible working to all parents of school-age children

INFORMATION

Duckworth, S. (2003) *Disabled Access to Facilities*. LexisNexis Butterworths Tolley, www.dataprotection. gov.uk.

3 Complying with Property Law

Nistha Jeram-Dave, Natalie Pecenicic and Nicholas Croft

Acquiring, selling and developing property for a company portfolio is a complex process. Company directors are increasingly calling on facilities managers to oversee this process and to report and advise on how to maximise the value of a property and minimise risks. Negotiating lease terms, ensuring any necessary planning permission is secured and procuring valuations all fall within the facilities manager's remit. Knowledge of the legal issues relevant to a company's portfolio is, therefore, a vital tool. It provides the facilities manager with a basis from which to navigate the complex legal processes encompassing property law, from acquisition and disposals, to planning, to liability for environmental clean-up – knowledge which will substantially impact on a company's costs.

ACQUISITION AND DISPOSAL

This section is aimed at those wanting to acquire commercial property in England and Wales for their own use. It outlines the procedures involved and highlights particular considerations to be borne in mind. It does not deal with the ways in which an acquisition could be funded or structured.

Scotland and Northern Ireland have different legal systems, and local legal advice should be taken.

How is property owned?

Leasehold and freehold

Property in England and Wales is owned freehold or leasehold.

Freehold ownership is absolute. An owner of a freehold property owns it until disposal.

Leasehold ownership is limited in time. Leasehold is when an owner (a landlord or lessor) enters into a lease with a tenant or lessee to transfer the property to the tenant for a specified period of time (the term) in exchange for payment. The payment will either be a lump sum (a premium) or a periodic payment (rent) or a combination.

The tenant may also, subject to provisions in the lease, create a further lease interest known as a sublease or an underlease. This sublease must be for a shorter

146

term than the original lease. In turn, further subleases can if permitted by the lease and sublease be created, provided they are for shorter terms.

The system of land registration

The owner of a freehold or leasehold interest in land must be able to prove ownership or 'title'. Title consists of either documents of ownership (deeds) where title is unregistered, or official copies where the title has been registered at the Land Registry. When title has been proved to the satisfaction of the Land Registry, the land is given a title number and a grade. Once title has been registered, a landowner need only show an official copy of the Land Registry entries as evidence of ownership. There are still parts of England and Wales that are unregistered.

Professional advisers: who does what?

The estate agent

When no independent broker is involved, the seller's agent usually acts as the broker. The agent introduces prospective buyers and sellers, will show a prospective buyer around various properties, and will often act as the go-between in the negotiations of the price and terms between the parties. In more substantial transactions the buyer may ask an agent to find a suitable property and to negotiate on its behalf. Offers should be made subject to contract and subject to survey. This means that the buyer will not be bound to buy the property if the result of the survey is not satisfactory or until the bargain struck between the parties has been properly and legally recorded and all relevant investigations made.

The surveyor

A surveyor will carry out a physical inspection of the property to check for structural or other physical problems. The survey result is sometimes used by the buyer to rene-gotiate its offer. A survey is not the same as a valuation. If a buyer obtains finance from a bank or other lending institution in connection with an acquisition, the bank will want to satisfy itself that the amount of the loan is sufficiently covered by the value of the property. It will therefore obtain a valuation. Unlike a survey, a valuation will not contain any detailed comments on the state of repair of the property or on any structural problems.

The solicitor

The buyer's solicitor will:

- negotiate and agree the terms of the contract by which the buyer agrees to buy the property (the first draft of the contract is usually prepared by the seller's solicitor)
- prepare the transfer document transferring the ownership of the property to the buyer
- negotiate and agree the terms of the lease, if a new lease is to be taken by the purchaser (again, the first draft of the lease is usually prepared by the landlord's solicitor)

- make all relevant title and other searches in order to build up a factual picture of the history and present status of the property
- arrange for exchange of contracts: this is when the matter becomes legally binding
- carry out 'completion' or 'closure'
- carry out registration and payment of taxes, following completion.

Buying property

Buying property in England and Wales is a two-stage procedure.

The first stage ends in an exchange of contracts, when the parties enter into a legally binding agreement for the sale and purchase of the property, but have not paid the full price or completed the transaction.

The second stage culminates in the formality of the purchase documents being signed and payment of the full purchase price, at which point the transaction is said to have completed.

Exchange of contracts

The first stage leading to the exchange of contracts is the most important. All legal issues must be settled and the buyer must be satisfied on all aspects of the transaction before exchange. All the searches and title investigations are done at this stage. Once contracts have been exchanged, the buyer cannot object to the terms of the contract or to anything about the title or the physical condition of the property. The contract is prepared in duplicate, a copy being signed by each party. The copies are then exchanged. It is at this stage that the buyer is committed to buy and the seller is committed to sell a specific property at a specific price. On exchange, a deposit of 10 per cent of the purchase price is usually paid. If a buyer fails to complete through no fault of the seller, it risks losing the deposit.

On exchange of contracts, the 'beneficial' interest in the property is passed to the buyer, so it will need to insure the property from that date.

Completion

Legal title does not pass until completion of the transaction. It is traditional to have some period between exchange and completion, during which the buyer's solicitor will:

- carry out final searches
- make practical arrangements for collection and transmission of purchase monies
- prepare the transfer document which transfers title to the property to the buyer.

The period between exchange and completion is flexible and may be for whatever period the parties agree, but is usually no more than four weeks.

On completion, the balance of the purchase price is paid and thereafter the buyer owns the property. If the transaction needs to be registered at the Land Registry, this is done immediately following completion.

If the property is being purchased with borrowed money, the mortgage documentation is usually completed at the same time as the completion of the purchase.

Costs involved

Apart from the professional fees, the following taxes and statutory fees are payable when acquiring property in England and Wales.

Tax

Stamp Duty Land Tax

Stamp Duty Land Tax is payable on all land transactions. The duty is normally paid by the buyer or the tenant and must be paid within one month of substantial completion (this usually means when the buyer takes occupation). Stamp Duty Land Tax is also payable upon the grant of a new lease and is calculated according to the amount of premium paid (if any), the annual rent and the length of the term. It is possible to mitigate Stamp Duty Land Tax in appropriate situations. Advice should be taken early in the transaction. For details of the current UK rates, see Chapter 4.

Value added tax

The current full rate of value added tax (VAT) is 17.5 per cent except for the period end 2008 to end of 2009, when it is 15 per cent. Where the price or rent is subject to VAT, the seller or landlord must pay it to HM Revenue and Customs, and will normally collect it from the buyer or tenant. A person selling the freehold of a brand new building will usually need to charge VAT, although this can sometimes be avoided by taking a long lease at a premium instead.

This apart, a sale or lease will usually be subject to VAT only if the seller or landlord has 'opted to tax'. Once a person has opted to tax a building, then in general they are irrevocably obliged to charge VAT on rent received from tenants occupying it and on proceeds of sale from it. The main advantage of a person opting to tax a building is to enable them to reclaim from Revenue and Customs the VAT charged on costs relating to the building, such as construction or refurbishment costs.

If VAT is charged on the purchase price or rent, the buyer or tenant may be able to reclaim it from Revenue and Customs, depending on the use of the property. If it is used for a business which makes many exempt supplies (e.g. as a bank or other financial business), all the VAT paid cannot be reclaimed, and so the VAT charge is a real cost for the buyer or tenant.

Many other businesses (e.g. manufacturers and shops) can reclaim all or most of the VAT that they pay on costs. Unless the buyer negotiates to defer payment of VAT, the buyer or tenant will have a cash flow cost in paying VAT to the seller or landlord before recovering it from Revenue and Customs. Where a buyer or tenant is entitled to reclaim, it can usually do so within one to four months. Exempt use of a property by a buyer within 10 years of its acquisition may oblige the buyer to repay to Revenue and Customs some of the VAT reclaimed on the purchase price.

For more information on the supplies liable to VAT, see Value added tax, in Chapter 4.

Other taxes

These include capital allowances (see Chapter 4), income tax and corporation tax.

Fees

Local land charges searches

There are various local authorities which maintain registers of planning and other matters affecting local properties. The buyer's solicitor will carry out a search at the local authority, for which a fee of around £200 is payable.

Land registry fees

In cases of dealings with registered land, a registration fee is payable. The fee is calculated on a sliding scale by reference to the acquisition value of the property, and currently runs from £40 for an acquisition of up to £50,000 to £700 for an acquisition of over £1 million. Fees are also payable upon first registration of previously unregistered land.

Other costs

Depending on the location of the land and whether it is registered or unregistered, additional search fees will be payable. For example, a property may be in an area for which it is necessary to carry out a coal or other mining search. It may be necessary to have a valuation carried out and it is often prudent to have a structural survey for which the surveyor will charge. It will also be necessary to have an environmental audit carried out to assess whether the land is contaminated and potential liability for the costs of a clean-up of the land (see Environmental matters, this chapter, for more on the law relating to contamination).

Acquisition for own use

Freehold or leasehold?

Before deciding whether to buy a freehold or a leasehold property, the buyer should analyse for what purpose the property is needed.

Generally speaking, a buyer with out-of-town site requirements needing to install plant and machinery that is not readily movable will be more attracted to a freehold interest: there are fewer restrictions and the level of initial capital outlay is such that the buyer is looking for a long-term site. By contrast, the buyer requiring town centre office use may be better suited to a leasehold property at a market rent; that is, a lease of 25 years or less for little or no initial premium, but subject to an annual rent based on the market value of the building.

Traditionally, market rent leases were for terms of 20 or 25 years, with five-yearly rent reviews, usually on an upward-only basis (see Business leases: Rent review provisions). In recent years, market forces have changed leasehold patterns, and it is now more usual for new leases to be granted for terms of 10 or 15 years and often for as little as five years. If the lease is for business purposes, the buyer as the tenant may have the right to renew the lease when it expires, subject to certain statutory restrictions (see Business leases: security of tenure, for more detail).

Leasehold: ongoing liability

A lease is essentially a contract between a landlord and a tenant. For leases granted before 1 January 1996, the rule is that the original tenant and the original landlord remain liable for their respective obligations under the lease for the duration of the lease term. This is the case even where the original tenant has sold its leasehold interest on to a new tenant, or where the original landlord has sold on its reversionary interest to a new landlord. This legal principle is known as privity of contract. In practical terms, it means that where the original tenant has sold its interest to another who in turn fails to pay the rent, the landlord may seek payment of that rent from the original tenant. The original tenant is not free from its obligations even though it has parted with its interest in the property. The landlord is thus cushioned from the effects of tenant default by being able to look to the original tenant to meet the liability.

The Landlord and Tenant (Covenants) Act 1995 (the 1995 Act) has abolished original tenant liability in respect of 'new' leases, that is leases granted on or after 1 January 1996. The 1995 Act provides that where a tenant sells its leasehold interest to the assignee, the selling tenant is automatically released from further liability. There is an exception to this rule. A tenant will not be released if it has sold its lease without obtaining any requisite permission to do so from its landlord. Also, it is possible for a landlord to require a selling tenant to give an authorised guarantee agreement in respect of the buying tenant's obligations under the lease. On a subsequent permitted sale of the lease, this guarantee falls away.

The 1995 Act means that a landlord loses the safety net that it had under the old principle of privity of contract. Landlords will need to be more concerned with the identity of proposed assignees. The 1995 Act has given landlords greater power to control assignments, although such controls are in practice limited, not only by the relative bargaining strength of the landlord over that of the tenant, but also by the likely impact of restrictions in the lease on the next rent review. The more restrictive the lease, the lower the rent that can be demanded for a property. Both landlords and tenants favour continuing the recent trend towards much shorter lease terms and are wary of qualifications for assignment which are likely to have an adverse effect on future rent reviews.

The 1995 Act also has implications for the landlord's liability. Although a tenant is generally automatically released from liability on a sale of the lease, the landlord is not automatically released from its liabilities on a sale of its reversionary interest. A landlord must request a release from the tenant. If the tenant refuses to give a release, the landlord may ask the court to grant one. If the court refuses, the landlord continues to be liable with the new landlord. If and when the new landlord chooses to sell its reversionary interest, the first landlord has a further opportunity to apply for a release.

BUSINESS LEASES

Businesses operate from the following types of premises:

- **freehold**: the occupier owns the premises outright for an unlimited length of time with few restrictions

- **leasehold**: the occupier has use of the premises for rent for a specified length of time
- **on licence**: the premises are used on flexible terms on a short-term basis.

This section sets out the main provisions of a lease of business premises.

Parties

The lease specifies the names and addresses of the landlord, the tenant and any guarantor.

Premises let

The technical term for the premises let is 'demised premises'. Both parties should understand exactly what is being let. A lease sometimes refers to plans which outline the area let. If there are plans, they should be checked for accuracy. If the lease is registrable at the Land Registry any plans must comply with the Land Registry's requirements for registration.

Fixtures and fittings

Ownership of the plant and equipment in the premises is important, as someone has to maintain them. Also, the tenant might want to remove items when the lease expires. Sometimes items fixed to the premises by a tenant can become part of the premises, in which case the tenant is not entitled to remove them at the end of the lease, even though those items belong to it. If, however, the items were put there for the purpose of the tenant's trade and can be removed without causing irreparable damage to the premises, the tenant can remove them.

Rights granted

In order to make full use of the premises, the tenant will need rights over adjoining premises or land. These could include rights to use common parts in a building, such as lifts and car parking areas.

Exceptions and reservations

Just as the tenant requires to have rights over common parts, a landlord will want to be able to enter the tenant's premises in certain circumstances, for example, to inspect the premises or to carry out repairs.

Term

This is the length of the lease. Traditionally, leases of business premises were long-term commitments, frequently of 25 years. Much shorter leases are now more common. Sometimes the lease allows the tenant to end it at an earlier date than the contractual expiry. This is an 'option to break' or 'break clause'. If a lease contains such an option, it is important that the relevant date is diarised and a solicitor instructed well in advance of the exercise date so that the notice is dealt with properly and on time.

Security of tenure

Business leases are governed by a statutory code in the Landlord and Tenant Act 1954 (the 1954 Act). Some business tenants enjoy security of tenure; that is, the right to have a new lease granted when their current lease expires, subject to the following of a detailed notice and court procedure. The landlord can only deny this request in certain circumstances. It is, therefore, a valuable right. The following paragraphs provide only an introduction to this very complex subject.

The 1954 Act applies to most leases where the property is occupied for the purposes of a business, unless the parties opt out of the 1954 Act before the lease starts.

Termination

A lease protected by the 1954 Act can only be determined by one of the methods prescribed by the 1954 Act. These include notice given by the landlord or the tenant. A lease which is not determined by one of the methods set down in the 1954 Act continues to run (despite expiry of the term) on the same terms until terminated by one of the methods detailed below.

Landlord's notice

A landlord's notice under the 1954 Act must be in a special form and must specify a date (not earlier than the lease termination date and not less than six or more than 12 months ahead) on which the lease is to end. The tenant must then notify the landlord whether or not it is willing to vacate the premises.

If the tenant wants a new lease, it must tell the landlord within two months of the landlord's notice and should make a special court application for a new tenancy, not less than two and not more than four months after the landlord's notice.

The landlord's grounds for opposing a new lease include:

- the tenant's failure to repair or pay rent, or its being in breach of other terms of the lease
- the landlord's intention to demolish or reconstruct the property
- when the landlord requires possession for its own purposes.

The time limits specified in the 1954 Act for service of notices and applications to the court are strict and in general no extension of those limits is permitted.

Tenant's request for a new tenancy

A tenant's request for a new lease must be in a special form and must specify a date (not earlier than the lease termination date and not less than six or more than 12 months ahead) for the start of the new lease. If the landlord wishes to object, it must tell the tenant within two months, stating on which statutory ground(s) it will rely. The tenant must then apply to the court for a new tenancy not less than two months and not more than four months from the service of its notice.

New lease

Unless the landlord can substantiate its objection to a new lease, a new lease must be granted. Usually this will be on similar terms to the previous one.

Compensation

If the landlord succeeds in obtaining vacant possession, based on a statutory ground which involves no fault on the part of the tenant, the tenant is entitled to compensation. In certain circumstances a tenant of business premises may also be entitled to compensation for improvements it carried out.

Rent

Rent control

There are no statutory limitations on the rent payable.

Payment of rent

The lease will specify the rent payable and when, usually quarterly or monthly. The usual quarter days are 25 March, 24 June, 29 September and 25 December. Different days apply in Scotland. A landlord will generally issue a rent demand several days before rent is actually due and will expect payment to be made punctually and by banker's order. Rent is usually paid in advance, not arrears. For possible action to be taken by the landlord in the case of non-payment of rent, see later in this chapter, Landlord's remedies for tenant breaches: Non-payment of rent.

Rent review provisions

Many leases contain rent review provisions. It is very common within the UK for leases to have provisions that protect the long-term investment of property from the landlord's point of view by enforcing 'upward only' rent reviews (meaning that the rent cannot fall).

Rent review provisions in leases can be complex. The dates for the reviews and the basis of the review are set out in the lease. Dates should be diarised and a solicitor and surveyor engaged to advise on the conduct of the review. The lease usually allows the landlord and the tenant to agree the new rent themselves for a certain period before the review date. If they cannot agree, the review can be referred to an independent surveyor to decide.

Rates and taxes

The tenant usually pays the rates and taxes imposed on the premises. A tenant pays VAT on its rent where the landlord has chosen to charge it (see Value added tax, this chapter).

Interest

Interest at a penal rate is payable on any late payments of rent and other sums due to the landlord under the lease.

Dealings

A lease will contain restrictions on a tenant disposing of its lease outright ('assignment'), subletting and sharing or disposing of occupation.

Assignment

A lease will usually say that the landlord's approval must be obtained for any assignment so that the landlord can be sure that the new tenant will be reliable. Usually, the landlord cannot take too long to give this approval and must not be unreasonable about it. The lease can impose conditions which must be met before assignment can take place.

Subletting

Similarly, a landlord will want to approve sublettings. Again, the landlord cannot take too long and must not be unreasonable.

Sharing occupation

A lease will usually contain restrictions on with whom the tenant can share occupation, but it will normally permit sharing with companies that are in the same group as the tenant. Other forms of sharing normally require the landlord's prior approval.

Use

The lease will specify how and for what purposes the premises can be used, such as a shop or an office. These restrictions can be enforced by an injunction (see also Planning: Contractually permitted use, later in this chapter).

Alterations

A tenant will often need to carry out alterations to the premises. The lease will set out which tenant's works are permitted and which are not. Most works will require the landlord's prior approval. When a landlord gives permission to a tenant to carry out works, the permission is often contained in a legal document called a licence for alterations. It describes the permitted works, attaching plans and specifications.

The lease will also allow the landlord to make the tenant 'reinstate' the premises at the end of the lease, undoing any work that it has done during the lease. A document called a schedule of dilapidations is prepared and negotiated by the parties, setting out the work that needs to be done. Sometimes, specialist dilapidations surveyors need to be engaged.

In certain circumstances, a tenant can claim compensation for improvements it has carried out to the premises during the lease, subject to a complicated notice procedure.

Repairs

The lease should set out clearly who is responsible for carrying out repairs to the premises. The definition of the premises actually let to the tenant is important. A tenant of an entire building will usually repair its entirety, at its expense. A tenant of only part of a building carries out internal repairs to its own premises and would contribute to the cost of repairing the whole building via a service charge.

New buildings

The tenant of a new building can find itself in the unenviable position of being required to make good 'inherent' or 'latent' defects in the design or construction of the building. Accordingly, when negotiating the terms of the lease of such a building, the tenant should:

- seek to shift liability for remedying inherent defects to the landlord
- require a duty of care agreement with the landlord's building contractor and professional team involved in the construction
- have the property surveyed and the plans and specifications reviewed by the tenant's advisers.

Where these proposals are not acceptable to the landlord, it may be possible to agree a compromise; for example, that the landlord will rectify inherent defects which appear during a specified period. The landlord should also be prepared to agree to pursue its legal remedies in respect of inherent defects against third parties, wherever it is reasonable to do so.

Old buildings

An old building may not be in a state of complete repair but the lease will generally require the tenant to repair any defects that exist at the time of the lease. The tenant should have a survey to reveal any problems. Ideally, the landlord should remedy existing disrepair or compensate the tenant. Alternatively, it may be possible for the landlord and the tenant to agree a schedule of condition (a detailed description of defects in the property accompanied by a set of photographs). The tenant's obligation could then say that the tenant will not be required to put the premises in any better condition than as evidenced by the schedule of condition.

Breach of repairing obligations

Where a tenant fails to repair, the landlord can either claim damages or, in limited circumstances, forfeit (terminate) the lease. Alternatively, the landlord could enter the premises to carry out the repairs itself and claim the cost from the tenant.

Rights of access

The lease will allow the landlord to enter the premises let to inspect their condition, to carry out repairs where the tenant has failed to repair the premises or to repair other parts of a building where such access is necessary. This is usually subject to the tenant being given advance notice and the landlord making good any damage caused.

Insurance

The lease will require either the landlord or the tenant (or both) to have various insurances. These relate to the premises, plate glass, plant, loss of rent and public liability. Consideration must be given as to which party should insure, whether the insurance should be in joint names and the sum and risks against which the property should be insured. Where parts of a building are being let separately, the landlord insures the whole building and then seeks reimbursement from the tenants for their share of the insurance premium.

The insurance should be repair on a full reinstatement basis, which means that the insurance cover should be sufficient to repair any damage or to rebuild.

What happens if the premises are damaged?

The tenant will be concerned with the following questions, which should be clearly covered by the terms of the lease:

- Does the rent cease to be payable until the damage is put right?
- Is there an obligation on one of the parties to reinstate?
- Are the costs of reinstatement covered by insurance?
- If reinstatement is prevented by factors which are outside the control of the parties, will the lease terminate and, if so, in what circumstances?
- If reinstatement does not take place and the destruction was caused by an insured risk, which party will own the insurance money?

Other financial considerations relating to insurance payable under the terms of a lease are covered in Financial Management: Insurance (Chapter 4, p. 224).

Service charges

Service charges are payments made by a tenant, in addition to rent, to cover the cost of works and services. The amount payable may be fixed or variable according to the costs incurred.

Management arrangements

The landlord will either carry out the management itself (with or without a managing agent) or pass it to a management company, which may be owned by the landlord, a third party or the tenants themselves.

The services

A tenant will wish to see the following services provided in most cases:

- repairs and maintenance of common areas
- redecoration
- provision of hot water, heating and lighting
- maintenance of plant and machinery
- cleaning and refuse disposal
- employment of staff.

Apportionment

The share of the landlord's total expenditure that will be reimbursed by a tenant will usually be a percentage or 'fair or due proportion' calculated by reference to the amount of space occupied or other factors.

Payment

The lease will require the tenant to make payments on account of service charge. There will then be a reconciliation procedure in the lease to deal with any difference between this provisional amount and the actual amount spent by the end of the year. Any shortfall is made up by the tenant and any overpayment is reimbursed or credited against future payments. The service charge accounts should be properly certificated by a suitably qualified surveyor or accountant.

Objections

The tenant will sometimes have the following objections relating to the service charge:

- that items have been included which should not have been
- that there has been a mistake in the calculations
- that the costs are excessive.

The way to challenge the charges depends on the lease. Many leases rule out any challenge at all by providing that the surveyor's or accountant's certificate shall be final.

Reserve and sinking funds

A reserve fund is an expenditure equalisation fund designed to avoid wide fluctuations in the amount of the service charges payable each year on account of relatively expensive and necessarily recurring items. As an alternative to asking tenants to pay every five years for expenditure such as the decoration of the exterior and the common parts of the building, the landlord makes an estimate of the likely costs in advance and collects this sum in five equal instalments over the years preceding the work.

A sinking fund is a replacement fund, by which a landlord aims to build up over many years a fund for the replacement of major items of plant and equipment, such as lifts and heating or air conditioning systems. The drawback for tenants is that these funds can be used to carry out improvements which ought to be carried out at the cost of the landlord rather than of the tenants.

A tenant will be concerned that reserve and sinking funds are kept separate from the landlord's own money to ensure that they do not form part of the landlord's assets on bankruptcy or liquidation.

Options to terminate

A lease may contain an option for the parties or one of them to terminate the lease at a stated time or times, or on the happening of stated events before the contractual expiry date. This is known as an option to break or option to terminate.

Any matters that are made conditions to the exercise of an option must be very strictly observed. The notice exercising the option must be given within the specified period and any other conditions, such as having paid the rent and complied with the terms of the lease, must be observed. Even the most trifling breach can prevent the tenant from effectively exercising a break option.

Landlord's remedies for tenant breaches

Non-payment of rent

For failure to pay rent, the landlord may carry out the following:

- **Sue the tenant** (or, if applicable, the tenant's guarantor or any previous tenant). Since 1 January 1996, a landlord can only recover arrears of rent from a previous tenant or guarantor of a previous tenant if it has first served an arrears notice on the previous tenant or guarantor specifying the amount to be claimed. That notice must be served within six months after the rent claimed first became due. Failure to serve the notice means that the right of the landlord to recover the rent arrears is lost.
- **Seize the tenant's goods**: known as 'distraint'. The landlord may seize goods on the property equal to the value of the rent due (and costs incurred). This process is subject to restrictions and can only be used against goods belonging to the tenant.
- **Take rent from the deposit**: if applicable, the landlord may take the rent due from a tenant's deposit held and then, if necessary, take action in the courts against the tenant for reimbursement of the deposit fund.
- **Forfeit or terminate the lease**: in England and Wales, the tenant has certain statutory protections against forfeiture of its lease. The degree of protection varies depending on the nature of the tenant's failure. This statutory protection is called relief from forfeiture. Usually, the action is dropped if the tenant rectifies the breach.

Other breaches

For other breaches, the landlord may:

- if the lease permits it, remedy the breach itself and charge the tenant for the cost incurred

- forfeit or terminate the lease
- sue for damages or seek from the court an injunction against the tenant (again subject to certain statutory restrictions.

CHECKLIST: BUYING A BUSINESS LEASE

A prospective purchaser of a lease of business premises should (in addition to the usual matters to be considered on a purchase) pay careful attention to:

- rent review
- repairing obligations
- service charges
- VAT implications
- whether the existing tenant is in breach of covenant, giving the landlord the right to determine the lease
- the likelihood of a lease renewal being opposed by a landlord on other grounds, such as redevelopment
- provisions relating to dealings with the lease
- permitted use.

PLANNING

When companies assess their occupation requirements they rarely give high priority to planning issues. Are these companies running a high-risk strategy? What if a company occupies new premises, only to find that planning restrictions prevent it from operating from that site? Consider, for example, if:

- the planning permission which attaches to the site permits warehouse use but not offices in conjunction with the warehouse
- the company wants to add an extension but local planning policy is such that planning permission for this development will not be given by the local planning authority (LPA).

The company would have wasted capital expenditure on acquiring the site and may have committed to a long-term lease, a situation which will not find favour with the board of directors.

What could the facilities manager have done to protect that company's position before acquiring the property? What development projections should have been made?

This section looks at the key planning issues relating to the use and development of business premises.

The law

The Town and Country Planning Act 1990 (the 1990 Act) contains most of current legislation controlling the application of planning laws. The 1990 Act has been amended by the Planning and Compulsory Purchase Act 2004 (PCPA). The effect of this is that certain amending provisions have been written into the 1990 Act, whereas other provisions remain as stand-alone clauses in the PCPA.

When is planning permission required?

One of the questions most frequently asked by facilities managers is: 'Do I need planning permission?' To determine this, it is necessary to establish whether what is proposed (building works or change of use) is classed as development for the purposes of s. 55 of the 1990 Act. If the proposals do amount to development, then planning permission will be required.[1]

What is 'development'?

There are two classes of development:

- physical works carried out in, on, over or under land: usually grouped together as 'operational development'
- development relating to how land is used: classed as a 'material change of use'.

Operational development

Any of the following will constitute development:

- **Building operations**: as defined in s. 336 of the 1990 Act, this includes 're-building, structural alterations or addition to buildings and other operations normally undertaken by a person carrying on business as a builder'. Demolition works were later brought within the concept of building operations and this is discussed below (see Demolition works).
- **Engineering operations**: as defined in s. 336 of the 1990 Act, this is generally interpreted to include operations usually undertaken by or calling for the skills of an engineer.
- **Mining operations**: this is not defined in the 1990 Act but will include the removal of material of any description from a mineral working deposit.
- **Other operations**: this is a catch-all category enabling unclassified development works to be brought within the planning permission net. These are judged on a one-off basis and will include such things as installing protective grilles on windows and raising the level of land by the addition of soil or other material. This is, then, a wide and potentially limitless category depending on the facts and circumstances of each case.

Use of land

A facilities manager is expected to deliver premises that operate in an efficient and commercially effective way. There must be no restriction affecting the use of the building which has a detrimental impact on the business. How then can facilities

---

[1] Section 57(1) of the Town and Country Planning Act 1990.

managers anticipate potential use problems? First, they must understand the distinction between 'contractually permitted use' under a lease or similar arrangement and the 'lawful use' for planning purposes.

Contractually permitted use

Contractually permitted use is the use allowed under a contract such as a lease or licence. For example, a lease may permit the premises to be used for the purposes of high-class offices. If the occupier uses the premises to run a shop then it will be in breach of the lease contract, even if the lawful use for planning purposes permits shop use.

Lawful use

Lawful use is the use permitted under planning legislation. The lawful use will override any use permitted under a lease or licence if they are inconsistent. For example, if shop use is prohibited for planning purposes and is permitted under the lease contract, the premises cannot lawfully be used as a shop.

Categories of lawful use are known as classes of use. The classes of use are set out in the Town and Country Planning (Use Classes) (Amendment) Order 1987, 1991, 1992, 1994, 1995, 2005 and 2006 (the 1987 Order). It consists of 13 classes of use arranged in four main parts.

Mention should also be made of uses which do not fall within the 1987 Order. It is relatively easy to identify the lawfully permitted use for planning purposes by examining the LPA's records. However, some uses do not fall conveniently within the 1987 Order and these are referred to as *sui generis* uses. Examples of a *sui generis* use would be where land is used for the purposes of a taxi business or for the sale of motor vehicles.

What is a 'material' change of use?

Material changes of use are a question of fact and degree. The change may be material because it constitutes a change in the type of use, or because it represents an increase in the intensification of a continuing use. The law determining what is a material change of use is not set in stone, although s. 55 of the 1990 Act does give guidance on which changes of use will constitute development:

No change of use

- use of land for agriculture or forestry and
- uses falling within the same use class of the 1987 Order.

Material changes of use

- deposit of refuse or waste material on land
- use of external parts of the building not normally so used for the display of advertisements
- change of use class within the 1987 Order.

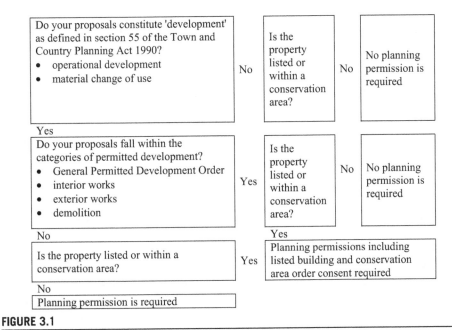

FIGURE 3.1

Do you require planning permission?.

Does section 55 apply?

The process by which facilities managers can make decisions as to whether or not physical works or anticipated changes of use constitute development within the meaning of s. 55 of the 1990 Act is illustrated in Figure 3.1. However, sometimes there will be grey areas, especially where changes of use are concerned. In such cases professional advice from a planning consultant or lawyer should be sought.

Listed buildings and conservation areas

Special controls apply if a property is a listed building or is situated within a conservation area. Listed building consent will be required for virtually all works carried out and a specialist consultant or architect should be involved at an early stage of any application process.

Conservation areas are designated as such to conserve the appearance of the locality. Therefore, works to the exterior of a property in a conservation area will usually require planning permission even if they are of a minor nature. It must also be remembered that trees growing in a conservation area are protected by preservation orders. Any works of lopping, topping or cutting to a protected tree require consent from the LPA.

Understanding planning requirements

Building development projects should be worked around planning requirements so that, in a complex situation, a planning adviser or consultant can be employed at an early stage to advise on compliance and also to instigate and maintain a dialogue between the company and the planning authority.

A facilities manager who understands the planning process will be able to advise the board of directors in respect of any proposed expansion or refurbishment plans. The board can then factor the relevant information into the company's budget and time schedule.

It will also be useful to find out what planning permissions already exist in relation to a property and this can be done by asking the local authority or arranging for a solicitor to carry out a search of the authority's records.

Applying for planning permission

So, if planning permission is required, what is the next step? Is this the time to bring in professional advisers? What role will they play?

Architects/planning consultants

An architect or planning consultant will guide the facilities manager through the application process. Such an adviser is well placed to advise on local planning policy. They will have experience in interpreting the local development plans.[2] Planning decisions by LPAs are plan led and it is important to interpret the development plan correctly. A planning application which is not consistent with the policy set out in the development plan is unlikely to succeed. Architects and planning consultants will also have local knowledge and previous dealings with the LPA. They will be able to streamline the application to give it the best chance of succeeding.

Solicitors

A solicitor can be involved in this process. Due diligence by way of a local search may reveal existing planning restrictions and adverse notices. The solicitor may also have confirmed the use class (see Lawful use). Legal advice will be needed to explain conditions and planning obligations attached to a planning permission.

Formulating the application

The following preliminary steps need to be taken in formulating the planning application:

- Clarify maps, diagrams and plans.
- Obtain copies of the relevant parts of the development plan and any other non-statutory plan which may affect the proposed development.
- Obtain an application form from the LPA, available free of charge. Each LPA produces its own form.

---

[2] Outside Greater London and the Metropolitan areas, the development plan consists of the structure plan and the local plan; within Greater London and the Metropolitan area the development plan consists of the unitary development plan.

- Check whether you are required to submit an Environmental Import Assessment (EIA) as part of your application. An EIA is the process by which the local authority will gather and consider information about the environmental affects of your proposed development. If an EIA is necessary, specialist advice should be sought.
- Consider a preapplication discussion with the appropriate case officer. This is particularly important if the development proposals are not consistent with the development plan or other planning policy statement.
- Decide whether the application will be for full planning permission or outline planning permission. Where the application is to erect a building, the applicant is entitled to apply for outline permission. In this case the application merely has to contain a description of the proposed development indicating its major features. An application for full planning permission will need to include full details of the proposed development. Full plans and elevations will need to be provided together with a considerable amount of detail setting out materials and construction methods.

Making the application

Figure 3.2 illustrates the application process through to the planning decision.

The planning application decision

The planning application will result in one of the following outcomes:

- refusal of planning permission
- failure to give a decision within specified time: a 'deemed' refusal
- grant of outline planning permission
- grant of full planning permission with conditions
- grant of full unconditional planning permission.

Refusal of planning permission

If the LPA refuses planning permission it must state clearly in writing the full reasons for the refusal. If full reasons are not given, this does not necessarily invalidate the decision itself, but it does mean that the decision can be challenged by judicial review and the failure to give reasons could be taken into account upon appeal.

In reaching its decision, the LPA must take into account the provisions of the development plan and also any representations received following publication of the application. Listed building and conservation area considerations will also be taken into account if applicable. The LPA will also consider any other material considerations, meaning considerations relevant to the application; they must be planning considerations and relate to the use and development of land.

There is a right of appeal to the Secretary of State. The appeal must be made within six months of the date of the decision.

Submit application form to LPA in triplicate. Include:
- fee
- drawings, plans
- supporting documents
- GPDO article 7 certificate where applicant not sole owner of land
- Agricultural Holdings Certificate

LPA lists the application in the public register

LPA publicises the application, usually by way of:
- site notice
- notification to neighbours
- local advertisement

Decision
Must be given within eight weeks of the submission of the application unless a longer period has been agreed in writing with the applicant.

Refusal of planning permission	Grant of outline planning permission	Grant of full planning permission with conditions	Grant of full planning permission	'Deemed' refusal: LPA fails to give decision within specified time

Appeal
An applicant can appeal against a refusal of a planning application in a limited number of circumstances. This right must be exercised within six months of the notice of a decision of failure to determine the application.

FIGURE 3.2

Planning application process.

A 'deemed' refusal

A deemed refusal is where an LPA fails to give a decision within the specified time. The specified time is eight weeks from the date of submission of the application, or longer if agreed in writing with the applicant.

There is right of appeal against a deemed refusal to the Secretary of State.

Grant of outline planning permission

By granting outline planning permission, an LPA is committed to allowing that development in principal, subject to approval of certain conditions. These conditions are known as reserved matters and require subsequent approval from the LPA. They deal with the siting, design, external appearance, means of access and landscaping of the development. Further applications will be required to determine reserved matters.

The outline permission cannot be revoked except on payment of compensation and additional conditions cannot be imposed over and above reserved matters.

Applications for approval of the reserved matters must be made within three years of the grant of the outline permission. The development approved by the outline permission must be commenced within three years of the grant of the permission or within two years of the approval of the reserved matters, whichever is the later.

Grant of full planning permission with conditions

The LPA may grant full planning permission but impose conditions. These conditions will relate to the development or use of any land under the control of the applicant, whether or not it is land in respect of which the application was made. The applicant can be required to remove buildings, to carry out additional works, to discontinue a particular use and to carry out subsequent reinstatement works if required. There are judicial restrictions upon the power to impose conditions to prevent these being abused by LPAs. These restrictions are intended to make sure that the conditions are:

- fairly and reasonably related to the development permitted
- not manifestly unreasonable.

There is a right of appeal to the Secretary of State against the grant of planning permissions subject to conditions within six months of the notice of the decision.

Subject to appeal, development work authorised under a grant of full planning permission with conditions must be commenced within three years of the grant of the permission.

Obligations

A mention should also be made of the obligations which can be linked to the grant of planning permission and imposed under s. 106 of the 1990 Act. An LPA can impose additional obligations in return for granting planning permission. Take, for example, an application to develop a new supermarket. An LPA grants planning permission for the supermarket. In return for the LPA's consent, the developer agrees to build a new roundabout in the vicinity. This agreement would be contained in s. 106 agreement.

Grant of full unconditional planning permission

Full planning permission may be granted without conditions. Development works approved by the permission must be commenced within three years from the grant of permission.

Appeals

Only the applicant may appeal against a decision by an LPA, even if the applicant is not the owner of an interest in the land. Third parties, including the owner of an interest in the land who is not party to the application, have no right of appeal. In certain limited cases, a challenge to a planning decision may be brought by way of judicial review.

An appeal can be made in the following circumstances:

- refusal to grant planning permission
- grant of planning permission subject to conditions to which the applicant objects
- refused approval of reserved matters on an outline permission
- refusal of application or grant of permission subject to conditions under s. 73 or s. 73(2) of the 1990 Act (these sections allow an application to be made for the implementation of an existing planning permission (before it has expired) where the new applicant is seeking to implement that permission without the conditions originally imposed)
- failure to notify the applicant within the prescribed period; in other words, deemed refusal.

In all these circumstances the applicant has a right of appeal to the Secretary of State within six months of the notice of a decision or failure to determine an application.

Challenging appeals to the Secretary of State

The validity of an appeal decision cannot be challenged in any legal proceedings. However, a 'person aggrieved' (s. 288 of the 1990 Act) may question the decision by appeal to the High Court if the decision was not within the powers of the 1990 Act or if any relevant procedural requirements have not been complied with. Again, in certain limited cases an application for judicial review may be made.

Human Rights Act 1998

The Human Rights Act 1998 came into force in the UK on 2 October 2000 and had an immediate impact upon domestic legislation. LPAs are now obliged to consider human rights issues when determining planning applications. In particular, the LPA must take into account the right not to be deprived of property (European Convention on Human Rights, Article 1) and the right to respect for private and family life (European Convention on Human Rights, Article 8). Facilities managers should remember that this will affect both their company's planning applications and the ability of the company to object to the applications of third parties. Although a company is a corporate body rather than an individual, it is deemed to have rights protected by the Human Rights Act 1998.

When is planning permission not needed?

In certain circumstances planning permission is not required, even where the proposed works or change of use fall within s. 55 of the 1990 Act. These circumstances are usually referred to as permitted development. It must be noted, however, that special rules apply to listed buildings and conservation areas. Professional advice should always be taken in these circumstances.

Permitted development includes:

- certain demolition works
- development permitted by the Town and Country Planning (General Permitted Development) (Amendment) Order 1995 (as amended)
- internal works
- certain external works.

Demolition works

Demolition of buildings has been brought within the scope of development. Section 13 of the PCPA 1991 introduced a new s. 55(1)(a) of the 1990 Act. 'Building operations' now include 'demolition of buildings; rebuilding; structural alteration of or addition to buildings ...'. Nevertheless, certain demolition works do not require planning permission.

Demolition works requiring planning permission:

- demolition of part of a building.

Demolition works not requiring planning permission:

- demolition of a building smaller than 50 m^3 such as a garage or shed.

Town and Country Planning (General Permitted Development) Order 1995 (GPDO)

The GPDO permits certain types of development which would otherwise require planning permission. There are defined lists of permitted development set out in the schedules to the GPDO. Schedule 2 is of primary importance. This has 28 parts setting out descriptions of developments that can be carried out without the need for planning permission. The most important are:

- Part 2: minor operations.
- Part 3: certain changes of use.

Internal works

Internal works which are not of a substantial nature involving major structural alterations do not usually require planning permission.

External works

External works which do not materially affect the external appearance of a building do not usually require planning permission. Again, what constitutes 'material' is a question of fact and degree. For the works to require planning permission, the change to the exterior must be capable of being seen by an observer outside the building. It must also materially affect the appearance of the building as a whole and not merely a part. The nature of the building must also be taken into account so that, for example, changes effected to a Georgian façade may have a greater visual impact than changes made to the external appearance of a factory. Special rules apply to signage and advertisements.

What if planning law is not complied with?

The facilities manager will now know that development work as defined in s. 55 of the 1990 Act will require planning permission unless it is permitted development. They will also know that operational development or a material change of use may have been undertaken in contravention of planning controls. If there is a breach of planning control, what action can be taken by an LPA? And what can the facilities manager do to protect the company's position?

The LPA will want to see that operational development and material changes of use have been undertaken within its planning district in a lawful way. In certain circumstances it may only need to view the exterior of the property. In other circumstances a more detailed inspection, including an internal inspection, may be required and it will need to gain entry to do this. Having collected evidence of potential breaches of planning controls the LPA will then want to remedy the contravention. Not surprisingly, LPAs are given wide powers of inspection and enforcement for this purpose.

Power to enter and inspect

LPAs may authorise any person in writing to enter land at any reasonable hour without a warrant where there are reasonable grounds for doing so. Where admission has been refused or if the matter is urgent the LPA may obtain a warrant to enter the property.

The penalty on any owner or occupier in these circumstances is that anyone who wilfully obstructs the person exercising a lawful right of entry is guilty of an offence.

The LPA can also compile evidence of a breach of planning control by using a planning contravention notice (PCN) or a breach of condition notice (BCN). A PCN is used to obtain information about any operations, use or activities being carried on at a property or on land. A BCN will be used primarily to establish whether or not conditions or limitations attached to an existing planning permission have been breached.

Enforcement action

An LPA has three options to enforce against a breach of planning control: injunction, enforcement notices and stop notices.

Injunction

An injunction can be used to prevent an owner or occupier from carrying out activities which are being undertaken in breach of planning controls. Whether or not an injunction is granted is at the discretion of the court.

Enforcement notices

The LPA can also issue an enforcement notice. An enforcement notice must contain the following information:

- a statement of the alleged breach
- steps necessary to wholly or partly remedy the breach
- a statement of the effective date of the notice, being at least 28 days from service of the notice

■ a statement of the time within which the steps specified to remedy the breach are to be carried out.

Non-compliance with an enforcement notice makes the owner or occupier liable on summary conviction to a fine of up to £20,000 or, on indictment, an unlimited amount. The LPA also has the power, in addition to prosecuting, to rectify the breach itself and recover reasonable expenses from the owner or occupier.

Stop notice

A stop notice can be issued. As its title suggests, a stop notice requires an immediate halt to the activities causing the breach of planning control. This is a drastic step for an LPA to take and will only be used in very limited circumstances. A breach of a stop notice is subject to the same penalties as a breach of an enforcement notice. However, the LPA will always be aware that if a stop notice has been incorrectly served or is subsequently quashed by legal action or withdrawn, the LPA may become liable to pay substantial compensation. LPAs are often unwilling to expose themselves in this way.

Protection for owners and occupiers

Faced with the impressive armoury of enforcement sanctions available to LPAs, the owner or occupier of the land subject to enforcement proceedings may be able to protect itself by appealing against enforcement proceedings. However, a more attractive and effective option may be for the owner or occupier to regularise its occupation of a property by applying for a certificate of lawful use or development.

Certificates of lawful use or development

Material changes of use and operative development are lawful if no enforcement action can be taken. This will include where the time limit for commencing enforcement action has expired. Enforcement action cannot be taken in the following circumstances:

■ where operational development carried out without planning permission has continued for a period of four years from the date on which the operations were substantially complete
■ where any material change of use and any breach of condition or limitation attached to a planning permission has continued for a period of 10 years from the date of the change of use or breach.

Facilities managers should remember that LPAs cannot impose further conditions when issuing a certificate of lawfulness. However, if, instead, an application is made for planning permission to regularise the existing use and development rather than an application for a certificate of lawfulness, this gives the LPA an opportunity to impose conditions.

Building Regulations

Often, little or no distinction is made between planning permission and Building Regulations. Although there is a degree of overlap in the sense that they must both be dealt with when carrying out building works, they are separate and distinct concepts. Planning permission deals with the obtaining of consent for the type or nature of work to be carried out. Building Regulations are necessary to ensure that the work is carried out with the correct materials and in accordance with appropriate health and safety laws.

The relevant building regulations are the Building Regulations 2000 which have been amended several times since coming into force. The most recent changes were made by the Building and Approved Inspectors (Amendment) Regulation 2007. Building Regulations are covered in more detail in The Building Regulations, in Chapter 1.

Complying with these regulations does not represent any form of consent to works and is not a substitute for planning permission. They ensure that the methods and materials used are of an appropriate quality and that they meet all current industry standards.

A Building Control Officer will be appointed by the local authority to inspect the works at various stages and must be notified before works commence. The officer will liaise with the various contractors to ensure that the works are compliant.

A certificate of satisfaction will be issued when the officer is satisfied that the works meet current industry standards. Conditions can be attached to the certificate and a final certificate will not be issued until these conditions have been met.

A fee is payable to the local authority. The level of the fee is calculated by reference to the cost of the works and the local authority will usually accept an estimated project cost as long as it is realistic. A final certificate will not be issued until the fee has been paid and the works have been completed to the officer's satisfaction.

Energy Performance Certificates

Facilities managers should be aware they may be required to produce or obtain an Energy Performance Certificate (EPC) for a building under the Energy Performance of Buildings (Certificates and Inspections) (England and Wales) Regulations. An EPC is a certificate containing information about the energy efficiency of a building. The regulations have four main requirements, which are:

- EPCs and recommendations for improvement of the energy performance of the building are to be produced whenever a building is constructed, sold or rented out. Since 1 October 2008 all buildings that are not dwellings require an Energy Performance Certificate on construction, sale or let.
- Display Energy Certificates (DECs) are to be displayed in larger buildings occupied by public authorities and by institutions providing public services to a large number of people.
- Advisory reports containing recommendations for improvement of the energy performance of such buildings must also be obtained.
- Air conditioning systems with an output of more than 12 kW are to be inspected at regular intervals.
- The energy assessors who produce EPCs and DECs and inspect air conditioning systems must be accredited.

EPCs are stored on a national register and are valid for a period of 10 years. Local authorities (usually by their trading standards officers) are responsible for enforcing the requirement to have an EPC on the sale or letting of a building. Failure to provide an EPC when required by the regulations means the seller or landlord may be liable to a civil penalty charge notice with a maximum fine of £5000. Listed buildings are a special case.[3]

ENVIRONMENTAL MATTERS

Environmental law, relating primarily to contaminated land, is a complex area of law. Little practical guidance can be given by examining individual statutory provisions. It is an area best dealt with on a case-by-case basis by an experienced environmental specialist, as the potential liabilities and remediation costs can be very high. Nevertheless, there are benefits in being well informed and facilities managers should be aware of the overall apportionment of responsibility for the clean-up of contaminated land.

The law

It is not easy to follow the relevant statutory enactments that deal with the clean-up of contaminated land. The Environmental Protection Act 1990 (the 1990 Act) was the first serious attempt at introducing detailed statutory provisions controlling this. Further provisions dealing with remediation and apportionment of liability were introduced into the 1990 Act by the Environment Act 1995.[4]

The provisions relating to the clean-up of contaminated land came into force on 1 April 2000. Note, however, that the regime is retroactive in effect: the regime will apply irrespective of when the land is or was contaminated.

What is contaminated land?

Land is 'contaminated' if, as a result of the substances in, on or under it, there is:

- a resultant significant harm[5]
- a significant possibility of such harm occurring
- pollution or risk of pollution to water.

---

[3] Internal works within listed buildings require listed buildings consent.

[4] Part II(a) of the Environmental Protection Act 1990. See also Contaminated Land (England) Regulations 2006 and DETR circular on contaminated land, dated October 2006, and the Contaminated Land (Wales) Regulations 2001.

[5] 'Harm' is defined as harm to health of living organisms or other interference with the ecological systems of which they form part and, in the case of humans, includes harm to their property.

Who is liable?: The 'polluter pays' principle

The 'polluter pays' principle is a two-tiered system which requires the identification of an 'appropriate person' who will be liable for the remediation of the contaminated land. An appropriate person is either the person who:

- caused or knowingly permitted the land to become contaminated (the original polluter): known as a class A appropriate person; or
- is the owner or occupier of the land for the time being: a class B appropriate person.

If a local authority identifies contaminated land, remediation works will be enforced against a class A appropriate person ('polluter pays' principle). If after reasonable enquiry the class A appropriate person cannot be found, remediation notices are enforced against a class B appropriate person.

Reducing liability for clean-up

Clearly, there is an obligation on facilities managers to ensure that no pollution is caused to a site during the period of their company's use of that site. However, they will also want to ensure that the risk of being held liable as a class B appropriate person is reduced. This means that they will want information about the environmental status of a property before that property is acquired for the company's portfolio. There is no public list or register of sites that may be contaminated. How then can the risks be quantified? The following options are available.

Desktop or phase I report

This involves a document-only analysis of the environmental aspects of a given site. Information for a phase I study is likely to be gathered by the following methods:

- specific enquiries of regulatory bodies
- specific enquiries of occupier
- researching planning history
- researching title deeds and preregistration documents.

Physical survey or phase II report

The phase II report is more extensive and requires the taking and analysis of soil, water and other material samples from the site. The phase II survey is expensive. However, where the potential risks of contamination are perceived to be high (if, for example, the site has previously been used for a high-risk use, such as storage of solvents or other chemicals), then the cost may well be justified in order to quantify the risks.

Occupier's liability

Where an occupier acts or fails to act and by doing so creates a dangerous condition which later causes harm to a lawful visitor using the premises, then the occupier may be liable under the provisions of the Occupier's Liability Act 1957 (the 1957 Act).

The grounds for making a claim against individuals or companies occupying or controlling premises are wide. They are an easy target, not least because it is widely thought that their insurers will step in to meet the claim.

Case study

Consider the following simple case study.

A company occupies office premises. The facilities manager has a strict cleaning and maintenance programme in operation. This involves polishing the client reception area to a highly polished finish. A client attends the offices, slips on the polished floor and is injured. Is the occupier of the office premises liable for a claim brought by the client because of this accident?

Is the claim valid?

The following key elements will determine the validity of the claim:

- Is the company an occupier?
- Has the accident occurred on 'premises' as defined in the 1957 Act?
- Does the company owe a duty of care to the client?
- Has the company discharged the duty of care owed to the client?
- Can liability be excluded?

Is the company an occupier?

The rules of common law (determined by case law) define who is an 'occupier'.[6] Exclusive occupation of premises is not necessary. The test is whether a person has some degree of control associated with or arising from their presence and use of or activity in the premises; it is possible to have more than one occupier of premises. Consider the example given above in the context of Figure 3.3. In this example, the company uses the premises for its business. It has control of the day-to-day running of the premises and is an occupier for the purposes of the 1957 Act.

Did the accident occur on 'premises'?

Facilities managers will be forgiven for thinking that premises means only buildings from which a company operates. For the purposes of the 1957 Act, 'premises' includes:

- any fixed or movable structure
- any vessel, vehicle or aircraft.

The definition is wide. Examples of what constitutes premises include scaffolding, ladders, grandstands, electric pylons and lifts.

If the 1957 Act definition of premises is applied to the case study there is no difficulty in establishing that the client reception area falls within the statutory meaning of premises.

---

[6] Occupiers Liability Act 1957, s. 1(2).

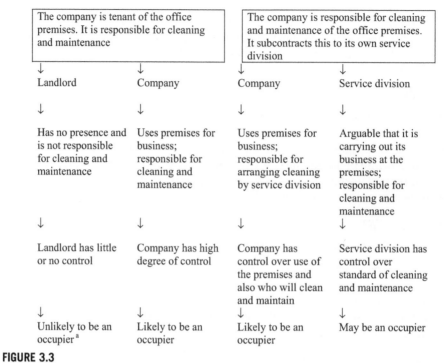

The company is tenant of the office premises. It is responsible for cleaning and maintenance		The company is responsible for cleaning and maintenance of the office premises. It subcontracts this to its own service division	
↓	↓	↓	↓
Landlord	Company	Company	Service division
↓	↓	↓	↓
Has no presence and is not responsible for cleaning and maintenance	Uses premises for business; responsible for cleaning and maintenance	Uses premises for business; responsible for arranging cleaning by service division	Arguable that it is carrying out its business at the premises; responsible for cleaning and maintenance
↓	↓	↓	↓
Landlord has little or no control	Company has high degree of control	Company has control over use of the premises and also who will clean and maintain	Service division has control over standard of cleaning and maintenance
↓	↓	↓	↓
Unlikely to be an occupier [a]	Likely to be an occupier	Likely to be an occupier	May be an occupier

FIGURE 3.3

Who is an occupier? [a] Duties are imposed on landlords who are under an obligation to repair by the Defective Premises Act 1972. The duty is wide and is even extended to trespassers. It is sufficient that the landlord merely has a right to enter to carry out repairs under the terms of the lease.

Does the company owe a duty of care?

An occupier owes a duty of care to its visitors, licensees and invitees ('visitors'). This will include anyone to whom the occupier gives any invitation or permission to enter or use the premises. Express permissions or invitations are relatively easy to identity.

An occupier must, however, be aware of implied invitations. In particular, it will always be difficult to argue against a claim that, for example, ungated pathways or a long-standing hole in a fence constitute an implied invitation, particularly where children are concerned.[7]

In the case study, the client has been invited to attend the company's office premises. The client attends the office and enters the client reception area. This is clearly an invitation to that client for him to be at the premises and he will be a visitor for the purposes of the 1957 Act.

---

[7] Cooke v Midland Western Railway of Ireland [1909] AC 229, HL.

Trespassers

What if the individual claiming against the occupier is not a visitor? What if that individual was a trespasser? Under the Occupier's Liability Act 1984 (the 1984 Act) an occupier owes a duty of care to those who are not visitors if:

- they are aware of the danger of a specific risk and have reasonable grounds to believe it exists
- they know or have reasonable grounds to believe that the trespasser is in the vicinity of the danger or may come into the vicinity
- the risk is one against which, in all the circumstances, they may reasonably be expected to offer the trespasser some protection.

Has the company discharged the duty?

The common duty of care owed by occupiers to visitors is a duty to take such care as is reasonable, taking account of all the circumstances. Visitors must be reasonably safe in using the premises for the purposes for which they were invited or permitted. As with any duty of care-based claims, such as a claim in the tort of negligence, the definition is legalistic and complex.

How then can the facilities manager judge whether or not the common duty of care has been discharged? Examine the case study below under 'Can liability be excluded?': surely it is sufficient for the facilities manager to have a rigorous cleaning and maintenance programme in place? If a client slips on the highly polished entrance floor, does that client have a claim against the company as occupier under the 1957 Act? Is this claim supported by case law? In determining whether or not the company has discharged its common duty of care as occupier to the client, the court will consider whether, in light of all the circumstances, the company has failed to achieve the standard of care required by the 1957 Act. Again this is legalistic and unclear. Facilities managers should consider the following standard of care guidelines developed through case law:

- Children must be expected to show a lesser degree of safety awareness than adults.
- Visitors attending premises to perform their particular skill are expected to be prepared for special risks associated with that skill.
- Where the use of independent contractors gives rise to a claim based on the contractor's negligence, an occupier may satisfy the standard of care if it has acted reasonably in employing that contractor, believing the contractor to be competent.

A lesser standard of care is owed to trespassers under the 1984 Act. The duty can be discharged if the occupier takes such care as is reasonable in all the circumstances to ensure that the trespasser does not suffer injury on the premises.

Can liability be excluded?

Consider the following in the context of the case study: the facilities manager arranges for the following warning sign to be placed in the reception area of the office premises:

NOTICE

All visitors are asked to take care when entering the reception area. The floors are highly polished and very slippery, particularly when wet.
No liability for personal injury or damage to property is accepted by the company.

Does this warning discharge the company's common duty of care? Once again the legal rules are complicated. If a notice or warning sign is reasonable in all the circumstances and is sufficient to allow the visitor to avoid the danger then the occupier may well have satisfied the required standard of care. To return to the case study: the client is warned of the slippery floor by such a notice. While they know that it is a potential hazard, they have no choice other than to cross the floor to reach the reception desk. The company has created a danger that cannot be avoided. The notice does not discharge the common duty of care owed to the client.

On the basis that the warning about the slippery floor is insufficient to discharge the common duty of care, can the company rely on the exclusion of liability for loss or damage to personal injury or property contained in the notice? It is possible to limit or exclude liability in certain circumstances; if, for example, a visitor has agreed to such an exclusion, then in so far as loss or damage to property is concerned it would be effective. Such agreement could be formal, by means of signing a disclaimer, for example, but may also be implied. A trespasser, for example, who enters premises uninvited, implies agreement to exclusion of liability. However, a decision to walk over a slippery floor to reach a reception desk does not constitute agreement on the visitor's part. Therefore, it is not possible to exclude liability for personal injury in these circumstances.

KNOWING YOUR PORTFOLIO AND MANAGING COSTS

There is no substitute for facilities managers knowing what properties are under their control. It is not enough merely to have details of the address. Building up a detailed set of information about the portfolio is time-consuming and potentially costly. However, the advantages of doing this cannot be overemphasised. Consider how the facilities manager can add value to their role in the company and participate in board decisions in the following scenarios:

- The company's board of directors wants to rationalise that company's premises requirements. A good knowledge of the portfolio will enable the facilities manager to answer questions such as:

  - Which premises are not cost effective?
  - Can leases be terminated early without penalty?
  - Are there title or planning restrictions which may impact on the timing or value of an open market disposal?

- The board is seeking to diversify the company's business. The facilities manager can advise on whether any of the existing premises can be redeveloped or undergo a change of use.

- Preliminary discussions are taking place with a potential purchaser of the pany. The facilities manager can address issues that may have an impact or price of the transaction, such as:
  - Can any of the properties be used in the negotiations?
  - Do any of them have unusual or special features?
  - Do any of them have onerous restrictions which should be brought into ___ open at an early stage?

- Lawyers or other professional advisers have been instructed to act on a matter relating to a particular premises. The factual details such as the landlord's address and the agent's name are missing. Various deeds and documents are also missing. The lawyers spend a considerable amount of time collecting basic information, which increases their bill. Clearly, it would be more cost effective for a facilities manager to provide factual information to the lawyer or professional advisor rather than to pay the lawyer or professional adviser to collect that information. A transaction is also likely to run more smoothly if the lawyer or professional advisor is provided with all relevant background information at an early stage. A well-informed facilities manager can use their knowledge of the property and the potential problems of a transaction to negotiate a realistic fee structure with the lawyer and professional advisors.

Moreover, knowledge of any occupational and/or use restriction enables the facilities manager to formulate solutions.

Premises requirements are rarely at the core of a company's key business strategy. This is likely to mean that the facilities manager will have little support from the board of directors for any initiatives aimed at scheduling or collating factual information about the property portfolio. Nevertheless, a full knowledge of the portfolio will enable the facilities manager to enhance their position by providing key information at important stages of the negotiations, anticipating problems, offering solutions and achieving cost savings.

Key knowledge requirements

The key information that any facilities manager should know about the premises under his control should include the following:

- identification of property
- planning consents and restrictions
- title matters
- terms of occupation.

REAL ESTATE VALUATION

The valuation of real estate is a broad topic, in terms of both the different categories of property and the methodology involved. Effective real estate management will often require a clear understanding of the value of the assets and liabilities held. A valuation is an essential tool for considering opportunities to release capital, realise development gains and achieve cost savings.

Valuation theory is best left to the valuer; however, an awareness of valuation procedure is of relevance to the facilities manager. This section provides guidance on procuring valuations, and considers certain practical issues which impact on property value.

The regulatory framework

The provision of valuation advice within the UK is normally prepared within the regulatory framework of the Royal Institution of Chartered Surveyors' (RICS) *Valuation Standards*, published in January 2008 and better known as the Red Book. The *Valuation Standards* have been updated to reflect the evolving requirements of the International Financial Reporting Standards. Further valuation guidance may also be found in the International Valuation Standards, published by the International Valuation Standards Council (IVSC).

The Red Book

Compliance with the Red Book is a mandatory requirement for chartered surveyors, who undertake the majority of property valuations in the UK, although certain categories of valuation are excluded. These include valuations undertaken in the course of litigation, arbitration and other disputes, valuations prepared in the context of negotiations and those undertaken by an in-house valuer solely for use within its own organisation.

As key requirements of any valuation, the Red Book's specifications include:

- the need to agree the valuer's instructions in writing prior to issue of the valuation report
- the need to adopt the correct 'basis of valuation', appropriate to the purpose of the valuation
- the minimum content of valuation reports.

Procuring valuations

When commissioning a valuation, it is important that both the client and the valuer have a clear understanding of the purpose of the valuation and the level of service required. This is unlikely to be achieved by a letter simply requesting a valuation of a given property for a specified fee, unless standard terms and conditions of engagement have already been agreed.

A more comprehensive definition of the task is normally required and the Red Book specifies the minimum level of detail needed. The main items that need to be agreed in writing before the final valuation report is issued include:

- the purpose of the valuation
- the address of the property and the legal interest to be valued, including the treatment of fixtures and fittings and plant and machinery normally regarded as part of the land and buildings

- the basis or bases of valuation
- any practicable assumptions to be made relative to the basis of valuation (e.g. to take into account any change in the value of the property if planning permission for an alternative use is being sought and is likely to be granted)
- the date of valuation (this will have to be either before or at the date of the valuation report)
- any restrictions on how or what the valuer may do in the course of providing the service
- the requirements for obtaining the valuer's consent to publication of the report
- the limits of liability to parties other than the client
- the nature of information to be provided (such as title documentation, lease details, trading accounts, planning permissions) by the client or its advisers and the extent to which the valuer is to rely upon that information
- the treatment of environmental issues, such as land contamination
- disclosure of any prior involvement with the property (or properties) by the valuer
- the fee basis.

It is always prudent when instructing a valuer to ensure that they have appropriate prior experience of the nature of the task, the category of property and the geographical location on which to base their opinion of value.

The purpose of the valuation

The purpose of the valuation and the nature of the property will determine the basis of valuation. The Red Book defines the accepted bases of valuation and among those most likely to be encountered in a facilities management context are:

- market value
- market rent
- depreciated replacement cost
- existing use value
- net realisable value.

The purposes and bases are discussed in more detail below.

Valuations for acquisition/disposal

Market value

If a valuation is required prior to acquisition or disposal of a property, the appropriate basis of valuation will normally be market value. This valuation basis is intended to represent the valuer's opinion of the best price at which the sale of a legal interest in the property would have been completed for cash consideration on the date of valuation (the full definition is set out in the Red Book).

As market value supposes completion of a hypothetical sale on the date of valuation, the valuer has to assume that, before the date of valuation, there had been a reasonable period to allow for the proper marketing of the property, agreement of the price and terms, exchange of contracts and completion of the legal formalities.

When considering a market valuation, specific factors to be aware of are as follows.

Special purchasers

Market value excludes bids from purchasers with a special interest in the property. A special purchaser might comprise another party holding a legal interest in the same property, such as a tenant, or the owner of a neighbouring property, who might require the property for expansion or to create an access to a nearby development site. In such circumstances, the special purchaser may be justified in submitting a bid in excess of the open market value, although the amount the special purchaser chooses to offer in excess of the market value will depend on how highly it values the property.

'Hope value'

Market value assumes an unconditional sale. This is particularly important to note in the case of development land. In many instances, development land is sold on a conditional basis, subject to achieving an appropriate planning consent, for example, the disposal of a factory site for residential redevelopment. The market value will reflect the circumstances of the property at the date of valuation. Therefore, if planning permission has not been obtained by that date, it will only reflect the so-called 'hope value' for an alternative use that exceeds the underlying value for the existing use.

The quantum of hope value will be determined by the probability, costs and timescale of obtaining planning permission. If a successful planning permission is unlikely, then the hope value will be limited. Alternatively, if achieving the required planning permission is a virtual certainty, the market value without planning permission may show little, if any, discount relative to the market value had planning permission already been obtained.

If a valuation is required of a property with the benefit of planning permission for an alternative use, before the planning consent has actually been obtained, then it will be necessary to request a 'market value of the property subject to the special assumption that planning permission has been granted'. The details of the planning permission assumed to have been obtained may also need to be stipulated; for example, the assumed density of development will often have a direct impact on the development value.

Further investigations

As market value is intended to represent the transaction price, ideally it should take account of all the factors that would be considered as part of the transaction process. This may include a detailed consideration of such matters as:

- the condition of the property
- the possibility of land contamination
- any onerous restrictions in legal title.

Often, for reasons of either cost or timescale, these detailed investigations are not undertaken when preparing a valuation, and as a result assumptions or caveats are included in the valuation instruction letter and report. Thus, if a property is actually acquired or disposed of and full and detailed enquiries are made, it should be appreciated that issues could emerge that cause the final price to differ from that included in an earlier valuation, prepared without the benefit of such investigations (aside from any issues of market movement). In this context, it should also be appreciated that it is difficult or impossible for a valuer to assess the impact of disrepair or land contamination on the value of a property without the costs of making good/remediation having been quantified. This would normally require additional expert advice from, for example, a building surveyor or an environmental consultant.

Market rent

Market rent is the estimated amount for which a property (or part thereof) should lease on the date of valuation assuming a willing lessor and willing lessee on appropriate lease terms in an arm's-length transaction after proper marketing (the full definition is set out in the Red Book). Appropriate lease terms refer to prevailing lease terms for similar properties in that location. If the market rent is normally accompanied by an incentive, such as a rent-free period, the assumed incentive and lease terms should be stated by the valuer. The rent paid, disregarding any incentive given, is normally referred to as the headline rent, while the rent paid, adjusted to reflect any incentives, is known as the net effective rent. Where incentives are given to a tenant, the headline rent (cost per square foot) will be higher than the net effective rent.

Valuations for incorporation within financial statements

Under International Financial Reporting Standards (IFRS), property is often to be held in the accounts at 'fair value'. Fair value is an accounting concept and is defined as 'the amount for which an asset could be exchanged or a liability settled between knowledgeable willing parties in an arm's-length transaction'. While this definition is similar to market value, fair value will not always equate to market value. The methodology for measuring fair value will normally depend upon the availability of market evidence. If market evidence exists, then fair value will usually be interpreted as equivalent to market value. However, if there is no market evidence because a property is specialised, a depreciated replacement cost or an income approach is likely to be more appropriate.

Financial statements are produced on the assumption that the entity is a going concern unless management intends to liquidate the entity or cease trading, or has no realistic alternative but to do so. This assumption underlies the application of fair value to property, except in cases where it is clear that either there is an intention to dispose of a particular asset or that option for disposal has to be considered.

In the context of real estate, the main bases of valuation used to determine fair value are as follows.

Operational property

Market value

This valuation basis is intended to represent the valuer's opinion of the estimated amount for which a property should exchange on the date of valuation between a willing buyer and a willing seller in an arm's-length transaction after proper marketing (the full definition is set out in the Red Book).

Depreciated replacement cost

'Specialised properties', such as an oil refinery, a museum or a power station, are those which are rarely, if ever, sold on the open market, other than as part of an ongoing business. The appropriate valuation basis for specialised properties is 'depreciated replacement cost', which by definition is not based on a market price.

In practice, the depreciated replacement cost is made up of the existing use value of the underlying land, together with the replacement cost of the buildings and building and site services, suitably depreciated to allow for age, condition, economic or functional obsolescence, environmental and other relevant factors. It is a prerequisite of depreciated replacement cost that every valuation assumes the adequate potential profitability of the business, taking into account the value of the total assets employed and the nature of the operation.

Existing use value

When valuing for accounts prepared under UK generally accepted accounting principles, non-specialised owner-occupied properties are valued on the basis of their existing use value, which closely follows the definition of market value, except that it assumes that the property can only be used for the existing use and that vacant possession is provided on completion of the sale of all parts of the property occupied by the business. It therefore excludes any hope value for more valuable alternative uses.

Non-operational property (surplus assets)

Under IFRS, surplus assets are to be initially accounted for at the lower of the carrying amount and the fair value less costs to sell, and subsequently at fair value less costs to sell.

Properties held for sale in the ordinary course of business

These properties are measured at the lower of cost and net realisable value, described in more detail below.

Net realisable value

Facilities managers should be aware of the difference between market value and net realisable value. Net realisable value is entity specific and while it would normally be the valuer's remit to provide the market value figure, the responsibility for quantifying net realisable value often resides with the facilities manager. The net realisable value will normally be lower than the market value, reflecting deductions such as professional fees on disposal and holding costs (including rates, security and building insurance), as well as any factors that emerge during detailed enquiries when a sale takes place (see above). In a stable market, clearly the longer a property takes to sell, the greater the reduction in net realisable value, reflecting holding costs over a longer period. When estimating net realisable value, it is therefore useful to obtain details of the disposal period envisaged by the valuer.

Investment property

Properties held as investments are normally valued on the basis of market value.

Valuation approach

Freehold interests

The approach to the valuation of freehold interests will depend on whether a property is owner-occupied or let as an investment.

Non-specialist owner-occupied properties will often be valued by the sales comparison approach, which involves obtaining details of sale prices on comparable properties and then making adjustments to reflect physical differences and market circumstances. Where there is limited evidence of comparable sales, but an active leasing market exists, an income approach may be more applicable. This entails estimating the rental value of the property and applying an initial yield to derive the capital value. The income approach is normally applicable to leased, that is, income-producing, investment properties. However, an owner-occupied property will be sold with vacant possession. It follows that the calculation will need to reflect the delays and costs of securing a hypothetical tenant for the property by means of an initial income void, allowance for letting costs and potentially the use of a higher yield.

Investment properties are also normally valued using the income approach, which in the simplest case involves capitalising the passing rent to arrive at the capital value.

Owner-occupied trading properties, such as hotels, public houses and private healthcare facilities, are typically valued as 'fully equipped operational entities, valued having regard to trading potential'. As properties of this type are usually sold as an ongoing business, the valuation methodologies normally capitalise the adjusted earnings for an individual property. Different types of trading properties often have bespoke valuation methodologies, each reflecting the underlying drivers of the trade.

Long leasehold interests

Leasehold interests fall into two main categories. The first comprises long leasehold interests where the rent payable is normally only a small proportion of the full market rental value and the lease term exceeds 50 years, for example, a 99-year lease at a nominal rent. Long leasehold interests will normally be categorised as assets, although these could be wasting assets as the unexpired lease term diminishes.

Short leasehold interests

The second category comprises short leasehold interests, such as a five to 25-year lease, typically with regular rent reviews. From a financial reporting perspective, the shorter leases will in many instances be categorised as operating leases and will therefore be off the balance sheet. However, the tenant's leasehold improvements will be capitalised and if a revaluation of these items is required, the value will often be measured on a depreciated replacement cost basis. Short leasehold properties that are surplus to operational requirements will often require a provision reflecting the holding costs to exit. In a market context, short leasehold interests are often perceived as a liability in value terms, particularly if the supply of a category of property exceeds demand. Often incentives will have to be given to an incoming tenant to take over the leasehold interest. In the UK, shorter leases and tenant's break clauses (see Business leases: Term, this chapter) have become more prevalent in recent years, reflecting tenants' requirements and government pressure on landlords to offer more flexible lease terms.

The value of a leasehold interest

The value of a leasehold interest is dependent on the relationship between the market rental value and the rent passing under the lease. If the rent payable is less than the market rent, then a profit rent is created and potentially a positive capital value may arise. This will often be the case with long leases where normally the rent passing is lower than the market rent. However, before reaching a conclusion as to the value of leases it is necessary to consider other factors, such as dilapidation liabilities and, in the case of short leases, what, if any, financial incentives may need to be granted to a new tenant to take on the lease. Such incentives could, for example, take the form of a rent-free period or payment of a reverse premium. With short leases, once these and other factors have been reflected, a positive value derived from a profit rent may in practice become a nil or negative value.

Negotiating lease terms

In the lease negotiation process, the facilities manager will often have a lead role in determining the heads of terms for the occupier (tenant). This will encompass defining the tenant's occupational requirements (e.g. the floorspace required, the length of lease needed and what level of rent is affordable). Once this has been done, the tenant will in effect have a shopping list of requirements. However, before entering the negotiating process, it is useful also to consider the landlord's motives. In most cases, a landlord will be seeking to maximise the income and capital return on their investment. This return will be very much dependent on the lease terms agreed.

Table 3.1 sets out some of the main terms of a lease, which will maximise or minimise the value of the landlord's interest. If the tenant has a particular wish to include a term or terms which reduce the capital value for the landlord (from the column on the right), it can in negotiating seek to counter the negative effect of this

Table 3.1 Terms of a lease which will affect capital value

Enhances capital value for the landlord	Reduces capital value for the landlord
Maximum lease duration, e.g. 25 years	Minimum lease duration, e.g. five years
A higher rental value	A lower rental value
Maximum security of rental income, i.e. tenant has strong financial covenant	Rental income at risk, i.e. tenant has a weaker financial covenant
No incentives for the tenant at lease commencement	Extensive rent-free period and/or capital contribution from the landlord
More frequent rent reviews	Less frequent rent reviews
No tenant's break clauses	Frequent tenant's break clauses

from the landlord's point of view by proposing the inclusion of a separate term or terms from the left-hand column.

In a buoyant property market, when demand exceeds supply, the features in the left-hand column will come to the fore. However, in a bear market, such as existed during the recession of the 1990s, the issues in the right-hand column will prevail. These factors contribute to the cyclical nature of property investment values.

FURTHER READING

Royal Institution of Chartered Surveyors (RICS). *Appraisal and Valuation Manual*. Also, RICS has produced A Guidance Note addressing service charges in commercial property; see either www.rics.org or www.servicechargecode.co.uk.

Managing your Business Effectively

Financial Management

4

Connel Bottom

Financial strategy and management are a core part of any business, supporting the achievement of the organisation's short, medium and long-term goals. For facilities managers, financial management refers specifically to the effective and efficient use of available finance through the use of planning and control mechanisms. This, ideally, is a proactive process which ensures that the right level of financial resource is available at the right times, enabling the required level of service quality to be delivered by a contractor, supplier or staff member.

This chapter aims to provide facilities managers with practical skills for managing finance, focusing on the theory and application of best practice management techniques associated with strategic planning, analysis and/or control of facilities, goods, services or projects.

BACKGROUND ECONOMICS

Financial management skills may be used and supported in different ways, according to the size, function, flexibility and complexity of an organisation. Before looking at the principles of issues such as cost control, benchmarking, procurement and value engineering, it is important to look at the economic structure of the organisation within which those principles will be applied.

Universal principles

Facilities management strategy, structure and scope within an organisation (public or private sector) may take many forms; for example:

- management of a direct labour team
- management through an intelligent client function (ICF) of both in-house and external service providers
- partial management through an ICF and assistance from an external managing agent or partnered provider, such as under the umbrella of total facilities management (TFM), Public Private Partnership (PPP) or Strategic Services Partnership (SSP) schemes
- management of facilities as a landlord/institutional investor or agent.

189

However facilities management is organised, the principles outlined here give basic and practical building blocks which can be replicated by any facilities management function in pursuit of sound, proactive financial management.

The significance of facilities management costs

Typically, the facilities manager is concerned with a wide variety of costs associated with the provision of premises or buildings, business and staff support services. Table 4.1 shows a typical classification of cost centres that may fall within the facilities manager's remit. It also illustrates audited annual costs for a UK office portfolio (as an example) and their associated percentage significance to the facilities management team.

Table 4.1 Typical classification of facilities management cost centres

		£	%
Property costs	Property: rent, rates, insurances, etc.	4,028,692	33.12
Premises costs	Building services maintenance	869,255	7.15
	Building fabric maintenance	125,595	1.03
	Grounds maintenance	80,180	0.66
	Alterations and fitting out	672,835	5.53
	Cleaning	922,083	7.58
	Security	602,217	4.95
	Utilities	1,189,355	9.78
	Internal décor	55,371	0.46
Business support	Archiving	18,990	0.16
	Reprographics	670,450	5.51
	Stationery	267,746	2.20
	IT communications	120,855	0.99
	IT computers	69,719	0.57
	Postroom functions	174,521	1.43
	Transport and fleet	92,744	0.76
	Porterage	94,719	0.78
	Travel management	8,569	0.07
	Furniture	14,667	0.12
	Business equipment	22,210	0.18
Staff support	Catering	1,947,575	16.01
	Gym	25,324	0.21
	Occupational health service	49,430	0.41
	Facilities management helpdesk	39,355	0.32
		12,162,459	100

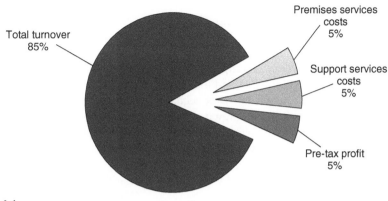

FIGURE 4.1

Typical facilities costs compared with organisational turnover.

To put these costs into context, Figure 4.1 illustrates a typical facilities management budget compared with organisational turnover and profit.[1] While the facilities management budget may appear small in comparison, it is worth considering the impact of facilities services on an organisation. For example, failure of building services owing to an inadequate maintenance strategy or the disruption caused by the absence of a disaster recovery plan following a terrorist bomb blast could have a significant effect on the core business financial performance. Furthermore, staff resources are a significant cost in most organisations and the workplaces that support them, managed by the facilities manager, are known to influence productivity and organisational performance.

Systems technology and financial management

All too often, an organisation's systems infrastructure is implemented without an understanding of the needs of the facilities management department. In some cases, this has the effect of encouraging individuals to set up or procure separate systems. It is quite common, for example, to see a facilities manager controlling and reporting budgetary performance using the organisation's enterprise resource planning (ERP) system and managing space-recharging or end-of-year benchmarking activities on separate spreadsheet files. However, information technology (IT) infrastructure suppliers now realise the significance and benefit of integrating facilities management information, and more and more facilities managers are able to use bespoke facilities management solutions. When setting such systems up, facilities managers should ensure that:

---

[1] Williams, B. (2001) Facilities Economics in the European Union. Building Economics Bureau, London.

- there is no duplication of data input or storage
- the system uses meaningful budgetary codes and classifications
- the classification is sufficient for conducting external benchmarking
- where there are several buildings or clear functional boundaries, costs can be collated and managed flexibly and logically
- the system computes costs fairly for recharging purposes
- direct comparisons can be drawn with contract performance information; remember that financial management is important, but performance information is arguably more important to facilities management customers and their productivity
- costs can be logically combined with key benchmarking parameters, for example, floor areas and headcount
- there are adequate measurement definitions for other staff to follow, for example, what should be included within a net internal area (NIA) measurement, or the definition of full-time equivalent (FTE) staff numbers
- descriptive units in automated reports are correct and full, for example, never show costs as £/sq m, always define the units fully as $£/m^2$ GIA p.a. (gross internal area per annum)
- taxes are separately identified and described
- facilities management functions in other buildings or countries adopt the same principles.

BEST PRACTICE FINANCIAL MANAGEMENT

Budgetary control

Budgetary control is concerned with ensuring that the financial management plan that has been agreed with the board of management is achieved. Control is effected through monitoring expenditure before and after commitment to prevent underexpenditure or overexpenditure. The principal considerations of budgetary control are:

- planning
- preparation
- coordination
- behavioural effects
- communication
- reporting and reconciliation
- change management
- evaluation.

Planning

The budget-planning process is concerned with the following issues:

- forecasting

- facilities managers' skills
- information on required service levels
- information on the organisation's future business strategy.

Budget planning is carried out in advance of the budget period; it is normal to plan for next year's budget two to three months before the start of a new financial year. In some organisations budgets are prepared on a rolling basis and in these situations planning will be a recurrent activity carried out at predefined intervals, say quarterly or half-yearly.

The planning process relies partly on the facilities manager's skills of forecasting, and partly on the quality of information available on previous financial and service quality performance and future business objectives. The facilities manager will normally be able to control the quality of information on the former through an efficient facilities management information system (see earlier in this chapter, Systems technology and financial management), and the presence of a strategic facilities management function with representation on (or at least good links with) the board of management would normally provide enough information on the latter. The absence of a strategic facilities management function in medium-sized to large organisations can hinder the budget-planning phase. If financial planning is not informed and is consequently not proactive then the facilities manager will run the risk of having to react to problems as they arise, sometimes called 'firefighting'. This scenario is clearly an inefficient use of resources in any organisation and is one that can be prevented through effective planning.

Preparation

Budget preparation is concerned with the following issues:

- the use of appropriate cost centres for management and benchmarking
- zero-based budgeting
- provision of data relating to service level quality and cost
- the declaration of all assumptions made
- capital and revenue planning
- treatment of depreciation
- tax planning.

The first three of the above list are discussed immediately below, and the latter four later in this chapter.

Cost centres

The facilities management department should have appropriate cost centres around which forecasts of expenditure and quality can be made. It is advisable to separate facilities services into functional groupings; for example, cleaning, services maintenance, furniture maintenance, security, catering and reprographics. These groupings can sometimes be subdivided further; catering, for example, can be divided into staff dining, hospitality catering and vending. It is generally accepted that the more detailed the classification the better the forecast, although this depends to some extent on the amount and quality of data available at the preparation stage. The classification adopted by the facilities management department may be more detailed than that required by the organisation's finance department, which is typically more interested in the correct accounting for the needs of the business than in the efficiency and performance of the facilities services. To maximise efficiency and performance, the cost centres must be logical, both in the context of service functions and in terms of retrospective performance appraisal or benchmarking (see elsewhere in this chapter, Benchmarking facilities costs). A suitable classification is effectively a benchmarking protocol, an example of which is given in the Appendix.

Zero-based budgeting

The facilities manager normally forecasts the activities necessary, the quality required and the associated cost of each service. It is poor practice to use the previous year's budget figure and add on a percentage for market inflation, as this approach does not always reflect an accurate forecast of future change. Detailed forecasting using a 'clean sheet of paper' is inherently much better practice, particularly where the facilities manager is not experienced or where it is expected that they will leave the organisation within the budget period. The construction of a new budget for each service from basic principles is referred to as zero-based budgeting. This approach has the effect of focusing attention on such issues as waste, unnecessary performance, leasing versus purchasing of equipment, and so on, prompting the facilities manager to ask such questions as 'Why do we do it this way?' and 'Why are we not getting better results?' – questions that are often asked by consultants after undertaking a facilities management review or a benchmarking study.

Data

Effective forecasting depends on the availability of good quality data for such issues as:

- business strategy or changes being proposed that affect facilities services
- key dates associated with business change
- service level qualities and costs for supporting the business through any planned changes
- the facilities management market, for example, price increases in labour, equipment and materials
- facilities management staff costs, for example, recruitment, salary and benefit costs associated with new appointments
- stability of budgets/cost predictions on ongoing capital projects, for example, a new building
- the effect of new legislation being enforced, for example, the cost of providing disabled access

- currency or purchasing power parity, if applicable, for example, budgets for a European office portfolio
- flexibility associated with contracts, for example, supply and payment of personnel in disaster recovery circumstances
- travel costs.

A zero-based budgeting approach will effectively accommodate all of these issues, assuming that the facilities manager has completed enough research into the potential for organisational or market change. The zero-based approach is in fact a step-by-step methodology that will ensure accurate forecasting, and the facilities manager should realise that it is important that all assumptions taken in relation to the above issues are well documented. Such information will be of use when reconciling the effect of change or if a new facilities manager is recruited during the budget period.

While zero-based budgeting is considered to be a best practice approach to budget preparation, there are other sources of data that the facilities manager may wish to use. These include the following:

- contract prices from existing or recent agreements
- a schedule of rates for a certain element of the supply chain
- pricing books published within the industry/marketplace.

In addition, contractors can often provide key budgeting information if requested and consultants usually record strategic and detailed cost information for many clients and buildings.

Typically, there are three levels at which data may be used for forecasting a budget figure (Table 4.2). The choice will normally depend both on the accuracy needed in the forecast and on the confidence and extent of the data available.

Table 4.2 Illustration of pricing levels within budget preparation

Level	Example	Price
High	Cleaning £11.00 per m^2 of NIA p.a.	
Elemental		$£/m^2$ NIA p.a.
	Office areas	6.00
	Toilets	2.00
	Restaurant	1.00
	Windows	1.50
	Pest control	0.50
	Total	£11.00
Itemised	Wash down partitions, 2000 m^2 @ £0.25 per m^2 of partition surface area = £500	

NIA: net internal area; p.a.: per annum.

Coordination

Facilities management budgets need to be coordinated with other departmental forecasts in an organisation, such as marketing and human resources. Furthermore, if the facilities management department is sufficiently large, it is possible that there are individual facilities managers charged with budgeting for the services under their remit. Best practice is for the ICF to prepare an outline budget for the coming year which coordinates with the wider business strategy for the short (annual), medium (next five years) and long term (over 10 years). Individual facilities managers should, within this framework, prepare more detailed budgets relating to their particular areas or 'bundles' (services that can be grouped and managed collectively).

Behavioural effects

The budget should provide tight financial constraints for managers to work within. The constraints must not be too tight or lax, however, as this can influence staff motivation to plan and control costs.

Responsibilities within the facilities management hierarchy must be clearly demarcated. For example, a benchmarking study may find that a facilities manager cannot control a particular budget because another department has management jurisdiction over the contractor. This may happen when there is a central procurement team letting contracts without full input from the facilities manager who will be responsible for managing service delivery.

Communication

The budget and its components, including objectives in relation to service level qualities, need to be communicated both to facilities management staff and to any core contract staff who need to understand the business requirements for the budgetary period. Effective communication will ensure that the facilities manager's plans are carried out as required.

Reporting and reconciliation

Reporting and reconciliation are mechanisms for control on a periodic basis, usually monthly. It is important that:

- the format used for reporting is clear, concise, meaningful and useful to the facilities management function
- the information is accurate
- the information is as close to real time as possible to promote proactive management
- the information is, ideally, conveyed in numerical and graphical formats (the latter will improve understanding and identification of potential problem areas)
- there is a written explanation of any numerical and graphical information
- the report is circulated to the staff responsible in advance of any control meetings: timely circulation will ensure that meetings are productive.

The information presented in the report normally includes the following:

- the principal cost centres, and where possible any subdivisions
- the monthly budget projection for each cost centre
- the actual cost incurred within each cost centre
- total projected and actual costs
- key resource-driver statistics, such as levels of staff resource and volumes associated with the actual costs incurred
- records of service level quality delivered in each main contract/cost centre.

The vast majority of events or services envisaged in the budget preparation period are likely to go according to plan. With this in mind, the facilities manager can concentrate on controlling specific cost centres that are, by their nature, likely to vary. This concept is commonly called 'management by exception'.

Evaluation

The presence of detailed budgets within facilities management provides a means of evaluating the performance of individual facilities managers or demonstrating the value of having an ICF within the organisation. Such a process is considered best practice, particularly where the performance of individuals is linked to skill sets and the facilities management staff training strategy.

The evaluation process can also be considered a useful protocol for financial benchmarking, the procedures for which are discussed later in this chapter, in Benchmarking facilities costs.

Change management

Once authorisation has been granted for a budget, any changes should be resisted; correct preparation and treatment of uncertainty should have eliminated many of the potential areas likely to cause a variation. Particularly dynamic organisations, however, carry an increased degree of uncertainty, and change in planned facilities management services is possible. If this change is significant and cannot be controlled within the normal reporting regime, then specific reporting within the management hierarchy may be necessary, as well as extra managerial meetings to control the effects of change.

Business change can drive significant expenditure within any facilities operation. Change management is therefore a very important process which should be supported, in organisational terms, with an effective and meaningful policy (normally this is part of an organisation's overarching facilities management policy). The following principles should be recognised and planned for proactively:

- Identify the forms and types of business change affecting the facilities management function.
- Understand and perhaps quantify the effects of the different types of change unique to the business. For example, will a new building need to be acquired? Will there be an impact upon existing floor space, its layout, quantity and quality?

Will service level quality change? Will there be a legislative dimension that needs to be addressed?

- New and existing service provider contracts should be designed to cover effectively and efficiently the potential for change. Change should not provide an opportunity for contractors to overcharge for supplies and services.
- A communications plan should be designed to cover both the participants of the change exercise and the facilities management customers whose expectations need to be satisfied throughout and after the change project.
- The identification and quantification process together with clarity of contract agreements will allow effective cost planning for budgeting purposes.

Procurement

Strategic considerations

The procurement of goods and services is a key function within facilities management, ensuring that goods and services are provided competitively and that they add value to the organisation's core business. In this respect, the choice, planning and implementation of procurement activities are a central tenet of efficient financial management.

Any procurement activity must be based around a sound understanding of the overriding strategic considerations. For most organisations these are essentially:

- the extent to which services are to be provided by contractors rather than in-house staff
- the scope/nature of the service that is to be provided
- the capability of the wider facilities management market to satisfy the service requirement.

Contracted-out versus in-house service provision

In deciding whether service provision is to be delivered in-house or by a contractor, the facilities manager should focus on keeping what the organisation considers to be its core business activities in-house, and contracting out non-core business activities (see What to outsource, in Chapter 7, for more on this topic).

There is a common presumption within industry that contracted-out service provision is automatically more economical than its in-house equivalent. In fact, there are numerous examples in the facilities management industry where the opposite has been shown to be true. What is certain, however, is that in-house facilities management departments have traditionally been seen as being uncompetitive and inefficient. This perception arises for a number of reasons, which include:

- **Poor communication**: facilities managers are notorious for not telling their customers how good they are (and for not having any performance statistics to prove it). In this respect communication is an important element of facilities management marketing.

- **Overprovision**: it is easy for the in-house team to be sidetracked into providing more than is required under their service level agreement (SLA), if there is one, thus adding to operating costs.
- **Organic growth**: many facilities management departments have grown organically with the parent organisation. As a result, procedures that worked efficiently in a 100-person environment are struggling to cope when the headcount reaches 500.
- **Poor governance**: controls over such issues as staff resources, including absence due to sickness.

Scope of services to be provided

Before initiating any procurement activity, firms should ensure that they have a thorough understanding of the scope and nature of the service required (see The tender document, in Chapter 7). This will enable an appropriate buying strategy to be put in place. The strategy adopted is likely to vary depending on whether the service required is:

- a single service (cleaning, catering, porterage, etc.)
- a group or 'bundle' of services (here the emphasis will be on the mutual compatibility of the services to be provided; thus it is common to find fabric, services and grounds maintenance bundled together)
- a 'total facilities management' package, whereby services across the three main facilities management subclassifications are supplied by one provider under a single contract.

Facilities management market capability

Within the global marketplace there is a plethora of service providers all advertising themselves as 'facilities management contractors', but in reality offering widely varying types and levels of services to their customers. This can have a significant impact on efficiency and quality of service provision, the level of support afforded to the core business units and thus the facilities cost/value model.

Any tender shortlist must therefore only be compiled after an exercise has been carried out to check that the core competencies of the prospective bidder companies are compatible with the scope and nature of the overall requirement. For example, there is little point in putting a facilities management contractor on a tender list for a catering contract if the company's core business revolves around the provision of maintenance services (for more on selecting contractors, see Choosing contractors, in Chapter 7).

It is possible to group facilities management contractors into broad categories which derive from the background of their parent organisation. These initial classifications can help to weed out weaker candidates from the selection process at an early stage. The categories are:

- **General contractors**: typically these organisations have moved into facilities management to take advantage of the improved profit margins (1–2 per cent for contracting compared to 3–5 per cent or greater for facilities management). Their core competencies tend to centre around the building-related disciplines.

- **Single service providers**: this group includes catering contractors, cleaning contractors and the like. In many instances such organisations have been encouraged to offer a broader range of services by their customers. Some have become facilities management contractors in the broadest sense, while others pay only lip service to the concept and may subcontract these bolt-on services.
- **Management buyouts**: Large property-owning/occupying corporates have sometimes opted to outsource their property departments. The majority of these ventures have proved extremely successful in a privatised environment; however, this was quickly recognised by the more predatory elements of the marketplace and only a few of them retain their independence. Typically, this type of organisation has strong managerial skills and operates at its best in a fee management/managing agent-type contractual arrangement.
- **Consultancy**: the recession of the late 1980s and early 1990s and the associated drop in the construction industry's workload forced professional consultancy organisations to seek alternative sources of income. Some of them moved into facilities management and again, on the whole, such ventures have proved successful. As is to be expected, such organisations also demonstrate strong managerial skills, but their preference is for the managing agent-type structure (this route is perceived as offering less business risk because the facilities managing agent does not have a direct contractual relationship with the service providers).
- **Consortia**: the advent of the Private Finance Initiative (PFI) and PPP has seen a growth of contracting organisations that have the capability of providing a total package of facilities and services to organisations in a long-term contractual arrangement.

E-procurement

The global facilities management market offers great potential for the use of business-to-consumer and business-to-business e-commerce, not least in terms of the creation of procurement supply chains. Facilities managers in large organisations will undoubtedly be aware of the drive behind the introduction of enterprise management IT systems (see earlier in this chapter, Systems technology and financial management) for the effective control of all business information, financial or otherwise. Internet technology is making it possible to integrate office functions with key suppliers of all but the most complex facilities management services.

Growth of e-procurement

The growth of e-procurement within facilities management is being driven by the following factors:

- The prequalification process can happen much more quickly, as e-technology enables the rapid transfer of key information.

- The production, copying and distribution of tender documentation can be significantly reduced when e-technology is used for procurement.
- Economies of scale can be created where individual packages of services are procured through the medium of a 'club' e-commerce site. The packages are essentially components of a large contract negotiated on the terms of the buying power of the club as an entity. They create a closed supply chain, linking a defined customer base with selected suppliers. A site would be based on a front-end browser which allows clients access to a global catalogue of business supplies and equipment from approved suppliers, helping them to search for the best value for money in their chosen product category.
- E-technology can support online tendering of commodities such as utilities, where each bidder is allowed to make offers within a certain time-frame and can view the result live in the form of a ranked league table.

Drawbacks

There are many problems associated with the movement towards e-commerce. Facilities managers should bear the following factors in mind:

- Portals can only create economies of scale if there are enough consumers and the volume of transactions is high. In other words, greater numbers of participants will bring greater economic benefits.
- Reliability of the technology and the supply chain information is of paramount importance if the system is to function efficiently. For example, service level performance capability information must be realistic and accurate if consumers are to remain confident in the e-procurement system.
- Many businesses or consumers need to streamline their existing purchasing and procurement processes if economies are to be expected. For example, poor business processes are often the cause when an organisation's travel department procures 'competitively' priced tickets that are in reality more costly than when individual staff purchase tickets directly from the airline. In such cases fundamental business processes may need to be re-engineered, and cultural changes must precede them.
- Many procurement departments will actually want to meet and study their potential service providers. In facilities management the 'people factor' is often an important issue when assessing the quality of proposals.
- Usability factors affect even the simplest of web sites, not to mention large and complex procurement portals.

PRIVATE INVESTMENT AND PARTNERSHIP

Private investment and partnership is normally associated with PFI, which was launched by the government in 1992 and reviewed in 1997 (The Bates Review).[2]

---

[2] Sir Malcolm Bates (1997) *The First Report on the Private Finance Initiative* for HM Government, June.

Now known as PPP, the central objective of the initiative is to involve private sector expertise directly in the procurement of new public sector buildings (or specific assets such as hospital equipment or IT equipment) and the operation of these facilities, including the provision of facilities management services, over a defined life-cycle period. The works and services are paid for over time by means of a single unitary charge. The methodology is equally applicable to the provision of buildings and services to a private sector organisation, sometimes referred to as a 'Private Sector PPP'.

The aims of PPP may be summarised as follows:

- to support the government's aim of reducing the public sector borrowing requirement (PSBR)
- to obtain cost savings (in terms of initial capital and operating expenditure) and efficiency gains by using private sector experts to construct and service buildings
- to enable the transfer of appropriate risks from the public sector to the private sector
- to bring the benefit of using and paying for the asset over time (rather than significant initial expenditure), so that more funds are available for other projects and public services can generally be improved.

Facilities management professionals should have significant input throughout a PFI/PPP project in order to ensure that value for money, particularly over the life cycle of the project, is optimised. The facilities manager should understand the procurement processes involved, how life-cycle economics play a significant role, the use of performance measurement systems and the application of payment mechanisms.

The PPP process

The recommended procurement programme methodology is to follow the Office of Government Commerce (OGC) Gateway Review process, which is a procedure for reviewing a project at key decision points using experienced professionals who are independent of the project team. There are five OGC Gateway Reviews during a project, three before contract award and two focusing on service implementation and confirmation of the operational benefits realisation. More information can be obtained on the OGC website (www.ogc.gov.uk).

In terms of the detailed procurement process, the European Union (EU) Procurement Directives provide for four main procurement procedures:

- **Open procedure**: all interested parties who respond to an advertisement in the *Official Journal of the European Union* (OJEU) [formerly the *Official Journal of the European Communities* (OJEC)] must be invited to tender.
- **Restricted procedure**: interested parties are invited to respond by submitting an expression of interest, which will be used to create a shortlist of tender candidates.

- **Competitive dialogue procedure**: applicable to complex procurements where more flexibility is needed to discuss the proposed contact with interested parties. Shortlisted parties are invited to participate in dialogue. This is now a standard procedure in most PPP projects.
- **Competitive negotiated procedure**: now limited to circumstances where the other procedures will not work. This used to be the standard procedure in PPP projects but is now only used in exceptional circumstances.

The following steps reflect activities normally undertaken in the competitive dialogue procedure:

1. **Establishing the business need for change**: issues such as poor building stock condition, lack of functionality and poor service levels are frequently key drivers for change.
2. **Appraisal of options**, for example, using techniques for financial analysis.
3. **The preparation of a business case**: establishing whether an investment option exists over the life of the project.
4. **Establishment of a procurement team**: the correct level of resource and skills are important factors.
5. **Select procedure**, that is, competitive dialogue in this example process.
6. **Issue OJEU notice**, that is, publication in the *Official Journal of the European Union*.
7. **Return of expressions of interest**: prequalification questionnaires, and so on, are returned for evaluation.
8. **Prequalification of bidders**: to identify suitable or competent participants and selection of a longlist for subsequent stages.
9. **Invitation to Participate in Dialogue (ITPD)**: longlisted candidates are invited to participate in dialogue.
10. **Dialogue**: used to refine the requirement by gaining the expertise of prospective suppliers.
11. **Call for final tenders**: participants/suppliers are invited to submit a formal tender. Tender documents will be well developed at this stage and will include a clear scope of work/services, schedule of requirements/output specification, proposed contractual terms covering such aspects as length of contract and performance measurement system.
12. **Submission and evaluation of final tenders**: assessments should be made against the criteria set out in the OJEU advertisement or tender documentation and should follow a predefined evaluation strategy. Weightings are commonly used to value responses in terms of financial and non-financial criteria.
13. **Appointment of preferred bidder**: award of contract and financial close, involving final drafting of all contractual/legal documents/schedules/mechanisms, and so on.
14. **Contract management**: throughout the life cycle of the project; commonly the client forms a strategic management function for sponsoring of facilities management policy and the monitoring of performance and payment.

Life-cycle cost models

A model, in this case, can be defined as a mathematical representation of the future cash flows associated with life-cycle costs for a given building. Life-cycle cost models can be extremely complex, but all will normally have a common theme, essentially a detailed classification of the elements or components making up the entire building, cross-tabulated with each year of the project. The level of detail depends on the particular objective. For example, it is common for the facilities manager to develop progressively more detailed models throughout the procurement stages outlined above. A good starting point is the classification known as the Standard Form of Cost Analysis, published by the Building Cost Information Service (BCIS) of the Royal Institution of Chartered Surveyors (RICS).

For a basic model the facilities manager will need to be able to write mathematical equations capable of forecasting future cash flows using the following information:

- the current (capital) cost of constructing the item (e.g. boiler) or element (e.g. space heating)
- the proportion of replacement (e.g. 100 per cent)
- the timing of the first and subsequent replacements (e.g. every eight years)
- additional on-costs (e.g. cost of an engineer working out-of-hours, scaffolding).

More detailed models can be developed which incorporate the following information:

- Data which explains the timing of replacements, for example, risk can be represented by a probability distribution of the mean life cycle for normal wear and tear conditions (or not).
- Facilities management service costs, for example, services maintenance, fabric maintenance and cleaning all have interrelationships that need to be mathematically represented in the models.
- Taxation and financial assumptions, for example, discounting rates and inflation rates.
- Facilities management strategies, for example, use of planned preventive maintenance instead of reactive-only maintenance.

Further related information can be found later in this chapter (Whole-life economics and financial analysis: Whole-life costs).

Payment mechanisms

Within PFI/PPP contracts, the unitary payment (paid by the client to the contractor/ service provider) represents financial consideration for the provision of buildings/ floor space and the provision of services. The payment mechanism is a mathematical formula for adjusting the unitary payment for issues such as:

- **Unavailability of floor space**, for example, in a hospital the mechanism would normally cover the availability of zones such as a ward or an X-ray room.

- **Performance**: the performance measurement score is normally one of the data inputs in the payment mechanism; a single score may be used or individual/ grouped service scores may be weighted and used in the formula.
- **Pass through costs**: where a cost borne by the contractor is passed straight through to the client; commonly utilities costs, postal charges, stationery costs, and so on.
- **Consistency**: sometimes a mechanism will only make an adjustment based on past 'performance' or track record over the past three to six months.

The formula normally includes weightings which are applied to the measurements listed above. The end result will normally be a financial adjustment figure for a given period, for example, the previous month, which will be deducted from the contractual unitary payment. Payment mechanisms frequently look very complex on paper, but a simple spreadsheet analysis can be undertaken to understand the true effects of the adjustments that may take place in the future. The facilities manager, armed with knowledge of the client's business, the particular building and the potential for service failures, should play a significant role in the practical testing of such systems and mechanisms before they are contractually agreed in the procurement process.

BENCHMARKING FACILITIES COSTS

Why benchmark?

In facilities management, benchmarking has been defined as 'a process of comparing a product, service process, indeed, any activity or object, with other samples from a peer group, with a view to identifying 'best buy' or 'best practice' and targeting oneself to emulate it'.[3] This definition effectively outlines one of the most important (but often misunderstood) aspects concerning facilities management benchmarking, that is, 'targeting' or taking action in order to release value to the organisation. Facilities managers should fully understand the reasons why they are embarking on a benchmarking exercise; they are often forced into following a market trend, an organisational mandate, or the potentially dangerous misbelief that at the end of the exercise costs can be cut. Of course benchmarking is about saving costs, where possible, but it is also about performance and value, and fundamentally, customer requirements.

The need for benchmarking within organisations can also be linked directly to the competitive environment in which they operate. Globalisation and information and communications technology advances inevitably mean that organisations must be increasingly dynamic in order to stave off the competition. Over the past 20 years there has been a business performance revolution which has been characterised by the introduction of methodologies and techniques

---

[3] Harrison, A. (2003) *The SANE Space Environment Model*. DEGW, London.

such as activity-based costing, the balanced scorecard, the business excellence model, the performance pyramid and shareholder value frameworks, all of which are approaches that many facilities managers will have experienced or even feared.

The techniques discussed above are useful when talking about the value chain that exists within organisations where the facilities management department or its activities could be described as a critical link. The techniques ultimately provide strategic management information through the use of performance measures (in various proportions and mixes) associated with the following issues/themes:

- stakeholders' (e.g. investors/shareholders, regulators, suppliers, employees) satisfaction/contribution
- leadership (e.g. experience, skill)
- company policies and strategies
- processes, skills, policies and procedures (e.g. time, quality, safety, waste efficiency)
- cost drivers (e.g. resource, productivity, supply chain)
- financial (e.g. cost of capital, share earnings, profitability, operating costs)
- capabilities (e.g. technology absorption)
- innovation and learning (e.g. training and development)
- customer satisfaction, loyalty and profitability.

Misinterpreting the value of benchmarking

It is common for benchmarking to be incorrectly mixed up with performance measurement techniques. The truth, however, is that benchmarking is a systematic process of evaluation; it should be a fluid methodology that *uses* performance criteria (among other measurements) in the search for improvement beyond best practice. Within the facilities management discipline this misconception is prevalent, largely as a result of facilities managers often viewing occupancy costs as their only output, rather than taking the wider view of adding value to the organisation by providing support through accommodation, workplaces and services. In addition, facilities managers always try to rely on general indicators that are typically available in the public domain. Many published databases provide information which is of questionable value; for example, a facilities manager may try to compare the costs of, say, security services against a published benchmark range of £3.00 to £15.00 m^2 GIA without any other information being available. Those who understand the nature of facilities management service provision will readily understand that this comparison provides no value at all, since different cost levels are driven by individual building/location characteristics and the quality of service provision.

The wider perspective

Facilities managers need to add value to the organisational value chain. There needs to be a realisation that the discipline of facilities management encompasses much more than cost alone. True facilities management benchmarking activities can be largely associated with evaluation of the following aspects:

- assets
- inputs
- processes
- outputs
- systems.

Buildings are significant assets to many organisations and the facilities manager would typically be concerned with such issues as physical, functional and financial performance. In this respect there are many methods of evaluating asset performance and feeding information into the benchmarking process. For example, building condition, postoccupancy evaluation, building quality assessment and investment appraisal techniques are all capable of providing data for comparative evaluation. Furthermore, the relatively simple analysis of space utilisation is often overlooked, although this in fact influences many of the other issues listed above.

Inputs can be associated with processes and outputs and can relate to many different circumstances within the remit of a facilities manager. For example, the procurement of a new building would be a discrete project with inputs and various activities interacting to make identifiable processes, the final output being the asset or building. The principal inputs in this case would be the labour and materials being used throughout the construction process. The provision of a catering service would normally depend on various inputs such as catering staff labour and raw food ingredients, the outputs being meals, and so on. Possible measures that a facilities manager may wish to use during the course of input/process/output benchmarking could relate to (examples given are as applied to security service provision):

- cost (e.g. hourly cost of guarding, cost of surveillance activities, total cost of security per annum)
- quality (e.g. employee skill/experience level, accuracy of intelligence reports, number of shoplifters apprehended or customer satisfaction)
- time (e.g. surveillance hours per annum, time taken to assimilate intelligence information, time to apprehend thieves)
- risk (e.g. health and safety breach by security staff, injury to third parties due to an unplanned activity, excessive loss to the business through theft).

Systems refer to the mechanisms that are in place to assist with the efficiency of processes. In the case of a new building project an example might be a web-based information system for sharing project knowledge. In relation to facilities management service delivery activities a computer-aided facilities management (CAFM) system should collect information to assist with the management of processes. Measurements relating to time, cost, quality and risk may equally be applicable in the case of systems.

Internal and external benchmarking

Benchmarking is about comparison with a best practice peer group, where the primary aim is not to copy but to emulate inputs, processes, and so on, with a view towards increasing output performance and hence value to the organisation. Peer group information can come from several sources:

- those internal to the organisation
- external to the organisation but similar in terms of business
- external to the organisation but dissimilar (at least in terms of principal business)
- first principles, sometimes referred to as 'should cost modelling'.

In facilities management it is quite common to find organisations operating in different sectors with broadly similar systems, inputs, processes and perhaps outputs. This is largely as a result of few facilities managers challenging the norm or because only a few suppliers routinely develop new and innovative services. This can be a detrimental problem when benchmarking as it is possible that 'peers' are broadly similar. Within the industry there is a growing realisation that the potential value of benchmarking increases as the comparison moves from within the organisation to a level that challenges current practice. Significant innovation can often be achieved by looking at the problem from first principles. However, the facilities manager should understand that this shift in comparative methodology will need to coincide with an increase in the level of commitment needed at all levels in the company hierarchy, the amount of resource dedicated to the project and an increase in skills needed within organisations.

The process of benchmarking

Depending on business priorities, the facilities manager will decide the nature and extent of the benchmarking process. In practice the objectives may vary widely, for example:

- benchmarking facilities services as part of a business process re-engineering or relocation exercise in order to inform business case analyses
- a complete review of all services within the facilities function to prove value for money
- a detailed service function audit covering analysis of costs and performance associated with, say, catering to test the adequacy of service processes and output performance relative to business group needs
- as part of a formalised contract, commonly called 'benchmarking and market testing' in PPP-type contracts.

The paragraphs that follow provide a generic step-by-step guide, illustrated in Figure 4.2, for the facilities manager to implement benchmarking practice and techniques. The methodology is applicable to cost benchmarking and is useful as it indicates the detailed processes involved in ending up with meaningful comparative data on which to base decisions.

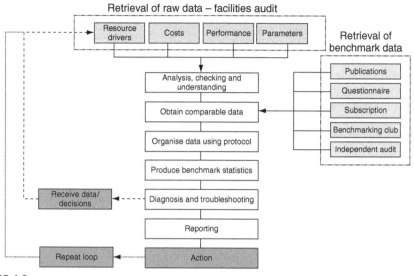

FIGURE 4.2

Methodology for benchmarking facilities management costs.

The facilities audit

The facilities audit represents a review of the costs of providing office space and services within an organisation. It is important to realise that the audit is not concerned with cost alone, but also includes analysis of the building and organisational characteristics that drive cost (resource drivers) and the associated levels of performance.

Resource drivers

A resource driver is a characteristic that influences the required levels and/or deployment of a resource. It is important for the facilities manager to understand that output performance (for example, how clean a building is) may remain at the same level even though the level of resource required (for example, the number of cleaners and frequency of cleaning operations) varies in different buildings (an otherwise comparable building may, for example, be located beside an area of pollution which increases the amount of dirt accumulating on the glazing).

Resource drivers have been classified as:

- **quantitative**: usually relating to characteristics of the building or organisation that can be readily measured, for example, floor area, window area, number of staff and contractors' staff, number of covers served in a restaurant
- **qualitative**: characteristics such as the location of the building or the specific preferences or aspirations of the organisation
- **economic**, for example, interest rates and market conditions
- **operating conditions**, for example, specific lease conditions and organisational aspirations.

Performance data

Performance characteristics are important within benchmarking in order to identify the level of output associated with cost. Unfortunately, many organisations do not record sufficient performance-related information (although this is changing in the advent of developments in CAFM and helpdesk software), and in any case it is difficult to make comparisons between organisations which measure performance metrics differently or not at all. In such cases the facilities manager has the difficult and subjective job of comparing and measuring performance. Customer satisfaction surveys can provide a quick means of procuring performance data.

Cost data

The retrieval of cost data will be a relatively simple process for the facilities manager who has developed and maintained facilities service budgets at a detailed level (see Budgetary control, earlier in this chapter). It is recommended that facilities costs are audited or collected at the greatest level of detail possible. This will ensure that the facilities manager understands what is included within an overall service cost. For example, from an accountancy point of view, a stationery budget may include reprographics supplies, whereas for the purpose of facilities management benchmarking it is often accounted for under the reprographics cost centre. It is often a lack of such understanding that leads to the failure of many commercial benchmarking groups or partnerships.

A spreadsheet is often the best way of assimilating cost information for analysis checking and ultimately for comparison with peer information. Figure 4.3 illustrates such a process for collecting and analysing the raw data.

Invoice description	Cost	Protocol categories	Services maintenance	Catering	Stationery	Mail/ distribution	Reprographics
Xy (lifts)	£7,612.00	Services maintenance	£7,612.00				
supplier → ABC (chillers)	£11,100.00	Services maintenance	£11,100.00				
Dp (sprinkers)	£1,010.00	Services maintenance	£1,010.00				
FG (CB cooler units)	£1,726.00	Services maintenance	£1,726.00				
HIJ (generator controls)	£46,000.00	Services maintenance	£46,000.00				
NOP (lighting controls)	£12,318.00	Services maintenance	£12,318.00				
EFG (kitchen equipment)	£6,840.00	Catering equip		£6,840.00			
P-touch tapes	£177.24	Stationery			£177.24		
stamps	£42.40	Distribution				£42.40	
April stationery	£3,028.25	Stationery			£3,028.25		
Xerox 3050 plan printer	£132.00	Printing & repro					£132.00
May stationery	£931.66	Stationery			£931.66		
July stationery	£782.49	Stationery			£782.49		
Staples for photocopiers (J23456Y)	£37.80	Printing & repro					£37.80
August stationery	£2,201.38	Stationery			£2,201.38		
Photos printed & framed for facilities	£151.00	Printing & repro					£151.00
		TOTALS	£79,766.00	£6,840.00	£7,121.00	£42.40	£320.80

FIGURE 4.3

Analysis of facilities service costs using a spreadsheet application.

Parameters

Parameters are the metrics that are used to express benchmarked costs in a meaningful way. In order to provide useful statistics, it is necessary to establish a direct relationship between the parameter and the cost of service. For example, it is unlikely that vending costs can be related to floor space, whereas there will, under normal circumstances, be a directly proportional relationship with the number of staff or occupants using the building in a 24-hour period. Similarly, it is common to express the costs associated with premises services on a cost per square metre of floor area, and support services such as catering, mail distribution and stationery are commonly expressed as costs per capita.

Parameters must be measured on a comparable basis between the organisation and its peer group. The RICS Code of Measurement Practice provides a standard protocol for floor space measurements that has been readily adopted within the industry. It should be noted that it is common for different countries to have slight deviations from this standard.

Comparing facilities costs using incorrect and incompatible parameters renders the benchmarking process ineffective. Services maintenance costs can be seen to vary significantly (in benchmarking terms) because of the use of GIA and NIA; it is a common error for facilities managers to use the wrong parameter by mistake. All too often, professional and managerial reports relating to premises and facilities are littered with incorrectly described floor area measurements. This is also a common reason for commercial benchmarking partnerships failing.

Obtaining comparable data

Obtaining good quality comparable data is probably the most difficult task that the facilities manager will experience during a benchmarking exercise. The primary sources of information are:

- **Publications**: professional journals frequently publish articles sharing information on facilities costs. However, usually this type of data has been desensitised and/or excludes knowledge concerning specific circumstances, performance characteristics, and so on.
- **Questionnaire**: from the very simplest to the most complex, questionnaires can provide good 'average' information. Individual respondents cannot necessarily be relied upon to adhere to measurement rules, standards of accounting, and so on. For these reasons the data may represent a very wide variety of circumstances including resource drivers, performance characteristics and, therefore, costs.
- **Subscription services**: the industry has many subscription ventures, particularly on the Internet. For a fee, facilities managers can obtain data relating to the costs and possibly performance characteristics of peer organisations. However, subscription companies must obtain the raw data in some manner, usually by way of questionnaires, so the quality of the data is constrained by the motivations of individual respondents and their accuracy of description and classification.
- **Benchmarking clubs**: a number of organisations can exchange best practice, cost and performance benchmarks and targets via a club. These are very

worthwhile as long as each individual is fully committed and a facilitator is present, that is, someone who can make decisions about measurement standards, accounting methodologies, reporting standards and the implementation of best practice. The absence of a facilitator usually leads to failure.

- **Independent audit**: this is carried out by an independent expert who should understand all the problems and pitfalls associated with collecting and comparing data from different organisations. It is common for such experts to own a consistent database containing service costs, performance measures and resource driver characteristics. Such an approach allows an efficient and informed assembly of true peer-group data.

The facilities protocol

A facilities protocol or standard method of accounting is one of the facilities manager's most important tools. An industry standard is given in the Appendix, and illustrates a classification of facilities management services using categories and more detailed subdivisions of activity, called sections. For novice users, a good protocol also provides definitions or examples of the types of cost that a facilities manager must include, as well as indicating which items should be excluded from the analysis (and where these items should in fact be included).

When conducting benchmarking, the choice of protocol is not as important as ensuring that the rules of what must be included and excluded are standardised, that is, ensuring that what the protocol represents is clear to all those using it. The spreadsheet analysis of an organisation's costs shown in Figure 4.3 illustrates how the facilities manager can label individual invoice costs with protocol category names. The methodology used is then clear to all parties participating in a club benchmarking arrangement.

THE IMPORTANCE OF CLEAR PROTOCOL CATEGORIES

The following example illustrates the importance of clear protocol categories. One benchmarking club was, on first comparison, misled as to the cost of building services maintenance owing to confusion over what should be included within this category. The original figures indicated that the total cost for maintenance was £50/m^2 NIA p.a., although an expert benchmarker would have known that this was too much expenditure for the relatively new facility. On further analysis the club realised that this figure incorrectly included costs associated with alterations and churn, which is a distinct protocol category in its own right. The correct building services maintenance cost was in fact £25/m^2 NIA p.a.

Production of benchmark statistics and graphics

Before interpreting benchmarking results, the raw data from the analysis work must be turned into information ready for comparison. The choice of correct benchmarking parameters will result in meaningful statistics, such as cost/m^2 GIA p.a. However, the computation of statistics at this level is normally called 'first strike benchmarking' – the most basic level – as opposed to a more detailed study using secondary or diagnostic indicators. These secondary indicators will help the facilities manager to understand the occurrence of variance between benchmark statistics such as cost/capita p.a.

The diagnostic indicators used by the analyst (generally in a more detailed study) usually include the following:

- resource driver characteristics; for example:

  - space utilisation/occupancy density
  - ratio of hard to soft landscaping
  - window to floor ratio

- levels of resource, for example, cleaners' hours/m^2 GIA p.a.
- consumption of volume figures; for example:

  - kWh p.a.
  - number of restaurant covers per annum
  - number of black and white copies per annum
  - number of mail items per annum

- labour rates, for example, cleaning operative cost/hour
- output performance levels; for example:

  - percentage of planned maintenance tasks completed on time
  - user satisfaction statistics.

Diagnosis and troubleshooting

Statistics are often meaningless when presented in tabular format. Graphical analysis is recommended for revealing the true relationships hidden in the data. An example of graphical analysis showing an organisation's costs compared with those of its peer group is shown in Figure 4.4.

A first strike interpretation of Figure 4.4 reveals that cleaning expenditure is high in relation to the peer group. This should prompt the facilities manager to ask simple questions such as:

- Is there a mistake in the calculations/analysis?
- Has the peer group reflected all resource driver characteristics?
- Is the peer group comparable in terms of the business performance requirements?

The above three questions would indicate a discrepancy in the benchmarking process. Questions should then be asked relating to the function itself; for example:

- Is the level of resource too high?
- Is the hourly labour rate too high?

All these questions represent examples of the diagnostic process that must be conducted before reporting any benchmark information. At this stage the facilities manager should review all analysis decisions on a 'repeat loop' basis until all

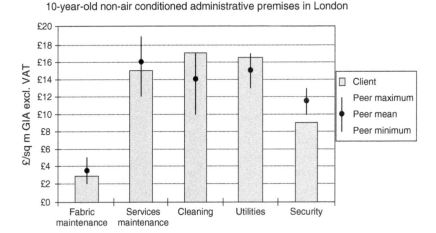

FIGURE 4.4

Facilities service cost benchmarking results.

variations can be explained confidently without the possibility of error within the process.

Reporting and taking action

Benchmarking is all about learning and doing things better. Just as it is important to understand resource drivers and to use a standard analysis protocol, it is important to communicate the results of the exercise to all concerned. A benchmarking report must always look to the future; it is not good value for money to conduct a fine level of analysis only to report current levels of expenditure and performance. Broadly speaking, a well-run facilities management function will show average levels of expenditure and performance. An 'average' performance, however, should not give rise to complacency, as there will always be ways of reducing cost and improving performance.

WHOLE-LIFE ECONOMICS AND FINANCIAL ANALYSIS

The role of life-cycle costing has been growing in momentum over the past 20 years and it is now widely accepted in the construction and property industries that the long-term operating costs and the initial capital expenditure on a building should not be considered independently of each other. More recently, with the advent of PPP-type procurement, life-cycle costs and the techniques used to analyse them have dominated many boardroom discussions.

Increasingly, the facilities manager is being recognised as an expert who understands the drivers behind life-cycle costs, whether relating to premises, business or staff support costs. Decisions made at the start of a building project, during the design stages, will have most impact on capital construction cost, but can also influence operating costs. Furthermore, there is an argument that if the design team proactively controls life-cycle costs (with a correspondingly small increase in design cost) then unnecessary costs can be avoided over the life cycle of the building. Knowledge about facilities management markets, processes, business requirements and the interaction of organisations with buildings is of paramount importance when planning the whole-life economics associated with a given project.

Whole-life costs

Whole-life costs comprise:

- **capital expenditure**, for example, the cost of constructing an office building
- **financing costs**: the cost of borrowing money to finance a particular project
- **occupancy costs**: subdivided into the following classification for ease of reference (note that these tie up with the protocol used for budgeting and benchmarking):
  - *premises costs*, for example, maintenance of building fabric and services, utilities, cleaning and replacement costs
  - *business support costs*, for example, mailroom, reprographics and archiving services
  - *staff support costs*, for example, catering, crèche and fitness centre
- **taxation**, for example, capital allowances on certain costs associated with new construction
- **residual values and disposal costs**.

Techniques for financial analysis

Financial analysis techniques are often considered difficult to apply and understand and are sometimes left to an accountant or financial analyst. They are, however, within the grasp of any facilities manager who can:

- accurately forecast expenditure (see earlier in this chapter, Budgetary control)
- decide which method of analysis is appropriate for the type of decision being made
- use spreadsheet software to conduct the analysis and interpret the results (although this is not a prerequisite).

In practice, the techniques used often centre around the analysis of future cash flows. The techniques are:

- payback period
- discounted cash flow
- net present value
- internal rate of return
- cost–benefit analysis.

Table 4.3 Payback analysis for new reprographics equipment

Initial cost of new reprographics equipment £10, 000		
Period	Income through recharge of occupants	Cumulative total
Year 1	£1000	£1,000
Year 2	£1500	£2,500
Year 3	£2500	£5,000
Year 4	£3500	£8,500
Year 5	£1500	£10,000
Year 6	£3000	£13,000

Payback period

This technique is applicable if the underlying question for the facilities manager is how long it will take to pay back the full cost of an acquisition. Table 4.3 illustrates the analysis in simple terms by showing that an initial expense of £10,000 is repaid at the end of year 5.

However, the payback analysis shown in Table 4.3 does not take into account the cost of borrowing the initial £10,000. Furthermore, the technique does not take account of time and income after the payback period; assuming that the equipment does not stop working at the end of year 5.

Discounted cash flow

In this technique, the simple payback period analysis illustrated above is subjected to a discount rate which takes into account interest or the cost of borrowing the £10,000. It is important to understand what is commonly called the time value of money before using this technique. For example, if the facilities manager decides to buy another piece of equipment in year 2, then what amount of money should be set aside or invested now (year 0) in order to match the anticipated expenditure? This can be illustrated as follows:

Year 2 expenditure (in one year's time)	£21,000
Real interest rate (net of tax and inflation)	5 per cent
Amount to invest now (year 0)	£20,000
	(i.e. $20,000 \times 1.05 = 21,000$)

This example may be extended to allow for subsequent elements of cash flows, such as that created by a planned preventative maintenance contract for the new piece of equipment. The simplest way to deal with this scenario is to work out the discounting factor to be used when analysing each year's expenditure in the cash flow. To calculate the discounting factor, you can use a spreadsheet with the following formula:

$$= 1/((1+i)^y)$$

Table 4.4 Example of discounted cash-flow analysis

Discount rate: 5%			
Year	1	2	3
Discounting factor	0.952	0.907	0.864
Net cash income	£1000	£1500	£2500
Discounted income p.a.	£952	£1361	£2160

where i = likely interest rate, and $^\wedge y$ = the year in which a certain cash sum is spent, for example, 4 for year 4 (NB. $^\wedge$ means 'to the power of').

A simple spreadsheet model using this equation to discount the income cash flows derived from the reprographics equipment example (above) is illustrated in Table 4.4.

The further away a sum of money is planned to be spent, the lower the present value of that future sum. This does not necessarily mean that expenditure in the future does not matter, rather the technique allows the facilities manager to compare between several options; for example, the technique could be used to good effect for evaluating options that differ in terms of specification quality and subsequent maintenance costs.

Net present value

The net present value (NPV) technique compares the sum of the discounted cash flows with the initial capital expenditure. When the sum total of the discounted cash flows exceeds the initial capital expenditure then the NPV is said to be positive and the project can be viewed as economic. When the initial capital expenditure is greater than the total discounted cash flows then the NPV is negative and is deemed uneconomic. Using the reprographics equipment example used above, Table 4.5 illustrates the use of the NPV technique.

Table 4.5 uses a formula to calculate the discounting factor for each year and is used in this document purely for clarity, to illustrate all the stages necessary to calculate the NPV. However, most spreadsheet applications have built-in functions that calculate the NPV using a single formula; for example:

$$= \text{NPV}(5 \text{ per cent}, -10,000, 1000, 1500, 2500, 3500, 1500, 3000)$$

In this formula, 10,000 represents the initial capital sum and the following figures represent the yearly net cash income (these figures would normally be substituted by spreadsheet cell references). It is worth noting that each spreadsheet application will have rules regarding the nature of the values being incorporated in the formula; for example, that the net cash income should be at the end of a given period and that each period must be equal. In addition, because of slight differences in approach, the result given by a spreadsheet calculation can be slightly different from the more explicit methodology shown in Table 4.5.

Table 4.5 Example of net present value analysis

Discount rate: 5%

Year	0	1	2	3	4	5	6
Discounting factor	1.000	0.952	0.907	0.864	0.823	0.784	0.746
Capital expenditure	£10,000						
Net cash income		£1,000	£1,500	£2,500	£3,500	£1,500	£3,000
Discounted income p.a.	–	£952	£1,361	£2,160	£2,879	£1,175	£2,239

£ 10,766: Present value of cash flows (sum of discounted income p.a.).
£ 10,000: Less capital expenditure.
£ 766: Net present value.

An important point to note is the fact that the payback period, with discounting, has increased to almost six years in this example. The higher the discount rate, the longer the (discounted) payback period. In terms of decision making, the facilities manager is likely to be comparing several options. The key criteria would therefore be that:

- The NPV is positive (discounted income is greater than initial capital expenditure).
- The required rate of return (5 per cent) is exceeded.
- The project with the highest ratio of present value to capital expenditure is providing the greatest return (in the above example this profitability index equals 1.08).
- The discounted payback period should be the least possible.

It would normally be considered best practice to present these statistics in tabular form, as they will vary with respect to one another according to the characteristics of the project. This will enable conclusions to be drawn easily and the best option for the particular company to be chosen.

Internal rate of return

The internal rate of return (IRR) determines the discount rate at which the present value of the cash flows equals the initial capital expenditure, that is, where the NPV equals £0. The IRR percentage should be greater than the company's cost of capital if a project is to be considered as economic. In the spreadsheet example given in Table 4.5, the facilities manager could use trial and error to find the discount percentage that results in an NPV of £0. However, most spreadsheet applications have an IRR function that automatically determines the correct percentage. The following equation is an example of how this can be set up:

$$= IRR(-10,000, 1000, 1500, 2500, 3500, 1500, 3000)$$

Again, the individual cash values in this equation would normally be substituted by spreadsheet cell references.

Cost–benefit analysis

The techniques described above essentially compare initial costs, ongoing revenues and the timing of transactions. The techniques do not necessarily (in their basic application) take account of costs and qualitative benefits that affect the business, third parties or the wider environment. Often these benefits are considered intangible, soft or subjective and difficult to measure; for example, noise, smell, image or even productivity. Cost–benefit analysis is a widely used technique where costs are directly compared with the benefits, usually through the use of a simple two-column table.

Applying the techniques

Because decisions made during the design stages of a building impact on capital occupancy costs, the technique of discounting is particularly relevant for comparing different design options. For these reasons a facilities manager should insist on the development of a life-cycle cost model at the earliest stage of a project. As the design becomes more and more detailed the model will grow and should be capable of assisting with ongoing decision making.

The methodology followed for the simple examples given above will assist in the development of practical financial analysis models. Indeed, the following are examples of some common uses of discounting future costs in relation to facilities management:

- **Maintenance**: a common example is the comparison of design options where specifications vary in quality, which impacts on the cost of maintenance and the life until replacement. The techniques are very important in this respect as there is a common misconception that high-quality specifications will always lead to lower maintenance costs.
- **Energy**: the specification of lamp fittings frequently causes a significant difference in energy consumption and hence utilities costs.
- **Cleaning**: cleaning costs are largely associated with time, degree of mechanisation and type of consumables required. Specification of building finishes, therefore, commonly impacts on overall cleaning costs. The problem may be complicated further by the relationship between the cleaning method and the life cycle of components.
- **Component and material selection**: the above example establishes that material/component specification may impact on cleaning costs. However, the effects are not likely to be limited to cleaning alone and can affect other facilities services, such as utilities, security, churn costs, and so on.
- **Investment appraisal**: it is common for facilities managers to be involved in portfolio or estate management decisions where disposals, acquisitions (both lease and buy), refurbishment and new-build options may need

comparison. In such circumstances all of the techniques described above can be used.

PROPERTY COSTS

As occupancy in a building rises, premises and support services costs will also rise, whereas property costs will largely remain static (assuming a short period for the increase in occupancy). For facilities economics, this means that the increase in facilities costs per capita (premises and support services) will be more than out-weighed by the corresponding reduction (as staff numbers increase) in property costs per capita.

Facilities managers are, therefore, principally concerned with space (driving property costs) and occupancy (driving facilities costs). In addition, however, the facilities manager should have a working knowledge of costs associated with the provision of property.

Property expenditure is normally associated with payments for rent, rates, insurance and service charges. These costs normally form a significant proportion of the facilities management budget; for instance 33 per cent in the example illustrated earlier in this chapter in Table 4.1.

Rent

Rent is an amount paid by a tenant in exchange for the use of a building and is usually regulated under the terms of a lease. Mortgage payments to a financial institution (commercial property loan) or 'internal rents' payable to a holding company may be encountered by the organisation if the property is freehold in title. Rent tends to be fixed for long periods, which means that the facilities manager will only be able to exercise an influence on cost at the agreed rent review date (usually every five years).

Facilities managers are often faced with making the strategic decision of whether to rent/lease or own the freehold title of a particular building. Analysis of the options should be related to the organisation's IRR on capital employed (IRR is covered earlier in this chapter, in Techniques for financial analysis: Internal rate of return). The answer to the question relates to whether the organisation can earn more on the capital which is otherwise tied up in the freehold.

For more on the legal issues concerning rent, see Business leases: Rent, in Chapter 3.

Rates

Within the UK, a uniform business rate (UBR) exists as a means of taxation based on the value of the property. All commercial properties are assessed at prevailing market rent levels by HM Revenue and Customs Valuation Office, and a rateable value calculated and fixed for a certain period (reviewed every five years; values become effective after an additional two years). For occupied property the UBR is added to this value to calculate the total sum payable for rates. The facilities manager should be aware of these rules, together with the following opportunities for effective financial management:

- Unoccupied offices and retail property attract half the calculated total sum.
- Unoccupied industrial properties, warehouses and listed buildings attract the full sum.
- Appeals may be made against rating assessments, for which an expert surveyor's services should be sought.
- Where the property assets have changed materially, through part demolition, for example, then rate liability may be reduced.
- External factors, such as neighbouring road works or building works, can reduce the rateable value.

Service charges

Service charges are normally covered under the terms of the lease and serve as a means for the landlord to recoup the cost of facilities services (building maintenance, grounds maintenance, fire protection, cleaning, security, vending, fitness suite, etc.) provided to common parts of buildings or estates. From the landlord's point of view, the investment needs to be protected through maintenance and cleaning so that there is no adverse effect on its value over the lease term. Until recently, service charges were largely hidden in so far as tenants did not contest or renegotiate the levels of cost.

In addition to the principles of benchmarking outlined here, the facilities manager should:

- Study the wording of the lease with care, as this will influence the method of remedy as well as interaction with other payments due in connection with dilapidations or costs of reinstatement at the end of the lease term.
- Request copies of supporting documentation relating to the financial calculations for analysis.
- Remember that the landlord's procurement processes and service contracts may influence the competitiveness of the service costs. The facilities manager is likely to have a better understanding than the landlord of what the expected cost of a service should be, and should press for the renegotiation of any uncompetitive contracts.
- Determine whether excessive or additional charges are being levied by managing agents.
- Determine whether excessive monies are being diverted to sinking funds for large life-cycle replacement costs, such as lifts, air conditioning plant, and so forth.

- Check the measurement of parameters used in the calculation and apportionment of service charges.
- Remember that the common approach for large service charges is for the landlord or agent to budget for the year ahead, recover actual sums periodically and reconcile the balance at the end of the year. In this respect the facilities manager has to rely on the other party's skills of budgeting for facilities management services.

The legal issues relating to service charges are covered in Business leases: Service charges, in Chapter 3.

Building insurance

Like service charges, insurances payable under the terms of a lease have also tended to be 'lost' among other charges levied by the landlord and therefore not generally contested or analysed until recently. It is worth considering the following:

- The exact scope (or comprehensiveness) of the policy and whether it is suitable for the business.
- Insurance policies may cover loss arising in connection with buildings (and foundations, but not the site), contents, rent abatement/loss of rent, contracts and loss of profits.
- The insured party should know the exact period of the insurance so that it is not allowed to lapse.
- In cases of indemnification (where the insured party is to be left in the same position after the occurrence as before), the insurance payment will be the cost of works less both depreciation and betterment allowances.
- In cases of reinstatement (where the insured party is covered for the cost of reconstruction or repairs to restore the property to its original condition), no adjustments are made for depreciation or wear and tear, although a deduction for betterment may be made.
- The effects of inflation may be covered in the insurance policy by way of index linking. This can have a significant effect if inflation has risen and a claim is made within the insurance period.
- Where statutory requirements have changed, the insured party may not be covered for the additional cost of complying with new standards.
- Insurers of commercial and industrial property are normally protected against underinsurance by an 'average clause' which works by adjusting a claim amount by the same proportion as the difference between the full reinstatement value and the sum insured. The remaining balance is normally paid by the insured.
- Remember that the insurance premium is affected by such issues as specification of materials and fire protection measures, which may be under the control of the facilities manager, therefore compliance with the insurance contract and/or legislation is important.

Flood insurance

While flood insurance has been common in the UK since the 1960s, the risk of flooding in certain parts of the UK and Europe appears to be growing. This risk can be translated into rising insurance premiums being passed on to businesses (risk-based pricing) and in some cases the risks are unfortunately deemed uninsurable. The facilities manager should be aware of the following issues that could influence the situation:

- Since the risks of flooding (and associated insurance and remedial costs) are increasing across many parts of the world, this should be one of the criteria used when making portfolio risk or location choice decisions.
- There is a direct relationship between flood insurance premiums and government investment in, or developer-led, flood defence initiatives.
- Insurance premiums can also be reduced if the buildings in question are designed to alleviate the risk of flood damage, for example, the incorporation of basement sump pumps or impermeable membranes/tanking.
- It would be prudent to check the obligations of any current or prospective lease.

Terrorism insurance

Insurance cover for damage caused by terrorist acts resulting in fire or explosion has been severely curtailed since the early 1990s. Costs for damage (for example, fire and explosion, business interruption and loss of rent) are limited in the UK insurance marketplace; however, additional insurance cover (for example, for biological and nuclear contamination and impact by aircraft) can be purchased which is covered by the government-sponsored mutual insurance fund known as 'Pool Re'. If, for example, there was a shortfall in these funds, then the government would provide the balance of cover (commonly called 'insurer of last resort'). The facilities manager should consider the following.

- Terrorism may or may not be a specified insured risk in a lease, therefore it is prudent to check the obligations of current or prospective leases.
- The definition of a 'terrorist act' can be important when weighing up the ramifications of a lease.
- Acts of war are excluded and must not be confused with terrorism.
- Pool Re cover must be taken out in respect of all properties in a landlord's portfolio; that is, a landlord cannot target buildings in high-risk locations. This will have a direct influence on a tenant's costs.
- Currently, cover does not extend to computer hacking or virus damage owing to the difficulty in preventing such attacks or proving their origin.
- Changes in legislation are occurring on an annual basis; therefore, it is advised that the facilities manager is familiar with current rules and guidelines.

The legal issues relating to insurance are covered in Business leases: Insurance, in Chapter 3.

Sustainable property

The global movement towards companies and individuals becoming more environmentally responsible is witnessed in the abundance of EC directives and regulations published on the subject. Sustainable development of property is founded on the principle that the processes necessary for the construction and operation of buildings do not lead to any environmental deterioration of natural world resources. Some of the primary issues which relate to property and impact upon development and operation costs can be considered as:

- **energy sources**: striving to use renewable sources, for example, wind turbines
- **water**, for example, recycling of rain water on site
- **construction materials**: using renewable sources, for example, sustainable forests; evaluation of embodied energy characteristics of different materials during design
- **Energy savings**: prevention of energy losses, for example, through sufficient specification of insulation
- **waste**: recycling of waste material where possible, for example, paper, aluminium and batteries
- **pollution**: reduction in emissions from building plant and service vehicles.

Distributed workplace strategies

Distributed workplace strategies can be directly related to the issue of sustainable use of resources in the urban environment (see SANE Project[3]). It is a common fact that organisations underoccupy and underutilise space that they own or rent. The problem is often exacerbated because of relatively short-term changes in the business environment compared with long-term duration of property ownership; for example, the prevalence of the 25-year lease. Organisations can address the situation by developing a formalised distributed workplace strategy which identifies the types of space necessary for supporting individual groups or teams of people. The key consideration will be to decide on the distribution of space and workplace throughout the city, country or region. For example, staff can be supported at different locations with different facilities services rather than assuming that all staff must be located in a city centre, headquarters office location. The mix of long-term leased buildings, short-term leased offices, serviced offices, professional clubs, business centres, homeworking solutions and other options will have a significant impact on the property and facilities costs described above.

PRINCIPLES OF FACILITIES FINANCE

The facilities manager should have a clear understanding of the background principles of business finance before embarking on any matters relating to the preparation and planning of budgets, the procurement of goods and services, benchmarking facilities costs, and the study of occupancy costs and whole-life economics.[3]

---

[3] Harrison, A. (2003) *The SANE Space Environment Model*. DEGW, London.

Taxation

The facilities manager can contribute significantly to reducing an organisation's overall tax burden. In many organisations, however, there are no processes encouraging the facilities manager to consider the effects of taxation on any particular service or product. Significant investments in facilities should be evaluated carefully through the identification of the correct tax burdens throughout their life cycle, using financial analysis techniques (see earlier in this chapter, Techniques for financial analysis). The effect of input and output value added tax (VAT) payments and recovery on cash flow is particularly significant in financial models. The facilities manager should always consult a tax specialist, as rules and regulations are complex and constantly changing. As a general rule, however, the following issues normally affect the availability of tax exemptions, reliefs and concessions:

- the nature of the company or legal entity performing business
- transactions or works being carried out
- location: different countries have different tax/allowance regulations
- the nature of any funding arrangements
- the nature of the industry or business: some organisations may attract significant tax benefits.

The principal types of taxation that the facilities manager should consider when budgeting are outlined below:

- Corporation tax is tax payable as a result of income calculated on a trading account.
- Income tax is payable on the portion of staff wages that are taxable.
- Capital gains tax (CGT) is payable upon disposal (sales and/or gifts) of a capital asset (where there is a gain in value), subject to exemptions, allowances and other reliefs.
- Inheritance and gift taxes are applied to the value of the inheritance or gift.
- VAT is chargeable on the supply or purchase of goods and services, subject to exemptions and whether supplies are standard or nil/zero rated.
- Stamp duty is normally payable in relation to legal documents associated with property transactions or company shares.
- Landfill tax applies to waste disposed of at landfill sites in the UK licensed under environmental law.

Capital and revenue expenditure

A business commonly has two types of expenditure: capital and revenue expenditure. The traditional (basic) accounting definitions of these are as follows:

- **Revenue expenditure** is associated with purchasing goods and services in the short term, including any associated business input costs: materials, labour, rent, rates, insurance, utilities, maintenance, and so on (common facilities management cost centres).

- **Capital expenditure** is the cost associated with acquiring fixed assets which cost a significant amount and usually provide economic benefits in the long term.

It is normal for a finance department to require capital and revenue expenditure to be budgeted and managed on a monthly basis so that the company will understand any peaks and troughs in cash flow. Such forecasting is intended to help financial managers to ascertain the impact of capital and revenue expenditure on the working capital of the organisation and on company borrowing.

Revenue expenditure (for example, that associated with maintenance and repair work) is allowable as a deduction from the gross profits of an organisation. In this respect, a full-rate taxpayer who pays 30 per cent corporation tax on taxable income effectively avoids any tax liability for facilities services associated with revenue expenditure (the equivalent of a refund of 30 per cent of the cost of qualifying works).

Depreciation

Fixed assets are normally discounted in the accountancy process in order to show the true (reducing) economic benefit of a capital investment over a period of years (the assets lose value owing to use, the passage of time and obsolescence). Another way of looking at the issue is that it would be 'unfair' if expenditure relating to assets purchased in a particular year were included as an expense in the profit and loss account for that year, thus significantly reducing profit levels. In facilities management terms, an effective building services maintenance strategy will normally extend the economic life of assets, which in turn reduces depreciation (thus profits will be higher). Furthermore, capital expenditure associated with improving an asset (for example, life-cycle replacement works) should never be charged as an expense on the profit and loss account. Using alternative accountancy terms, such expenditure should be 'capitalised' and not 'expensed'.

There are two common methodologies for calculating depreciation:

- **The 'straight line' method**: this divides the capital sum (less final salvage value if applicable) by the useful economic life of the asset. The resultant figure is the annual amount of depreciation that reduces the company's profit measurement.
- **The 'reducing instalment' method**: this is considered closer to reality in that the asset is treated as if it loses its value more quickly in the early years compared with later years. A fixed percentage is applied to the reducing balance year on year.

Capital allowances

The law and regulations surrounding capital allowances are particularly complex, with the primary difficulty being how to determine 'qualifying expenditure', that is, capital and revenue expenditure, described above in basic terms. The facilities manager should understand that allowances may be available for certain categories and elements of land, plant, machinery, buildings, and fixtures and fittings.

FURTHER READING

Kelly J, Male S. (1993) *Value Management in Design and Construction*,The Economic Management of Projects. 1st edn. E & FN Spon, London p. 17.

Lucey T. (1996) *Costing*, 1st edn. DP Publications, London p. 387.

Robson W. (1997) *Strategic Management and Information Systems*, 2nd edn. Pitman Publishing, London.

5

Risk Management

Chris Taylor and Frank Booty

Some risks are worth taking; others are not. We identify the hazard: the possibility that harm or distress will occur and assess the risk: the chances of the hazard playing out to our disadvantage. All along, we balance risk and reward.

Managing risk goes somewhat further than simply assessing the risk/reward ratio. It acknowledges the possibility of a situation where you forgo the reward because of the risk. Or where, by removing the hazard, you remove, or at least significantly reduce, the risk, as with stopping smoking for instance. Naturally, there is the middle course whereby, having formed a view about a risk, the likelihood of its occurring and its potential impact, you may be able to reduce it to an acceptable level. Many of the same principles apply in the business world.

Risk management has assumed an ever-greater importance over the past few years with the growing feeling of unease and concomitant risk of collateral damage from urban terrorism. But how many facilities managers have included risks from a terrorist's source in their workplace risk assessments? Health and safety legislation, contrary to popular belief, does include security risks.

THREE PHASES OF RISK MANAGEMENT

Risk analysis

This is the process of working out the many and various things that can go wrong, the likelihood of something happening, and what the consequences might be. Do not simply focus on the major crises: companies in business or physical areas where such things are appropriate may have coherent plans in place to deal with terrorism, bomb attacks including to nearby buildings, and so on, but nothing to counter the effect of a fuel strike, an extended national rail strike, and the like. Risk does not only cover major events.

Risk assessment

In this phase, the appropriate action to minimise the probability of the risk occurring is determined, and the cost and resource allocation to manage the impact of the harm, in the event of occurrence, are established.

228

Risk mitigation

This phase involves the preparation of plans, the identification of management teams, and the assignment of responsibilities for managing the process.

RISK MANAGEMENT IN BUSINESS

Business is an inherently risky enterprise, and good managers will seek to minimise risk through removing or reducing hazards but, in the business as opposed to the domestic context, what exactly does this mean, and how do we set about making it so? The aim of this chapter is to cast light on the subject and, by doing so, to offer facilities managers an understanding of the issues sufficient to understand and recognise hazards which fall within their remit and to establish, then subsequently minimise, the level of risk through a planned and managed process.

Definitions from the Health and Safety Executive (HSE) tell us that 'hazards' are those things which have the ability to cause harm, while 'risk' is the likelihood that harm will occur. But what is 'harm'? In its purest form, in a business context, 'harm' occurs when an organisation, or any of its constituent parts, has to divert from a predetermined plan to react to unforeseen circumstances. If the goal of business is to make money or create value, anything that interrupts or slows that process could be defined as harm.

Practically speaking, a business can be diverted from its desired course for many reasons, some within our control, some without. Those without our control call for business continuity measures and insurance policies to be in place (see Chapter 6). Some within our control are legislated to determine minimum levels of risk management (see Chapter 1, section on Risk assessment), but there is always room for improvement from statutory minima, and there are plenty of situations which are not subject to legislation.

Different types of risk

To consider risk in its proper context, it is important to consider the different types of risk, sometimes called 'pure risk' and 'business risk'.

Pure risk

This includes:

- physical effects of nature, such as fire, storm and flood
- technical events, such as equipment failure
- personal issues, such as sickness or injury
- social deviations from norms of behaviour, including theft, violence and negligence.

Business risk

The key components of business risk are:

- the impact of new technologies or changes in technology
- social impact, such as changes in customer expectation, or increasing litigiousness
- economic impact, such as inflation or budgetary constraints
- political impact, such as the imposition of government ideology, policy or philosophy.

For example, consider a self-employed sole trader who wishes to take an overseas holiday, perhaps a skiing trip. Pure risk here includes everything from flight delays, accidents and injuries. A prudent traveller would carry insurance to reimburse the direct cost of these events. But business risk goes further. What is the business cost to this individual of a week in hospital, or the inability to make that important meeting with a significant prospective customer? Again, key worker insurances and the like can reimburse some of the direct cost, but rarely is this sufficient to account for loss of time or loss of opportunity.

On a larger company scale, think of a fire destroying a retail store, for example. Pure risk covers the structural elements – and here we are very much into business continuity areas – but business risk includes loss of business attention while management time is necessarily diverted elsewhere to deal with the business disruption. Insurance covers the simple rebuilding and loss of earnings, but as the opportunity is almost always taken in such circumstances to bring a building up to a different specification, the costs of this may well not be covered under a straightforward rebuild 'like-for-like' policy. Loss of earnings cover, too, is always a time-consuming negotiation, of which the best that can be said is that if both parties, insurer and insured, are unhappy, that is probably as good as it gets.

The point, however, is that business risk represents the damage that can accrue to a company in these circumstances, in terms of both lost income and lost opportunities. Despite the best insurance cover, many businesses never trade again after such events: that is the real business risk.

There is also business risk in success. A winner of the US Malcolm Baldrige National Quality Award – America's highest and most prestigious business quality award – suffered significant losses in trade after its management team shook hands with the US President who named the company as one of the 'best of the best'. The problem? The directors took their eye off the ball and rested on their laurels for too long. They accepted every invitation to describe at conferences and seminars how they had reached the heady heights of Baldrige but, while this was happening, the company was falling into trouble. Quality – and risk – needs constant attention.

An area of risk of particular interest to facilities managers is the very real risk that can accompany the outsourcing of vital services. Few would argue that the chances of problems are increased when day-to-day control is delegated beyond the immediate structure of the company. The business risk here is a diminishing of productivity, that is, cash loss in real terms, for a variety of reasons. These include loss of reputation through those unversed in company culture delivering poor messages about the organisation, poor service rendering the host company relatively unattractive as an employer, unreliability of the contractor, and even security issues such as theft or industrial espionage. These should all be of concern when placing outsourcing contracts.

DEVELOPING A RISK MANAGEMENT STRATEGY

Awareness of risk in company boardrooms is rising across many areas. A survey by the Institute of Chartered Accountants of England and Wales and the Risk Advisory Group found that directors across the FTSE 500 companies claimed a significant increase in business risk over the preceding 12 months.

This increased risk spread across a wide spectrum. Those surveyed cited as threats accounting issues including the growing detection of financial misstatement, non-compliance with the fast growing number of regulatory and legal frameworks, and criminal and terrorist activity. The risk that caused the greatest concern was damage to corporate reputation. The threats were real, and the concerns high, yet the FTSE directors acknowledged that many companies reviewed risk too infrequently, once every six months or less.

The need for a formal strategy is not an option. It is required by the Turnbull Committee recommendations (see section on Risk planning and responsibilities, below), which place responsibility for risk management squarely on the boardroom table.

A risk management strategy would provide the framework determining an individual company's response to risk, including who would undertake the work involved at a tactical level. It would determine the priority of different risks to the organisation (see section on Prioritising risk, below), and state the frequency and scope of reviews, and the trigger levels of risk that would lead to individual areas being upgraded for more frequent, or more stringent review, or to the stage where remedial action needs to be taken in terms of risk mitigation involving investment or other work.

Where to turn for help

A major concern for companies is the question of information security, and a British Standard (BS 7799 Pt 2, or ISO/IEC 17799) now exists to outline best practice in this area. Much information about this standard can be sourced on the Internet.

Most companies are aware of the risks they face, but the task of bringing them all together 'under one roof' may seem daunting at first. There are specialist companies (e.g. the Risk Advisory Group) or dedicated sections of the major consulting firms (e.g. KPMG's Enterprise Risk Management practice) that can offer help, but most professional bodies and functional organisations are now keyed up to offer useful advice, if only to recommend an endorsed supplier of the appropriate advice.

Prioritising risk

A problem for companies when they start to assemble their different and dispersed elements of risk into a single area for attention is that there seem to be so many, they don't know where to start to address the problem. A sound corporate risk strategy will help, in that it should determine exactly what the company aims to achieve through such a strategy, and actions can be measured simply against that yardstick. But there will always be a call for a list of priority elements, and that comes down to someone making decisions.

The key to effective decision making is that it should be a process. There is a right and a wrong way to do it, and often it is too important to work on the lottery basis where you hurl all the alternatives in the air and go with the one which you catch first.

One process for prioritising decisions is described below.

- **Work on the right decision problem**. In this specific situation, are you aiming for finite security, or confidence building, or something else altogether? Pose the right question, and the answers will come more easily.
- **Specify your objectives**. What do you most want to achieve through action? Which of your predetermined concerns are best addressed by any specific action?
- **Create imaginative alternatives**. Consider all the options. The more alternatives you have, the better the solution may be. The best answer is not always obvious.
- **Understand the consequences**. Consider all the knock-on effects. Sometimes what seems like the best alternative may, under scrutiny, not work so well.
- **Grapple with your tradeoffs**. You will need to strike a balance between time, cost and long-term effectiveness, and maybe other parameters. If you take decision A, will that make B impossible for any reason? Is that acceptable?
- **Working through all the alternatives** and ranking them by their performance in the above criteria will help to prioritise your to-do list effectively. You may use more, for example, you may wish to include uncertainty (including, not least, future requirements) and risk tolerance (that of your company, or the individuals involved).

Outsourcing risk

Many companies take the outsourcing route with some elements of their exposure to risk, effectively transferring risk from the company to the contractor. For example, a company may elect to outsource the physical security of its premises and hold the contractors responsible for the safety and compliance of the services they provide.

Transferring risk is a wholly normal manner of dealing with the problem, after all, insurance is little more than transferring the risk from insured to insurer, but companies need to be aware that however stringent the terms and conditions, they cannot transfer the complete business risk through outsourcing, so even these areas need to come under the scrutiny of those responsible for risk monitoring and review.

Contractually, the client company will be likely to have recourse to the supplier in the event of inadequate or improper service provision and, to a degree, the financial implications of any unsatisfactory work. But as the survey mentioned above describes, corporate reputation is a major cause of concern, and the implications of loss in this area are considerable, as are those of consequential financial loss through unsatisfactory behaviour on the part of a contractor.

So long as the contractor operates on the company site, health and safety law holds the client company responsible for many of the possible risks (see Chapter 1). Risk transfer is a comforting theory, but the trade is more usually between different types of risk, rather than reducing their number or intensity.

RISK PLANNING AND RESPONSIBILITIES

Turnbull recommendations

Who, in a company, is responsible for identifying, monitoring and reviewing risk? The Turnbull Committee on Corporate Governance places risk management squarely on the boardroom table.

Turnbull was given teeth at the start of this century, when the London Stock Exchange required that company results from 2001 would be required to comply with five key recommendations, namely:

- Listed companies are expected to have in place a robust system of internal control to protect both the assets of the company and the investments of shareholders.
- Such controls and their effectiveness should be reviewed at least once a year.
- An evaluation of the risks facing the business should be carried out at regular intervals.
- Risk management is a corporate governance issue, and hence requires the collective responsibility of the board of directors.
- The board, while the responsible authority, may delegate aspects of the role.

For the first time, this gave company directors a personal incentive to fulfil their responsibilities as custodians of their company's assets, including its data and other digital assets (see section on Digital risk, below). Every listed company – the principal but not exclusive target of the Turnbull guidelines – is now expected to publish, in its annual report, the processes it has established to protect the shareholders' interests as well as the company's assets. This includes every area of risk management: protection and survival are unequivocally within the shareholder interest.

Furthermore, if the company does not have an internal audit function and the board has not reviewed the need for one, the rules require the board to disclose those facts. The point and purpose of these recommendations are to shame directors into taking responsibility for adequate risk management. If they do not do so, they must disclose the fact to their shareholders, and risk the wrath of the market.

Essentially, this means that the legal case is clear: if a company suffers through failing to manage risk, including digital risk, as best as possible, the path is open for shareholders to sue the board of directors individually and personally. It is a move that will surely gain the attention of most people around the boardroom table.

Roles and responsibilities

Practically speaking, who will do the work, and how will the risk management role be structured?

A growing number of companies are finding that there is benefit in coordinating risk management under a chief risk officer (CRO). Key drivers for this development have been the September 11 attacks, the invasion of Iraq, and the internal affairs of Enron and others, and the fear that scandals of this scale and scope will cross the Atlantic to Europe. The CRO role first started to appear in US financial services companies, notably GE Capital as far back as 1993, but is now spreading worldwide. The number of CROs was estimated at some 1000 in 2007, and is expected to grow steadily.

The presumption and rationale for the new role are that risk is a quantifiable factor, and best addressed in an integrated manner across the entire organisation rather than piecemeal. The CRO acts as central coordinator.

Whether or not a CRO is in place, someone within the organisation must have overall operating responsibility for risk management. Logically, bearing in mind the weight of responsibility, this should be the chief executive officer (CEO), but that may not be practical bearing in mind the other pressures on his or her time and attention. Nevertheless, it is important that the person responsible at a tactical level has the ear of the CEO and can visibly demonstrate support from this quarter; in practical terms, this means the appointment (or additional duties, depending on the situation) must fall no lower than the executive senior management level.

Risk management plan

The CRO should then lead a risk management team, comprising responsible members from the different functional areas of the organisation where significant degrees of risk have been identified. The role of these team members will be to produce an integrated risk management plan to cover the risks of the company as a whole, incorporating elements such as project planning, health and safety requirements, environmental considerations, contracts and general business procedures.

The plan must do more than identify risk, it must prioritise it (see Prioritising risk, above) and devise plans for implementation in the event of the hazard occurring, and means to minimise the risk exposure. Implicit within this are visible top management support and company-wide communication.

This plan requires regular review in the light of changing concerns, changes to the company exposure for whatever reason and changing personnel; the team itself may be dynamic, with members changing to reflect the areas of highest priority within the organisation.

DIGITAL RISK

How many people are content that their computer networks are secure against criminal attack, accidental abuse, or the results of flood, fire or theft? How many people take it seriously, and view the cost as a corporate requirement as opposed to an expensive option?

Examine the facts. Statistics and experience show that one company in five will, over the coming five years, suffer major disruption to its business through fire, flood, storm, power failure, terrorism, or hardware or software failure. Following the event, four out of five of those who fail to have a comprehensive business continuity plan in place will be out of business within 13 months (see Chapter 6).

There are, however, risks that fall short of going out of business: consider stock price, corporate reputation, the ability of the company to attract and retain the best staff, and more, and consider how these can be impacted by problems caused within the organisation, often by accident rather than deliberate wrongdoing.

Consider how the following human errors reflect on the host organisations:

- An IT consultant accessed credit card details of 7000 customers of a utility company.
- A similar occurrence with Barclays Online, following the introduction of a new security infrastructure designed specifically to strengthen the bank's security.
- Microsoft stored details of 18 million customers on a public server.
- Network Solutions (a US domain name registration company) emailed confidential details of 86,000 web sites to its entire customer base.

Such events are not uncommon, and are widely reported. Consider the following facts:

- In 2000/01 one email in 2500 contained a virus, in 2005 one in 44 emails contained a virus, principally because of spam (defined as unsolicited emails) and in 2007 it was near one in 30.
- Businesses are not prioritising compliance issues related to email.

The increase in spam is dramatic. In June 2002, 0.1 per cent of emails scanned (by Message Labs) were spam. A year later, this had risen to 38.5 per cent and in June 2004 volumes reached 86.3 per cent, while 2007 peaked at 98 per cent. The impact of such volumes on email infrastructures is significant, costing businesses in unnecessary bandwidth consumption, storage and technical support. The impact on productivity is equally alarming, with Gartner Group indicating that at least an hour is consumed daily by an average employee in managing email.

Types of digital risk

There are different levels, and categories, of risk. Operational risks include staff and human error, a lack of preparedness and information technology (IT) security hazards, while those under a data protection heading include the legal responsibilities of the board, the wide risk management responsibility of the board, compliance issues with the UK's Computer Misuse Act 1990, internal risks (such as sending the wrong material to the wrong person) and compliance with the Data Protection Act 1998.

How to minimise digital risk

Most companies make almost no effort to manage their level of digital risk. Four out of five companies with external electronic links do not use any type of firewall protection whatsoever; with microbusinesses (those with up to 10 employees) one in three has no virus protection. One InfoSecurity Europe study found that the favourite business computer password was just that: the word 'PASSWORD', followed by the user's first name or child's first name in second place, and a favourite football team in third place. While many of the solutions are simple and inexpensive, companies with the need to adopt the highest levels of security are looking at biometrics, such as palm prints, voice recognition and iris scanning, to manage their digital risk.

LINKING RISK MANAGEMENT TO THE QUALITY PROCESS

Managing risk is inextricably linked to quality. If we understand business risk as the chance that an organisation will have to divert from its chosen path to attend to unforeseen circumstances, the link is even clearer. While quality programmes seek to remove from the process all direct causes of interruption and delay, such as faulty products or components, human error or inefficient processes, risk management seeks to minimise the interruption due to circumstantial or situational factors including the effects of anything from a rail strike to terrorist action.

Both quality programmes and risk management programmes require continual reassessment in the face of changing parameters inside and outside the company: both require highly visible senior management endorsement and involvement; both require their actions to be communicated and understood across the organisation.

Both quality and risk require a 'champion' within the organisation and neither can be undertaken and then left aside. The review of both risk and quality is continuous, as the business and the environment, both micro and global, in which it operates, are also constantly changing.

It is hard to avoid concluding that risk management and quality are two sides of the same coin. Reduced risk can lead to improved quality, for practical reasons – there will be fewer unexpected diversions from the company's chosen path, so productivity will be higher – and, less quantifiably, because the presence of a risk plan itself will heighten the workforce's ability to spot early warning signals of potential problems. The corollary also holds good: reduced quality can lead to heightened risk.

RISK MANAGEMENT IN PRACTICE

Risk management: the facilities management workload

Right now, on your desk you have the following: the board wants a report for its next meeting assessing the overall security risk to the company with recommendations for action; the union wants to know what plans exist to counter the perceived and possible threat of biological contamination; the mailroom staff have specific concerns about anthrax and the chairman is due to fly to the Middle East next month for a conference, then spend a few days as a tourist with his wife, who will be travelling with him. Is it safe to travel, he wants to know, or should he cancel?

No one has mentioned it, but you cannot get it out of your mind. You occupy a high-rise, high-profile, building close to an international airport. How safe is that? What about cleaning staff, even security staff, and those others whose work allows them to roam the premises with a high level of autonomy, often at times when few people are around? How much, if anything, do you really know about them? For that matter, how much do their employers, your service providers, know?

You start to think of all the other issues that have seemed relatively unimportant for years, but in this atmosphere of heightened security now seem increasingly urgent. Questions of data security, including everything from privacy to disaster recovery, by way of intellectual property theft; questions of travel safety, not just by air, but no less by rail or road and insurance: do you have enough cover, and are the right things covered? Things like key personnel, for example, and loss or damage as a result of terrorist activity. How is your business interruption cover looking?

You cannot fix everything overnight, but you do need to identify, and quantify, the risks for the board report, allowing for both time and financial constraints. At the same time, at best it is a matter of employee morale, and at worst, very much worse than that: that clear and visible steps are taken to protect the sense of physical safety perceived by your people, and of the premises in which they work.

Define the problem and make immediate changes

The first step is to define the problem. When faced with many apparently equal and urgent tasks, how does one prioritise between them?

Pick the low-hanging fruit first. These are those actions which demonstrate commitment, build confidence, and buy time for the essential planning, talking and meetings that will be needed before major changes can be made. Examples are those tasks that are low cost and high visibility; for example:

- Give the mailroom an unlimited supply of disposable gloves. Make it known that anyone who handles mail can ask for, and rapidly receive, a similar supply.
- Ensure the medical centre (or nursing team, etc.) is up to speed on both the realities and the perceptions of the situation and ensure the workforce understands that this is the case, and that if anyone has concerns, they are to feel free to ask for a check-up, which will be available without delay. Their concerns will all be taken seriously.
- Suggest the union liaises with colleagues in neighbouring offices and buildings to share concerns and solutions, and promise senior management attendance at any meetings called to coordinate efforts.
- Do the same with your own opposite numbers in neighbouring organisations.
- Ensure your heating, ventilation and air conditioning plants have been recently serviced and are in good order. If not, undertake this right away.

- Review your outsourcing agreements with companies whose staff enter your premises, and establish whether those companies have an adequate policy regarding security screening. Look at your contracts and service level agreements.
- Review total site security. Ask colleagues where the weak points are.

You will think of more that can be done in a morning's work and with marginal cost: do them all, and communicate widely that they have been done. While your actions may prove in retrospect to have averted a crisis, even saved life, your principal cause at this stage is to reassure, and to build confidence. To this end, any action taken is a waste of time if it is not widely and effectively communicated.

Tackling the harder issues

For many or most of the other risks you face, or at least want to discount, you need specialist help. Again, it is a decision making problem, but one that can be addressed by dividing the challenge into bite-sized chunks. Some issues, including insurances, data protection, disaster recovery, and the like, may call for a professional review of the status quo. Reports from the appropriate organisations may be worth commissioning, as they will almost certainly cost a great deal less than the loss of the functionality you seek to protect. Only when the reports come in, and the recommendations and costings for action are compared, will decision time come and this may well turn out to be a board matter: the tougher issues are likely to take time and/or cost money.

Risk assessment is in itself a specialist task, and it may be that you will want to recommend to the board that you invite a team in to review and undertake a thorough cost–benefit analysis across the entire spectrum of your operations, addressing your specific concerns and any other issues they feel appropriate.

Addressing the internal issues

There will always be issues that you know will benefit the organisation if you undertake them internally, and which need no consultancy report to initiate. In this category, for example, you may wish to review and improve internal computer security, looking to make a start on intellectual property and knowledge management issues. You will almost certainly need help to end up with the optimal system, but there are many ways in which a beneficial start can be made using internal resources.

Or, for example, if air contamination or pollution is a reasonable concern, you may wish to consider the relatively straightforward process of turning your workspace into a positive air pressure zone, through the installation of a fan to blow air into the building, passing through a high-efficiency particulate air filter (HEPA). If a single input, of sufficient capacity, is blown in by fan, every other 'leak' in the building will then leak outwards rather than inwards; appropriate filtering on that input will provide a high level of protection against airborne contamination.

In conclusion

Finally, what to do about your chairman and his wife? It has to be his decision, but he has asked for your views; so give them, highlighting that it is your subjective opinion.

What would you do in the circumstances? Would you be willing to travel to that destination? Would you be happy to extend the trip and take your partner? A small amount of time on the phone or by email will give you information you can pass on about any added security arrangements at the airline and the hotel and the conference organiser will have views too. But the chairman will almost certainly have made his mind up already and you will be unlikely to change it. So do not waste time on this one.

Note, a threat assessment is only a snapshot in time. As with workplace risk assessments, a threat assessment should be carried out regularly. Activity should not be confused with achievement: do not just think about counter-terrorist measures and then do nothing. The one risk assessment you do not undertake could be the one that catches you out.

According to security specialists, there are five stages of effective counter-terrorist planning: threat assessment; design, including risk assessments; build, install and introduce; operate, maintain and test and go back to stage 1 and start again. You must remain vigilant.

Building better: case study

Once viewed as a symbol of corporate power and importance, a company HQ is now often perceived as a target, especially with high-profile global operators. One leading bank, no stranger to activist action over many decades, occupies a new 33-storey head office in London Docklands. The building has been designed from the ground up to minimise every form of risk, not least from terrorist attack.

The building has been described as one of the most robust buildings in the world, and a blueprint for future tall buildings. Among other elements, the following features have been designed in as part of the company's risk management strategy:

- **Structural redundancy**: the design allows two adjacent structural columns to fail without the building collapsing. Floors, too, have been designed so as not to collapse progressively, as seems to have happened in the World Trade Center towers.
- **Strength**: the central core's concrete walls are 30 per cent thicker than standard, and contain additives to reduce fire risk from, for example, aviation fuel.
- **Stairs**: the standard two internal flights of staircases have been augmented by two concrete encased stairways at the edge of the building and stairway widths have been increased to boost capacity, hence speed evacuation.
- **Fireproofing**: as well as the core concrete (see above), the building has substantially upgraded fireproofing aimed at increasing fire resistance from 1.5 hours to 2 hours.
- **Biological threats**: all plants, including intakes for air and water, are on the roof 34 floors above ground.
- **Glass**: all glazing in the building incorporates a laminate layer to prevent damage from flying shards of glass; most injuries occur in this way after a bomb blast.

SAMPLE CHECKLIST FOR BUILDING IN RISK MANAGEMENT

- Give specific consideration to protecting structural elements that are fundamental to the survival of the building.
- Ensure existing fire protection has not deteriorated or become damaged over time.
- Ensure existing fire compartments are gas tight.
- Ensure any modifications to a building or its services have not compromised fire compartments.
- Make provision for simultaneous evacuation of a building's occupants.
- Consider the use of lifts to speed simultaneous evacuation.
- Ensure easy access for fire and rescue teams.
- Have emergency response plans in place for different scenarios.
- Train and select staff to manage evacuation.
- Ensure building services are secure against unauthorised access.
- Ensure air intakes can be isolated.
- Ensure vehicles cannot get near to the building.
- Secure the building against unauthorised entry.
- Monitor cleaning and support staff and mineral water supply contractors.
- Ensure guards have been trained to react appropriately if faced with a suicide bomber.
- Remove unnecessary hiding places for explosive devices, including litter bins, window ledges and planters.

Business Continuity

6

Frank Booty

There is a lack of understanding of business continuity management (BCM), particularly among the small and medium-sized enterprise (SME) community. Too many companies believe that the phrases business continuity and disaster recovery are synonymous. However, BCM is concerned with the whole business whereas disaster recovery tends to focus on information and communication technology (ICT) recovery and back-up plans in response to major physical incidents such as terrorism, fire and flood.

There is a business continuity standard, British Standards BS 25999, a two-part publication that describes the activities and 'outcomes' of establishing a BCM process. It also provides a series of recommendations for good practice. Part 2 defines the requirements for a management systems approach to BCM. The Business Continuity Institute (BCI) had introduced PAS56 – British Standards guide to BCM, Publicly Available Standard, in 2003, subsequently seeing it adopted by Sainsbury's, the Metropolitan Police, the Post Office and the government.

Disaster can strike at any time, and often from the least expected source in the least expected area, as was so painfully evident with the New York 11 September 2001 terrorist atrocity. Disaster can encompass fire, flood, theft, a crash, a bomb or any event occurring against the odds which stops a company operating at its expected level. There are over 30,000 commercial building fires each year. Information technology (IT) and telecom, particularly mobile phone, theft increases continually; the Association of British Insurers reckons that over £600 million worth of IT equipment is stolen each year in England. According to the Home Office, 42,000 laptops were reported stolen in 2007, an average of over 800 a week, and research by the Association of British Insurers (ABI) suggests that a further 16,000 a year (300 a week) go unreported. Of those reported stolen less than 1 per cent are currently recovered by the police. With over 50 active terrorist organisations in the world, terrorist activity continues to pose real threats and, as any company dealing with commercially sensitive information knows, it need not come from any particular terrorist organisation. How does a business continue?

241

THE TURNBULL REPORT

The Institute of Chartered Accountants of England and Wales published the *Turnbull Report* in 1999 to the Stock Exchange, recommending that all quoted companies should have a risk management strategy. Companies not planning for interruption stand to lose more than just cash flow, customers and confidence: eight out of 10 companies cease trading after a major incident. Partly because of that, many organisations stipulate that their suppliers must have business continuity plans in place to minimise the knock-on effects any catastrophe would have.

Under 50 per cent of companies who depend on computer systems and other technology have formal business continuity plans, and only 12 per cent of those plans are considered to be effective outside the IT department and across the business. One of the most valuable contributions *Turnbull* has made is to help companies to see risk management not just as a means of staying afloat, but as a way of embracing business continuity in order to thrive in the future.

Business continuity demands total commitment at board level, the dedication of key individuals in a company, assistance from business continuity specialists, and an enthusiastic and informed staff to carry out all the necessary processes. Remember, even if your building is not the target, it could be impacted by a nearby target.

DEVELOPING A STRATEGY

Business continuity management can best be defined as:

> '*The ongoing process of ensuring the continual operation of critical business processes through the evaluation of risk and resilience, and the implementation of mitigation measures.*'

Business continuity management had its roots in disaster recovery planning, but as the market and the players within have matured, so too has the belief that risks to business can be mitigated as well as recovered. Disaster recovery set out with the intention of providing business with protection further to that originally provided by standard maintenance and insurance contracts. Companies have accepted that while an investment in business continuity directly affects the bottom line, and often with no immediate tangible benefits, it is the most important investment they will make. Crucially, companies have woken up to the difference between business continuity and simple insurance, which will at best replenish the value of the equipment lost, and that at a time when it could already be too late.

Business vulnerability

Each business has a different threat portfolio, as no two businesses are exactly alike. In the event of a disaster, at best an organisation can expect to suffer damage to its principal reason for operating. This will mainly be financial, but may also mean damage to reputation, share price, image and customer support. At worst it will result in the damages being irreversible, and the total failure of the business.

There are many statistics published about UK business interruption:

- power failure accounts for over 10 per cent of all business interruptions
- 57 per cent of disasters are IT related
- 61 per cent of companies do not publish their business continuity plans throughout their organisation
- 84 per cent of companies do not identify risks in the supply chain.

Average outage times of business interruption in the UK are:

- fire: 28 days
- IT failure: 10 days
- lightning: 22 days
- flood: 10 days
- theft: 26 days
- power failure: 1 day.

Another non-quantifiable aspect is job loss.

Arguments in the business continuity industry suggest that, rather than being the forum for doom-and-gloom merchants to excel, business continuity can be rightly seen as an investment which, rather than just reducing the impact of disaster on business, can also increase employee, investor and shareholder confidence in the company. Note that a good business continuity services provider can be measured by how regularly it is asked to speak to executive boards about the corporate requirement for continuity planning.

An effective strategy

A business continuity strategy has to meet several criteria to be effective:

- It is crucial that the board leads, and is seen to lead; the strategy must be constructed and implemented as an integral part of the company's structure.
- There can be no parts of the organisation that the strategy does not touch. As well as becoming a key and integral part of each department, the strategy has to cross all departmental and functional boundaries.
- The strategy must reflect the way the company interacts with its environment, and it must equally anticipate future demands from that environment and from within the company.
- All necessary resources (vital information such as contact names and numbers) must be available, appropriate to the nature of the company's business.
- Through a greater awareness and better management of risks, the strategy should seek to add value to the company. The reduced uncertainty in the implementation of business strategy will aid this process.

The ultimate aim is for a cost-effective and focused business continuity infrastructure.

An organisation's exposure to operational risk is a measure of its susceptibility to unwanted, unplanned events. The implication is that if the two dimensions of 'unwanted' and 'unplanned' can be managed effectively, the organisation's exposure to risk is de facto under control.

Conducting a business impact analysis

At the core of continuity planning is an understanding of the organisation's unique threat profile and subsequent exposure to operational risk. Central to this is the business impact analysis (BIA). Note, neither management nor the board will ever truly buy into the concept of business continuity until they can see the *quantified* risks of failure associated with serious business interruptions. Unless board members can see risks broken down into pounds and dollars they are unlikely to appreciate the issues.

A BIA can be defined as 'a management-level analysis which identifies the impacts of losing company resources'; it 'measures the effect of resource loss and escalating losses over time to provide senior management with reliable data on which to base decisions on risk mitigation and continuity planning'.

Analysis of potential risk

The first step is to look around you:

- **Study historical data**: what types of emergencies have occurred in the community, at the company's premises and to others in the area?
- **Look at the local environment**: is the site near major transportation routes, airports, flood plains, dams or power plants? Is the company close to other companies which produce, store or transport hazardous materials (which may be a terrorist target)?
- **Analyse the technological risks**: what could result from a process or system failure? What emergencies can be caused by an employee error? Human error is the single biggest cause of workplace emergencies and can result from poor training, poor equipment maintenance, carelessness, misconduct and fatigue.

Analyse each potential emergency from beginning to end. Consider what could happen as a result of:

- prohibited access to the facility
- loss of electrical power
- communication lines going down
- ruptured gas mains, water damage
- smoke damage
- structural damage
- air or water contamination
- explosion
- building collapse
- chemical release.

Once all the risks have been considered, estimate probability and plot the results on a graph. The x-axis should indicate probability and the y-axis should show the amount of probable downtime (the outage) in terms of days. So, for example, while a power failure is highly probable, it causes minimal downtime, whereas flood is unlikely but will have a serious long-term effect.

Analysis of potential business impact

To assess the potential business impact, consider the loss of market share. Assess the potential impact of:

- employees unable to work
- customers unable to reach the facility
- company violation of contractual agreements
- imposition of fines and penalties or legal costs
- cash flow slowdown
- damaged reputation and image
- job losses
- lack of delivery of critical supplies
- slowed or halted product distribution
- damage to the environment
- drop in shareholder and stakeholder confidence
- compliance with statutory obligations.

Once a company appreciates the impact an interruption could have on its business, it will be in a position to minimise the risks on a priority basis. If it decides reputation is more important than profits, it should first minimise those risks that could affect its reputation.

If the BIA is done objectively, it will have a stimulating effect on the business. Employees will understand how seriously the whole continuity issue is taken. Suppliers, who have to be approached as part of the analysis, will realise that they should conduct a similar exercise, which in itself will help to protect the company's business. In addition, the board will start to take the matter of business continuity seriously.

IMPLEMENTING THE STRATEGY

Drawing up a plan

Planning is the next logical step and procedures need to be in place for staff, recovery of IT and buildings, evacuation procedures, media management and dealing with trauma, as well as any other issue pertinent to the company's operations. A business continuity plan need not be a sprawling document, and if there is confidence in the staff's capabilities, the smaller it is the better. This, together with ongoing training, will also encourage staff at all levels to back the project; any plan is useless unless the people operating it both support it and are capable of implementing it. The plan must be amended with the emergence of new staff, buildings, procedures and any technological upgrades, in conjunction with any third party disaster recovery supplier.

Plans – operationally oriented – should address or include:

- up-to-date record drawings
- asset registers (addressing those critical to the business)
- how staff or visitors on site will be identified and accounted for
- safety procedures
- mirroring of facilities if these are critical to operations (such as duplicate dealing rooms in the City)
- fall-back communication systems
- essential utility supplies in case of mains failures
- triggers that will initiate the plan or component parts
- prioritisation of action
- any necessary training requirements
- recovery process
- stockpiling of critical supplies: do not forget dependence on single small items which are used once in a blue moon but without which a service folds
- regular updates: remember that staff change, buildings develop, and so on

Plans can be developed by considering exactly what aspects of your operation the business is dependent on. It is important to create a vision of what the business capability will be in the event of an incident: will it be business as usual or will output be reduced?

Roles and responsibilities

Most companies will have a management team to continue running unaffected or undamaged sections of the business, consisting of existing personnel carrying out their normal duties. A second team – the emergency management team – has the task of accomplishing the recovery or continuance of the damaged sections of the business. Such a team is supported by the business continuity manager, business continuity teams and other teams involved in the recovery effort. Mobilisation only occurs in the event of an incident.

The constitution of these teams will have to cover many key areas:

- damage assessment and salvage
- IT recovery
- telecom and data communications recovery and call centre operations
- premises restoration
- procurement of disaster recovery resources (such as sandbags)
- media management (covering corporate communications, public relations and marketing)
- operational/production recovery
- business unit/departmental recovery

- support and coordination for such areas as personnel, transport and administration
- welfare support for staff affected by the disaster
- dealing with insurers.

There needs to be a team leader and a deputy for each business continuity team. All staff must be trained. In the event of a disaster, staff need clear communication informing them of what to do. One way of effecting this is to issue everyone with a telephone number and/or web site where up-to-date details of the status of the business can be communicated (if there is a disaster). This system should be tested regularly.

Roles and responsibilities must be assigned and understood. Details must be rolled out to all sites for multisite operations (an obvious but often overlooked fact). Back-up sites and locations have to be planned. The location of all staff members at work must be known, and full details of home addresses and phone contacts available.

Rehearsing the plan

The work does not end once the plan has been drawn up and roles and responsibilities have been recognised. The final and critical step is to carry out rehearsals and maintenance. The BCI reckons that only 18 per cent of companies rehearse their plans; but non-rehearsal is the biggest weakness in a business continuity plan. Rehearsal identifies weaknesses in both the plan and the people in the designated crisis management team. The plan has to be kept relevant and up to date, and rehearsals should be carried out at least twice a year. Testing the plan until it runs smoothly and everyone knows exactly what to do is the only way to stay prepared.

OUTSOURCING DISASTER RECOVERY

Disaster recovery is fast becoming a mission-critical service (see Chapter 7). Customers need the best the industry has to offer to keep their businesses operational whatever the circumstances. As part of their business-wide disaster recovery policy, businesses almost without exception have chosen to invest in one or a series of third party disaster recovery agreements. Disaster recovery companies can now help to protect almost any vital asset a company may own. This may be a critical server or personal computer (PC), and may include standby facilities, buildings or even people.

Disaster recovery companies work by responding to a company's invocation, in a precontracted time-frame. This is achieved through taking the actual technology and engineering support into the client or recovery site.

Risk management should be central to disaster recovery provision. Companies must first understand what they stand to lose (in real terms) in the event of an interruption to business. This will hopefully have been calculated at the BIA stage of the planning cycle. The investment in the disaster recovery project will only be a valuable one if it is outweighed by the potential loss in the event of an interruption. It is because of this that the industry has only recently seen the SME and public sector markets invest in disaster recovery planning.

Organisations need to make sure their disaster recovery contracts minimise their risk exposure by catering for any eventuality. A thorough under-standing of IT and information systems is essential for a disaster recovery compa-ny, but business continuity and disaster recovery are primarily risk management tools.

Selecting a disaster recovery supplier

Although a disaster recovery contract is much more sophisticated than an ordinary insurance policy, it is nevertheless based on a similar principle: risk assessment. A dedicated disaster recovery company should have a comprehensive system of risk analysis that considers every type of risk with probable occurrence rates. This system should include site inspectors and a detailed evaluation of the customer's own risk management plans.

Ideally, customers should be grouped in small syndicates based on contrasting risk levels, including geographical location and industrial sector. The risk assessment should also consider factors such as:

- proximity to transport infrastructure
- electricity supplies
- telecom networks.

Each grouping should minimise the risk to individual members and ensure there are adequate resources to cope with multiple invocations.

The supplier of disaster recovery services should sell contracts on a risk, rather than unit numbers, basis. There is no point in buying a contract that covers 200 PCs only to find that when the contract is invoked, the service provider cannot supply 200 computers because it underestimated or miscalculated the risk exposure. The likeli-hood of invocation, rather than the number of machines covered, should be the key consideration in a disaster recovery contract.

Invest wisely

As more organisations turn their attention to disaster recovery and start investi-gating the market, it is only natural that some choose to select their supplier on the basis of cost. But disaster recovery, like insurance, is one area where buying on price could prove to be a mistake. Cheap insurance or disaster recovery cover can soon turn out to be expensive if the provider cannot honour its policies.

If a disaster recovery contract is very cheap, there is usually a good reason. The supplier may:

- have failed to make an adequate risk assessment
- have deliberately underestimated risk exposure to keep prices down
- be overselling its PC recovery stock to make a profit.

Demand the best

Today's competitive economic environment dictates that companies must move towards best practices and procedures. They should demand the very best the disaster recovery industry has to offer, including complete risk assessment to highlight areas of vulnerability, dedicated disaster recovery stock in secure storage facilities, and reliable up-to-date equipment. Above all, customers should insist on a disaster recovery company with a successful track record of providing total business continuity solutions for companies of all sizes.

Specifically, a good disaster recovery firm should:

- **Assess and respond to change**: it is important to remember that risk, like all business continuity issues, is not static. It changes as political, economic and social trends develop. Disaster recovery firms must understand these changes and appreciate their impact on risk calculations. This fact makes it essential for a disaster recovery company to review each customer's circumstances regularly and to advise on appropriate changes where necessary. Active response to risk assessment should ensure there are no nasty surprises if a company's disaster recovery cover has to be used in a real emergency.
- **Keep abreast of changes in technology**: not just in terms of staff training but also in their hardware profiles. Dedicated disaster recovery stock should be continually updated to keep pace with technology. As customers regularly update their IT infrastructure with more powerful and efficient equipment, they should expect their disaster recovery provider to do the same.
- **Provide regular testing for recovery plans**: a company should not have to ask for this; the disaster recovery firm should insist on it itself.

Finally, the term 'disaster' should be applied to any incident that prevents businesses operating as usual, from microchip theft to computer virus outbreaks, not just to large-scale incidents such as fires, floods or terrorist bombings. Ideally, it should be defined by the client.

DEALING WITH STAFF

Care of staff is the final piece of the business continuity jigsaw. A company's people constitute its best (and potentially worst) asset. It is not enough simply to include them in a business continuity plan: they must be at its core.

Remember that human error is the single largest cause of workplace emergencies with, for example, 95 per cent of all commercial building fires started by employees. People also cause disasters without malicious intent if:

- they have not been trained properly
- they have not been adequately communicated with
- they are tired or unhappy
- they are put under extreme pressure.

A company's staff are central to its success. If a whole department were to win the National Lottery as a syndicate and decide to leave, business would undoubtedly be severely disrupted. While IT would still be operational, there would be no one there to use it. Some insurance companies are offering advice and premiums against the possibility of this happening, although this will obviously not cover the potential loss to credibility and image that such an interruption would bring. Planning for such interruptions should be as central to BCM as the potential loss of IT.

Training

Key to this must be dedicated training of staff, where politics is discarded and everyone is suitably prepared for the shock factor an interruption can bring.

Training will:

- raise the awareness of business continuity and its importance within the organisation
- lend ownership of the business continuity plan to employees of the organisation
- ensure the business continuity plan will be effective (even the best laid plans will not work unless people are aware of their role and how to carry out that role)
- motivate: individuals will feel that the organisation is investing to meet their needs
- help to streamline the business continuity plan (in time) as key employees become business continuity experts.

Awareness raising for senior management is often most effective when carried out by an outside agency. Outsiders will not be seen to have any ulterior motive for promoting business continuity, whereas an in-house presentation may be seen as 'empire building'. Outsiders may also be seen to be more credible. They should have had first hand experience of crisis situations and their repercussions and be able to convince senior managers of the real risks faced by the company.

Training the business continuity planner

Practical training for the business continuity planner should also be outsourced from a reputable training supplier. Often, the most effective training course is a bespoke course held on the organisation's own premises. The trainer would normally meet with the business continuity planner, look at existing business continuity plans and procedures, and from there design a course around individual training needs, addressing any information gaps identified.

Communication

Avoiding disaster

Equally important should be an effective channel of communication to ensure staff know how best to avoid causing a potential disaster. This needs to be implemented as early as induction, and continue to be regularly updated and communicated. Failing to communicate clearly and regularly not only threatens recovery should a disaster occur, but also demotivates individuals who interpret a lack of board-level endorsement of business continuity issues as an evaluation of their own personal function at the company as not 'mission critical'.

Immediate disaster response procedures

After any full-scale interruption, such as fire or flood, management must be confident that its staff are suitably prepared to act in a controlled manner and will know exactly what procedures they need to follow.

If a business evacuates its building, it must be to a location that is safe for its staff to group and far enough from any neighbouring businesses that may have been forced to do the same. Typically, businesses should avoid underground car parks, or areas too close to offices. Many people may leave personal belongings in the building and need transportation to return home, for which money should be readily available. There should be enough mobile phones for staff to check others are safe and to reassure them of their own welfare.

Staff welfare after an event

Intense emotions, numbness, flashbacks and severe reactions can be triggered by seemingly unrelated events. Professional trauma counsellors for post-traumatic stress disorder (PTSD) argue that an ideal PTSD system should include both an external consultant and system facilitators within the organisation, as well as a debriefing team and occupational health staff, all tailored to the structure and specific nature of the business. A company should immediately issue all personnel with a generic statement in the aftermath of a disaster to let them know that the company respects their feelings and actions, and continue to support them long after the event by including trauma counselling as an integral part of its disaster recovery plan.

Businesses cannot cope with this on their own, but must make provision for it in a contingency plan. This is where dedicated business continuity groups are able to offer advice and help, through training, seminars and other similar media. Again, dealing with trauma should not be restricted to company-specific incidents, but should encompass any severe change in mood in an employee.

DEALING WITH THE MEDIA

In the event of a disaster, the media will want immediate responses. The crisis management function needs to focus on the media interface: one individual must be appointed through whom all communications will be made or managed, who will report directly to the BCM team. No individual member of the company should talk to the media; instead a unified agreed message must be given out. The crux of this vitally important area should be viewed as turning a negative scenario into a positive situation.

Media management does, of course, vary according to the type of business affected. The financial sector, for example, is very coy about releasing information, whereas companies in the retail and distribution sector generally address the situation openly.

It is widely regarded that media relations are bottom of the list where the SME market is concerned. This is indeed their Achilles' heel. Large companies are typically much better prepared.

A UNIFIED AGREED MESSAGE

A good example of the success of this strategy is a situation where a fire started at a company in part of its manufacturing facility and administrative complex. The business continuity plan was invoked and the fire was brought under control. The crisis management team agreed a message which was broadcast through a glossy in-house publication with pictures of the fire, key messages and pictures of the facility after the event. The company survived. The publication was then sent to suppliers and customers; not one of them had been aware that there had been a disaster at the company. The continuity plan had ensured it was business as usual.

INVENTORY AND RESTORATION

Business continuity is not a complicated science, but to carry it through successfully it requires:

- dedication from a number of key people in an organisation
- total commitment at board level
- assistance where appropriate of dedicated business continuity specialists
- enthusiasm and understanding of staff
- inventory control.

Inventory control

Inventory control is a key but difficult area. The biggest problem by far is management, and management of assets in particular. Strategy-planning software is recommended to cover suppliers and inventory. Everything is relevant – all IT, for example – and the complete asset list must be incorporated into the contingency plan. The software allows companies to run different scenarios of disasters to see what would be lost, what the company would still have, and where the company stands. In a disaster situation, a company might lose 100 PCs but still have 300 or so elsewhere within the company. The plan would be to allow the reallocation of existing resources, as well as having another company ready to be able to supply 100 PCs at short notice.

Restoration of the business

Many business continuity specialists will have relationships with specialist restoration companies. This will cover such reactive services as cleaning and drying, providing advice and counselling services, through to loss adjuster specialists. Environmental crises, which will not be anticipated by most people, may arise as one result of a disaster. Few are aware that in the event of a fire, mixing water with PVC results in the production of hydrochloric acid, as happened once with a chemicals and plastics supplier. The smoke was bad enough, but acid contamination caused widespread environmental damage. Specialist decontamination companies are required to deal with acid spills and damage.

Restoration is not generally the province of disaster recovery providers. It is a specialist service, which is typically outsourced. Disaster recovery providers would usually carry out an environmental survey prior to any incidents to provide a breakdown of what can be covered. Health and safety and environmental issues are key, and are core responsibilities of the facilities department.

Technology restoration

The main restoration challenges that facilities managers are likely to face after a disaster relate to the integration of technology. The following are all vulnerable:

- data centres, which will typically have their own rigorously enforced plans
- Internet telephony, also referred to as voice over Internet protocol (VoIP)
- computer telephony integration (CTI)
- 'smart' telephone switches
- call centres.

Re-establishing matrices across such networks is automated, and in so doing poses problems for the business continuity industry. With data transmitted over voice cables, wide area networks and flexible bandwidth networks spread across Europe, single points of failure are difficult to track.

Call centres are seen as being particularly vulnerable owing to the need for rescripting the complex computerised switches deployed across the IT and communications infrastructures of these sites, should a disaster occur. Companies with one or more centres would usually need to implement a plan involving fall-back or mirroring policies.

BEST PRACTICE PROCEDURES

Business continuity issues are present every day in the working environment. There are many procedures that companies can implement at once to improve their plans and help to create a business continuity ethic across their organisations:

- Ensure visitor parking is as far from the reception area as possible, reducing the risk of suspect cars containing dangerous devices being left close to the building.
- In a multitenancy building, check the evacuation procedures of the other companies.
- Cover all windows in protective film to minimise damage caused by exploding glass.

- Back up all data regularly and store off site in a secure environment.
- Use a password-protected screen saver so that no one can use a PC if the building has to be evacuated in a hurry.
- Perform a weekly restore test to check back-ups: restore failure is the most common reason for total data loss.
- Agree with the company's insurers now on the policy for paying out for bomb and explosion-damaged PCs and servers, as loss assessment is lengthy and often unrewarding.
- Make sure that safe areas are away from windows, stairwells, lift shafts and rooms with suspended ceilings.
- Keep hard hats, goggles, and so no, in every office.
- Ask insurers now for the policy on injury claims, especially if the police advise on evacuation of a building in a disaster and there is a decision not to do so.
- Make a video of the property to help with insurance claims, finding things in a hurry and loss adjusters.
- If answering machines are relied on to take messages during a building evacuation, check how many minutes of recording time there are: it is no good evacuating for four hours if there is not sufficient storage capacity to cover that period.

SIMPLE STEPS TO EFFECTIVE BUSINESS CONTINUITY

- Train all key staff to the criteria already outlined.
- Identify all mission-critical processes, people and applications.
- Perform a risk and business impact analysis (BIA).
- Write the plan, preferably using an experienced third party consultancy and a recognised software package.
- Take out any necessary additional disaster recovery cover on any mission-critical equipment.
- Ensure the plan is kept current and up to date, in accordance with changes in technology, staff and company direction, and test it regularly: it can never be tested enough.

INFORMATION

British Standard 25999 www.bs25999.com, www.bsi-uk.com/business_continuity
British Standard 7799 – UK standard for information security management, section 9 of which deals with business continuity management: www.bsi-global.com
Business Continuity Institute, for PAS56, Tel.: 0870 6038783, www.thebci.org/pas56, www.pas56.com
Cabinet Office, Civil Contingencies Secretariat, detailed information for practitioners, www.ukresilience.info

Centre for Crisis Psychology, Tel.: 01756 796383, www.ccpdirect.co.uk
Cranfield School of Management, Tel.: 01234 751122, www.cranfield.ac.uk/som
Information for general public: www.preparingforemergencies.gov.uk
Joint Treasury/Bank of England/FSA: www.financialsectorcontinuity.gov.uk
National Infrastructure Security Coordination Centre: www.niscc.gov.uk

7

Outsourcing

Chris Taylor and Frank Booty

Outsourcing today is an increasingly common way of doing business. There are sound financial reasons for not employing permanent staff – who must be fully paid regardless of work flows – when you can buy in outside expertise which may be able to do the job better, cheaper and more quickly.

Facilities managers' time is, therefore, increasingly dominated by managing contracts with external suppliers. This has required facilities managers to develop new skills. Whereas before, facilities professionals had to be multiskilled, multifunctional and good managers of people from a variety of backgrounds, now – on top of all that – comes the responsibility for continuing to motivate those over whom they have no immediate line management authority, while managing outcomes through a third party employer with their hands tied by legal contracts.

Successfully working with contractors on a strategic level is another challenge facing the facilities manager. Today's outsourced contractors are 'partners' who 'add value' to a company's operations. Command and control, as a management style, would be as self-defeating here as anywhere else. Partners are looking for long-term relationships, where trust means more than mere on-time delivery. It means sharing cost bases, profit ratios and business objectives. To a degree, it also means sharing information you might prefer to keep in-house.

This chapter aims to explain the nuts and bolts of outsourcing, with guidance on the whole process from choosing what to outsource, to writing and managing the contract, to maintaining a dynamic working relationship.

WHAT TO OUTSOURCE

When outsourcing first entered the management arena in the 1980s, it was all about saving money on essentially manual tasks; premises cleaning was a typical and early example. The focus is now on access to skills, with outsourcing expanding to include areas closer and closer to the centre of business; companies are now looking to buy in outside expertise so that they can concentrate on their own core activities, and contract with external suppliers to provide many or most of the tactical elements.

256

The shift from those early manual tasks has seen us move through other administrative and infrastructure areas, towards innovation: how can the company move further ahead faster than the competition? Outsourcing today, and the expertise that comes with the specialist contractors, has now embraced research and development (R&D), design, product management, marketing, communications, even personnel supply and management.

That development has not finished. Already into strategic functions, it is not unlikely that outsourcing of core business areas, and even the strategic direction of the company, could follow.

Outsourcing should not be seen as just a money-saving management device or tactic, but considered in terms of the potential value it can bring to a business. It may cost more to outsource, but the job may be better performed, the company image may be enhanced, and it may release expensive management time for core activities. However, 2005 saw the possible critical turn of opinion regarding outsourcing with some high-profile industrial activities. Some companies are reviewing how they define 'core' activities, which begs the question, 'are we seeing a start of the move to bring stuff back in-house?'.

Notwithstanding that development, what, in your specific business, should you consider outsourcing and what should you retain in-house?

Core and non-core activities

A first assessment in considering whether or not to outsource is defining what you consider to be your core business activities. Current practice dictates that companies retain control of these areas. The thinking behind this is that organisations should be left to concentrate on their core business activities without the distraction of providing and managing non-core activities. These are better provided by an outsourced commercial organisation whose personnel can be better motivated and rewarded in an environment that recognises their special training and skills.

Of course, what is defined as 'core' will vary significantly from organisation to organisation. A good example here would be the law firm that outsources its conveyancing work on the basis that its core skill is commercial litigation and it is from the latter that it earns the majority of its income.

Consider next the degree to which any selected function is routine and well defined. If it can be easily defined, then it can be more easily measured, and managed at arm's length; and remember, you are not devolving responsibility, just functionality. So, if a non-core function can be defined and measured, it may be worth considering for outsourcing.

Anticipating pitfalls

Before embarking on any outsourcing initiative, there are several potential problems which should be anticipated and evaluated. Many of these are concerned with the people affected by the change in operation, others are to do with changes in the business itself and the manner in which the business is undertaken.

People issues

It is unfortunate that outsourcing is rarely welcomed by a workforce, especially that section responsible for the function being outsourced. Often it is perceived as downsizing by other means, and strenuous efforts are made to oppose it, diverting business time and energy from development to firefighting. Most of the problems that can arise can be pre-empted through intensive discussions and planning. Talk continuously with staff; be certain that they understand what is happening, why, how and when, and understand that simply telling them is not the same as being certain they understand.

Do note that when you transfer staff to another company, you send them a very strong message. You are telling them their work is a commodity, which they are welcome to continue performing under new management. You are informing them that your business needs to concentrate on what is really key, and that does not include them or their work. Also, you are telling them you expect to obtain more for less under a new regime.

From that point, the messages they receive will devolve from the vendor's management team, and it is unlikely that these people will be pushing for, say, a healthy information technology (IT) section. In place of the professional environment based on common goals the team once shared, or the partnership with an experienced and knowledgeable organisation you were seeking, you will find you have a commodity-based service with goals contradictory to your own. That darker side of outsourcing was experienced at a major global US company which outsourced its IT function.

Before proceeding, ask the following questions:

- Is there a convincing case for outsourcing? People and unions will generally accept an argument in which the logic is unassailable, in which all the numbers, from every source, have been reviewed and checked, and where every aspect has been openly examined. Secrecy is not a good strategy here.
- What are the legal requirements? Does *TUPE* apply? (See Chapter 2.) How much will it cost to be generous, rather than just to stick to the letter of the law? Is the cost worth it?
- How have the levels of knowledge and skill that exist within the organisation been established, especially within the area to be outsourced? Is the loss of some of these skills acceptable – most probably as a gift to the contractor – or should some or all of these be kept in-house as a supervisory or management resource?
- Is everything possible being done to involve and inform all staff – not just in the affected areas – of the progress towards outsourcing? Be aware, however, that different circumstances apply when an internal bid is on the table alongside external competitive bids; managing information in these circumstances

is especially challenging. The function must continue, but one bidder has an effective stranglehold; tell them too much, and external bidders will claim favouritism; too little, and the existing team claim secrecy. The only option here is honesty and openness, supported by a policy, clearly stated, of what is acceptable and what is not, and what information is to be shared and what is not.

Business change

All businesses change. Before entering into an outsourcing contract, you need to consider the following questions so that you are not hampered by a restrictive contract as you pursue your development goals:

- What happens if you are bought out, or buy out another company? They may have a state-of-the-art in-house provision for what you have outsourced; or they may have none, and you need to look at rapid expansion of the service.
- What happens if you have a fundamental shift in business focus: if you decide, for example, to close plant, move headquarters of strategically important offices, or pursue a new but promising business direction which affects the need for the outsourced service?
- What happens in cases of force majeure: in the case, for example, of a significant market downturn which leaves expensive equipment and facilities underused? What provision has been made to allow for a ratcheting down of the service, and the associated costs?

Process change

The way business is done changes continuously. Think how long you have been using email as a key communications channel. How many standalone fax machines are still used regularly in your business? Do your younger staff even know what a telex machine looks like? Such technological changes can significantly alter the manner in which an outsourcing contract will work over time. For example, if outsourcing had been a significant factor a generation ago, the typing pool would almost certainly have been a prime candidate for it; now typing pools simply do not exist. It's not just the technology: working and management styles can alter considerably over the duration of a contract.

Consider the following:

- What structural changes could affect the outsourcing agreement? Consider, for example, the effect on the catering service of introducing flexible working; or how devolving autonomy from head office to regional offices could impact on supplier strategies, from transport to telephony or utilities.
- Bearing in mind the growth in power of computing (remember the old axiom which still holds good: every two years double the power/speed and halve the price), what could you be doing for free which used to require specialist input? Think of display media, and how that function has been usurped with presentation software packaged into most laptops, or of the production of newsletters, posters or other promotional material.

CHOOSING CONTRACTORS

The tender document

Time spent preparing the tender document is a requirement, not an option. Although you may feel there is only one logical choice for contractor, it is essential that you approach the tender process with an open mind, and ensure that the process is competitive. Do not rule out the possibility of bids from angles you may not have considered, not least the existing function team, and do not assume you know all the potential bidders.

The process involves three stages:

- detailed specification of requirements
- invitation to tender
- preferred contractor selection.

The first stage, the detailed specification of requirements, is the most time consuming. It involves a microscopically close examination of the function concerned, and a thorough definition of that function. This is necessary because few if any functions of a business can be said to be completely self-contained; all have areas of overlap with other functions, and these must be resolved before you can spell out unequivocally exactly what you are asking a contractor to take on. It may be as simple as defining reporting procedures from company to contractor, or it may involve changing internal processes, between, say, the area being outsourced and the accounts department, to enable an unambiguous command line.

The outcome of this stage should be a lengthy document. It will describe the function in detail, in terms of actions as well as outcomes, and how it dovetails into the organisation. If changes are planned or being considered that will affect the function, they should be spelt out here, but not to a degree that compromises company confidentiality. While the document should state that the contents should be treated as commercially confidential, it is essentially a public document over whose circulation you have less than perfect control.

Invitation to tender

To whom should the document be sent? The second stage, the invitation to tender, can be as broad or as narrow as you choose. One option is to advertise in an appropriate trade journal or newspaper, send out multiple copies of the tender document, and accept bids from all and sundry; more than one company has discovered the perfect outsourcing partner in just this manner, from businesses they would not have heard of otherwise. Alternatively, you may prefer to limit the number of bids at the outset through a preselection process, either through advertisements asking for 'expressions of interest', or through a strict invitation-only process where you involve only a shortlist of contractors already known to you, or discovered through advertising or word of mouth recommendation. The 'expressions of interest' route ultimately comes down to a similar selection: all manner of relevant information can be asked of prospective bidders to enable an informed choice of candidates to invite to tender.

Final selection

The next challenge is to draw closer to a final selection. You will have received a wealth of information, and almost certainly business references, from several companies, and three or four will stand out as meriting deeper investigation. A systematic comparison is important, as is involving people close to the function to be outsourced; but clearly, this can be sensitive if the in-house team are among the bidders, whether they have made it past the first 'cut' or not. Be sure to establish what each candidate has to offer. You are not just buying another pair of hands, but a functional expert with a brain. If this is to be the partnership you want and expect, it is reasonable to expect them to bring ideas as well as skills with them, and you should listen carefully to their suggestions as to how the function can be tailored more closely to the objectives spelt out in the tender document.

After discussions, and sometimes more formal interviews, it is likely that you will be able to reach 'preferred contractor' status with one candidate. This is the goal of the exercise so far and, arguably, where the real work begins.

Final negotiations

Next in the process is the interval between the selection of a preferred bidder and the contract signing. This is when 'proceed with caution' is a wise advice; this stage can be compared with the period between choosing a new home and exchanging contracts. Both parties have reached an understanding and have a period of time, which their enthusiasm may seek to shorten, to make certain that the decision is the right one; mistakes can be costly.

Sometimes a bidder will fall at this final hurdle. Through a detailed examination of needs and abilities on either side, your goal is to agree a service level agreement (SLA) which will be your working blueprint for the day-to-day operation of the contract. These final negotiations may reveal incompatibilities, and it is better to discover this before signing contracts than afterwards. Remember that the selection of a contractor is not necessarily a one-off process; you may almost get to the 'altar' and then have to start all over again with a different partner.

Creating a synergy

What are the areas you should be discussing? It is important for any company to start with questions of its preferred bidder:

- What skills and abilities do both parties have that will provide a synergy, creating added value?
- How will intellectual property and expertise be managed to the benefit of each?
- What cultural factors will make working together easier or more difficult?

Good channels of communication and working towards shared goals are critical from day one, and objectives must be clearly defined. Ensure that you discuss:

- The establishment of unequivocal and open lines of communication between named individuals at all levels of operation
- The level of priority that customer satisfaction should hold: the final consumer of your goods or services will notice, and be confused by, conflicting priorities.

Flexibility

Look, above all, for flexibility to adapt to the changes that will come. Think of the future, and look for special skills that a partner can bring to the table:

- Can the outsourcing contractor offer a strategic plan?
- Can the outsourcing contractor offer potentially value-adding concepts?
- Can the partner demonstrate commitment to a long-term relationship?

A two-way process

Remember that to turn the supplier into an asset, they need to be part of the organisation, and this means flexibility on both sides. The choice of supplier may come down to how much risk a company is prepared to take. Things will change, and companies should be willing to adapt:

- Be willing to change your corporate culture if necessary.
- Focus on business objectives.
- Consider making your outsourcing partner a risk-sharing partner: a recent trend is to offer the supplier shares within the client company so they have a direct interest in profits.
- Establish and agree metrics to reward risk sharing.

Also, of course, accept that all parties need to feel that what is offered and received represents value for money. The specific procurement process particular to Public Private Partnership (PPP) projects is outlined in Chapter 4.

The supplier market

When researching possible contractors, it may be useful for client companies to understand that there are four pedigrees within facilities management service providers, each with different experience, priorities and business focus. They are:

- Single service background: these companies started by providing one service and expanded into others to meet client demands.
- Construction background: these companies diversified into facilities management during the 1990s' recession.
- Professional construction background: these are managing agents, quantity surveyors, architects and engineering companies which have moved into facilities management.
- Client background: many companies have spun off from former employers and expanded to offer the same or expanded services.

The well-informed client will consider which approach and background may be most appropriate for its requirements, as each background style will have its own traits, skills and experiences, and not all will suit every market.

SERVICE LEVEL AGREEMENTS

An SLA is not a summary document, nor is it an outsourcing contract. It is, quite simply, a detailed memorandum specifying the outcomes from many elements of the outsourced function. SLAs need to be defined in detail, as they are an important way of measuring an outsourcing supplier's performance. The most common metrics of quality, speed and accuracy clearly enable each partner to assess the current level of service the buyer is receiving. If performance slides, the SLA may trigger penalties. Putting these metrics in writing provides a legal basis (in the worst case) for contract termination, and gives the client the ability to influence the supplier's performance.

Negotiating an effective SLA that will provide value in the outsourcing relationship need not, and should not, be a one-sided process. The willingness of a preferred bidder to enter into the negotiation of the SLA design speaks volumes for their future acceptance of it, and indeed, they may have as much or more experience in the task as the client. Furthermore, while it is important to specify the desired outcomes in detail, the SLA must not be so prescriptive in the inputs side that it prevents the contractor from seeking ways to do a better job.

An effective SLA:

- identifies certain service levels or performance standards that the outsourcing contractor must meet or exceed
- specifies the consequences for failure to achieve one or more service levels
- includes credits or bonus incentives for performance that exceeds targets
- establishes the level of importance of key service areas by a weighting system.

SLAs are not easy to design or negotiate. But a comprehensive, fair and effective SLA is critical for a successful outsourcing relationship. In the course of negotiation, outsourcing clients and suppliers have the opportunity to learn much about how their future partner will approach important issues in the outsourcing relationship, which can only help in the smooth running of a contract.

Penalties and incentives

The most important factor at this stage is for the outsourcing supplier and client to agree on credible measurements, and establish what would be classed as above and below acceptable levels of service. By weighting the impact on key areas of exceeding or falling short of these measurements, to total 100 per cent, then penalties or payouts can be tailored to fit. If the outsourcing supplier then fails to achieve some of the key service levels, the percentage missed for the month can be applied as a service level credit against a proportion of the invoice.

Often the parties identify a subset of the key service levels as critical. For these elements, the parties will agree that more extreme penalties will apply, even that the outsourcing contract may be terminated if the levels are not reached to the frequency specified. Contract law generally entitles one party to terminate a contract if the other party 'materially breaches' the contract; defining critical service levels and providing specific conditions for termination eliminates ambiguity by determining what is, and what is not, 'material'.

Reasonable clients will avoid overmeasuring and trying to include every imaginable service level. They should agree to fair credits for failures in meeting service levels. Suppliers should be willing to understand that the client requires significant protection in the SLA and to acknowledge that there are certain levels of performance that would justify termination of the contract.

Similarly, exceeding the levels may be seen as a potential trigger for bonus payments or other incentives, payable when a quantifiable benefit to the client can be seen, far beyond any expected performance as laid down in the SLA. It is important for the contractor to know that they will be properly rewarded for success. If the supplier can add real value to the client's business, clients should be willing to share the value gained as a result of superior performance. In this way, service level objectives become highlighted as a critical parameter for both parties.

Force majeure clauses

Force majeure clauses excuse a party's failure to perform if the failure resulted from a natural disaster. In outsourcing contracts, negotiating the provisions of excused performance in the context of the outsourcing contractor's responsibilities and liabilities can be challenging and time consuming. Examples include failures resulting from the client's non-performance, failures of third parties, and failures in hardware and software.

Outsourcing suppliers tend to seek a broad definition of force majeure, while clients seek a narrow, tightly defined provision. Most organisations accept that a fair agreement lies somewhere in the middle, without entirely absolving the supplier from the responsibility of correcting and mitigating the effects of an excused performance failure.

WHAT TO INCLUDE

Every SLA will be different, tailored to the specifics of the contract and its application, but they will usually have several sections in common, including:

- Introduction
- Definition, boundaries and parameters of the service being outsourced
- Minimum acceptable service levels for all aspects of the job
- Detailing of improvements from the status quo (if sought)
- Agreed cultural norms of, for example, appearance (if contractors are to be perceived as company employees)
- Quality criteria (where contractors have to supply materials)
- Who reports to whom, how, when, and why

- How service levels are to be monitored, when and by whom
- Measures for rewarding exceptional performance, if appropriate
- Measures for penalising underperformance, if appropriate
- General payment terms and conditions
- Conflict resolution procedures

CONTRACTUAL ARRANGEMENTS

Having established an agreement with a single supplier, the next step is to finalise the contract. Only when all the details are agreed is it time to sign. It can take well over a year to reach this stage, and should not be rushed: it is a contract you ideally never want to terminate, so it is important to get it right first time.

The function of the contract

The contract should define both the work itself and the manner in which it is to be undertaken. If the task to be undertaken is ambiguously defined, in terms of both scope (the work to be done) and style (when and to what standard), then there is room for individual interpretation; and one interpretation will almost certainly differ from the next. The contract process is the best system to define unambiguously what is needed and at the same time to lay down performance criteria so the execution of the contract can be monitored.

The SLA is arguably the most important document in the tactical day-to-day management of the relationship. It provides the key performance criteria by which success or failure will be assessed. While in industry practice the SLA is a separate addendum to the outsourcing contract, in law it is not a separate agreement, but merely a set of terms and conditions of the substantive contract itself. In other words, the function of the SLA is to specify the goals of the outsourcing relationship, while the contract is the administrative document which outlines all the practical arrangements necessary to ensure these goals are met.

What should the contract include?

As well as terms and conditions, renewal dates and criteria, payment terms, and arrangements for rewards or penalties in the case of overperformance or underperformance (with definitions of these being spelt out), a contract should build in the following considerations.

Flexibility

If you know what is going to happen, you make plans. If you do not know, you make contingency plans. It is inevitable that circumstances – markets, technology, supply chain, and so on – will change, and that some of those changes are likely to be radical. A contract which allows no flexibility, therefore, is a bad contract. Unless the contract allows for change, the client may find himself tied down to the letter of the contract by the supplier, even though it may be obvious that a failure to change will be detrimental to the business. That will clearly lead to distress, or worse, dispute.

Disputes and exit strategy

Be aware, as indicated above, that there are likely to be disagreements – every relationship has them – and there needs to be an agreed methodology for dealing with them, configured in such a way that the work continues while the disagreement is resolved. This methodology should be written into the contract.

In the worst possible case, disagreement may lead to a complete breakdown of the relationship, so an exit strategy should also be agreed. In many cases contracts are expensive and difficult to terminate prematurely. Without careful management, and the right contract in the first place, the outsourcing contractor can easily gain the upper hand, which can be disastrous for the prospects of a long-term relationship. No client should be obliged to continue with an unsatisfactory contract, simply because the implications of breakdown are worse than maintaining the status quo.

The client should ensure the existence of a reasonable exit route in case the relationship becomes unmanageable. Terminating a contract is the worst case scenario in the relationship; ideally, the contract should define less final consequences for any lack of service or failure to provide to agreed service levels.

Decision making

Outsourcing relationships should be true partnerships, but one party must take the lead. There are decisions where there may be no right or wrong in qualitative terms, but where someone has to call the shots. It should be made explicit in the contract that while the partnership recognises the expertise brought to the relationship by the supplier, the ultimate call comes from the client.

Contract 'management'

Anyone can sign a contract, but unless it is adhered to, monitored and driven to its optimum, it is simply paperwork. Managing the contract is indisputably the clever bit. It is not the same as managing the same process when handled in-house. Once contracts have been signed and the supplier is providing the service, you have less direct influence on how the job is undertaken, as you have delegated away the tactical element of execution, except in so far as the contract allows. You can no longer hire and fire, or change emphasis or priority; you have a set of rules laid down that must be adhered to, no less by you than by the contractor.

Yet at the same time you have a team to inspire and motivate, but this team does not report to you. You have to be seen to be interested and involved, to be monitoring and checking; it is important that the supplier shares your enthusiasm for continuous improvement. Unless you are seen to be involved in the contract, without treading on the toes of the supplier, demotivation will follow, leading to declining standards and ultimate contract failure. These issues are covered in more detail in this chapter's upcoming section on building the relationship.

Consequences, positive or negative, can take the form of financial penalties or additional payments for high levels of performance by the supplier (see previous section, Penalties and incentives).

KEY PERFORMANCE INDICATORS

A balanced scorecard

If an outsourcing relationship is to be successful, clear measurements are needed to manage achievements and expectations on both sides. One of the best ways to establish this is through the balanced scorecard approach. A balanced scorecard involves goal setting, target setting and an information collection process. It includes a number of categories that represent the most general level of expectations, usually around cost, service and quality. Categories are divided into a number of subsets, known as attributes, defined through a joint buyer–provider process, with the exact composition and number depending on the goals of the relationship and the service in question. Choices are made about an appropriate measure for each attribute.

When establishing the scorecard, some of the key areas to consider are:

- Does the overall scorecard balance long-term and short-term goals, and financial and non-financial goals?
- Do the metrics and measures compensate for each other's blind spots? Consider balancing subjective and objective measures, and qualitative and quantitative measures.
- Are multiple perspectives taken into consideration? Does the scorecard balance the perspectives of the strategic and operation managers, the buyer and the provider, and so on?

What will a balanced scorecard achieve?

Having balanced scorecards in place during an outsourcing relationship accomplishes at least three things, namely:

- a client-defined, mutually agreed performance management system to reward exemplary service and to discourage below-par performance
- an established set of metrics by which performance is measured with the opportunity to make ongoing changes in service levels and expectations
- the provision of historical information to help decide the future of the relationship when it comes to contract renewal.

Client satisfaction

Signing an outsourcing arrangement should not mean the abdication of responsibility for a business activity to an outside provider. While one of the true benefits of outsourcing is the transfer of responsibility to an outside expert who is well equipped to handle the task, clients must put systems in place to ensure their own satisfaction with the service. Below are some guidelines for measurement and evaluation.

- **Measure what you want to manage**: you can only manage what you can measure and you can only measure what you can see. However, it is possible to come up with creative and useful solutions to measuring and managing activities that were previously beyond reach: cost savings and customer satisfaction to name but two.
- **Change what you can control**: not measuring and, in turn, holding providers accountable for things beyond their control is counter-productive and frustrating for all. The activities and measures must be within the service provider's control and having them at the table defining the scorecard is the best way to understand that.
- **Recycle and reuse, do not repeat**: the last thing anybody wants is competing measurement systems; a single 'good enough' system will do, defined as providing the information you need to manage with. The balanced scorecard can absorb earlier measurement systems; do not reinvent the wheel, but use the best measures that already exist.
- **Set the systems early**: construct a balanced scorecard as early as possible, customise it once the provider has been chosen and then jointly set target levels. A provider's ability and willingness to live up to the criteria set in a balanced scorecard can be a consideration in the selection process.
- **Timely and efficient measurement**: the measurement system to support a balanced scorecard should collect only what is useful and not duplicate material. Unnecessary and overlapping measurement processes are wasteful. Collecting information to support balanced scorecards should also happen as soon after the fact as possible.
- **Measure realistically**: be realistic in what can be inexpensively, quickly and easily measured. Certain things should not be 'scorecarded'. For example, do not measure the number of failed state inspections – such events do not provide for ongoing management of the relationship, they represent significant and immediate problems.
- **Use all the measurement tools and data sources at your disposal**: there are at least three tools which buyers and providers have at their disposal:
  - **Surveys**: these can be used with external customers, internal customers, and employees of both provider and buyer. Surveys are effective at collecting behavioural information or opinions but the measures tend to be subjective. Relying on this method leaves the process open to serious questions and undercuts the ideal of balance. When they take the form of checklists, surveys provide for more objectivity.
  - **Management information systems**: these can provide valuable information on costs, inventories and other accounting-related data; however, they are predominantly financial.
  - **Audits**: these avoid the subjectivity of surveys, while collecting non-financial and other specific information not present in management information systems designed for other purposes.

BUILDING THE RELATIONSHIP

Surveys have consistently shown that companies are, in general, dissatisfied with the overall results of their outsourcing agreements. So how can you make the outsourcing relationship one of value, providing ongoing benefit to your business?

Continuous improvement model

In order to have an outsourcing relationship that works, it is important to consider the supplier as a source of value that needs to be constantly realigned if the contract is to succeed. Levels of service need to be continuously improved if they are to result in long-term relationships. This calls for flexibility, which starts in the planning stage and should be the result of candid communication, frankness and a willing approach to working with the aim of creating a 'win–win' situation. With a good plan, there should be flexibility to meet new opportunities and redefine the relationship on an ongoing basis. That means measuring the value and rewarding the supplier on their ability to deliver that value. Then the bar can be raised continually and the relationships can be moved and aligned with business objectives.

Defining added value

Decide early on how to define added value. Look for one or more of the following:

- the ability of the supplier to come up with the initiatives that reshape the relationship to meet ongoing objectives
- the ability and willingness of the supplier to set and meet concrete and measurable service levels
- the ability of the supplier to commit to and meet specific financial targets.

Clear channels of communication need to be mapped out which encourage the supplier to be proactive and bring ideas to the strategy of your business. The supplier should also be rewarded for ideas and revenue opportunities, with rewards reflecting the benefits brought by the implementation of ideas.

Good communication

Ideally the relationship should start with each organisation appointing an informed and empowered point of contact, to act as contract administrator. These two individuals are responsible for making the relationship work, as opposed to the site managers who will oversee the day-to-day running of the function on site. This is an important first step in implementing a successful outsourcing relationship. Other factors which help to build and sustain the partnership include a strong management team; a dedicated account manager; a good working relationship between management on both sides; a consistent communication chain; a single point of contact, possibly the administrator (see above), to resolve queries and remove duplication; and a database of written communication between parties to track commitments.

Trust is essential in developing a successful outsourcing relationship, and communications need to be based on this. An outsourcing contractor should be instrumental in the client's success, therefore they need to have access to the information they need. It is necessary to drive home what the business is all about; highlight objectives; be frank in disclosing issues facing the company or its industry, both short and long term; give the supplier the tools they need to achieve or maintain status as an industry leader; and be clear about the roles in the relationship.

Coordinating standards and budgets

Relationships need to be developed at multiple levels throughout the organisations involved, to create the trust and understanding required for long-term success. Both parties need to coordinate standards, such as protocols and business processes, and perform joint budgeting exercises to understand the key cost driver information inherent in the other's infrastructure. In addition, there needs to be qualitative information associated with the reliability and performance of the outsourced services.

Anticipating problems

Most problems arise because the client underestimates the future work they will require the supplier to undertake. In addition, the client often fails to say up front what work is in the scope of the contract and what work is additional. This can lead to a misunderstanding of responsibilities. For example, the manager of a client's data centre is frequently retained in-house to help manage the supplier. That manager is familiar with the way the centre used to be operated in the past but, of course, the supplier may choose to do things differently. This may cause problems all round.

As this example shows, most management problems are not actually 'people' problems, more a result of a lack of clarity over how key objectives are to be implemented. This highlights the importance of the contract. Without a suitable contract the two parties enter their 'marriage' and may go into the honeymoon stage, where everyone is enthusiastic and focused on the objectives of the agreement and the potential benefits that are to follow. At this stage, everyone talks frequently, gets involved with fine-tuning objectives, works enthusiastically towards service levels and has great expectations. However, too often time erodes this initial flurry of excitement and a sense of anticlimax develops which can lead to exasperation. At this point, communication breaks down, key people leave, the original mission is lost, the client and supplier forget original objectives, and frustrations grow.

Ensuring that the contract is clearly defined, that flexibility is built into it, and that there is constant scope for communication and assessment and reassessment of the tasks and responsibilities in hand will help to prevent misunderstandings and frustrations.

Team dynamics

Most outsourcing agreements represent long-term, dynamic relationships in which unforeseen opportunities and conditions will appear and will have to be addressed. This means there is a need for flexibility, which should be built in to the first stages of the relationship. However, keeping on an even keel can be fraught with difficulties. Day-to-day decisions have strategic impact, while lack of planning and misunderstanding can create difficulties in moving forward. Personality clashes can also cause rifts, and managers on both sides need to have good people skills.

Facilities professionals should consider the following guidelines:

- The client should ensure that the supplier's best people are working in the relationship.
- There should be contractual terms and conditions spelling out certain processes and disciplines.
- The client should have the first right of refusal for members of the supplier's team.

Regular reviews

Many problems can be avoided if the dangers are anticipated and dealt with in the planning process. Communication and incentives for success are key to nurturing long-term relationships, with the company's outsourcing team focusing daily on the mission in hand. Outsourcing specialists often suggest that companies should communicate clearly every month, discussing performance and anticipating any changes for the next month. These discussions can provide short-term measurement to monitor ongoing behaviour, and benefits expected to be achieved over the length of the contract can be reaffirmed.

SUCCESSFUL CONTRACT MANAGEMENT

The key messages are:

- Avoid relationship deterioration by planning for the pitfalls.
- Schedule monthly meetings to reaffirm objectives and measure performance.
- Give the supplier financial incentives to succeed.

MANAGING THE END OF THE CONTRACT

A key area, often overlooked at the outset of any contractual relationship, is what happens when the contract comes to an end; and it will, eventually. Here the talk is not necessarily about a contract ending in dispute, rather what happens to a contract when its agreed term has run. Few companies can afford to go 'on hold' while they sort themselves out; competitors are waiting to swoop in and take advantage of the gap in the market.

What are the alternatives on contract termination? Assuming that the function remains necessary, they are limited. You can:

- Continue with the same contractor.
- Find a different contractor.
- Bring the function back in-house.

Many companies stay with existing providers through inertia, and fear of change. They may be unsatisfied with the existing arrangement, but they suspect that it would be easier to obtain improved contract terms, and tighter SLAs, from a new contractor, than by renegotiating the existing arrangement. This has dubious logic. The possibility, necessity, even, for change and continuous improvement should have been built in from the start.

Changing contractors can take a year or more to organise, and is an exercise that is fraught with hazards, including the possibility of delaying tactics from the outgoing contractor, removal of key staff and run-down of efficiency, and possible confusion over ownership issues of dedicated equipment. Many of the same reservations apply if attempts are made to bring operations back in-house, since key assets, in terms of people, knowledge and equipment, have been relinquished, and you should also ask why you chose to outsource in the first place.

Key lessons

First, too few contracts have adequate provision concerning termination or expiry, and many fail to take account of the way business needs will change over the relationship term. Second, usually, it is 'better the devil you know'; that is, it is preferable to stay with the existing provider, who knows your business. If you want new SLAs, renegotiate. Asking incumbents, with others, to retender can focus everyone's attention on the key aspects of the task, as well as pinpointing whether there really is a better option out there.

OUTSOURCING FACILITIES MANAGEMENT: CASE STUDY

A global financial and travel services organisation, with its headquarters in the USA and European operations centred in the UK, had a facilities management contract in the UK with the incumbent supplier for seven years. Up to the end of 2003, everything had been prescriptive, being based on an input specification. Retendering was handled differently using teams from the procurement and real estate departments and external advisors, and a new five-year contract was awarded.

In the US market, all matters to do with facilities management are input driven generically. But in Europe, most organisations either are moving towards an output specification or are there already. With input specifications the risk stays with the client, whereas with output specifications the risk is transferred to the vendor. This is fundamentally different from what happens in the USA.

The agreement covers all the property portfolio in the UK (nearly 200,000 m^2), which includes the HQ, financial services operation, bank property and many travel offices. The supplier handles all facilities management operations within these properties, except for security, which is also outsourced but overseen by the company's own security department. That is everything from grounds maintenance, cleaning and lift maintenance through to energy provision.

The continuing operation of the contract relies on SLAs and key performance indicators. To use the cleanliness of a meeting room as an example, the company expects the room to be clean at all times; how the supplier achieves that is up to them. If it required a person practically living in the room, then so be it. Under the previous system, the company would tell the facilities management provider when and how often it would want the room cleaned: twice a day, except on Fridays when it would be three times, for example. The risk would be taken by the company. Now, the onus that the room is clean is transferred to the facilities management provider. There are monthly key performance reviews with facilities management teams and the supplier. Any negative input is redressed by the company with the supplier.

The European approach to a facilities management contract has been based on close links with the European real estate department. The output specification has been produced in consultation with local facilities managers in several countries. Local countries' asset registers and operational specifications were passed to the UK and fed into the overall tender documentation, which included designing the contract and also covered plans of all buildings. Any company seeking a pan-European solution has to be aware of potential local solutions, and local providers need to be part of the bidding process. For example, France and Germany could be handled by one provider with the remaining estates handled by small local operators. To have an operational centre in a pan-European context, you really do have to have full support at a local level or the contract will not go anywhere.

Some e-tendering processes involved UK and European vendors in France, Germany, Italy and Spain. The process involved physical meetings, reference site visits and commercial meetings with bidders. While the UK operation is leading the way with its procurement-led facilities management outsourcing deal, it is not leading with e-tendering (the USA is showing the way forward).

Now, the facilities management contract for Europe and the UK is to be retendered at the same time, or the European contract will be let 'co-terminus with the UK contract'.

This was a joint effort of global procurement and real estate on the UK contract, and the same thing with local colleagues individually on the pan-European contract. People issues are pertinent to this work, which continues to demonstrate the key need for, and crucial importance of, personal communication.

8 Transport Policies

Frank Booty

Transport costs remain one of the biggest factors impacting on the bottom line. For some businesses, transport is the second largest cost after personnel. With the government's tax changes now impacting on companies' transport budgets, this situation is unlikely to change. However, with businesses constantly evolving towards a more global working culture, the need for transport is greater than ever before. How can facilities managers seek to resolve this conflict of interests?

Careful travel planning can bring huge benefits to a company. The environmental imperatives remain: transport-derived air pollution, carbon dioxide emissions and traffic congestion still need to be reduced. But the reasons for developing travel plans go beyond environmentalism. Other benefits include time and cost savings, greater flexibility and accessibility for the workforce.

It is easy to overlook the parking issue. The car park represents the largest untapped asset in the property portfolio, with brand names waking up to the potential it offers for floor graphic advertising. But the car park is also becoming the target of both government regulators and taxmen. How can facilities managers ensure they get the most out of their car parks, while taking new government guidance into account?

TRAVEL PLANS

Legislation

A travel plan sets out to combat overdependence on cars by boosting all the possible alternatives to single-occupancy car use. By reducing car miles it can not only benefit the environment but also produce financial benefits and productivity improvements, saving businesses and their staff time and money. A travel plan is an effective site management tool that also meets the need for continuous improvement in environmental management.

After the government's *Planning Policy Guidance Note 13: Transport* (PPG13)[1] was approved earlier this century, all businesses were encouraged to try their utmost to encourage staff to take up car sharing, use public transport and work from home

---

274 [1] See www.planning.detr.gov.uk/ppg/ppg13/index.htm for further information.

wherever practicable. Good travel plans have typically succeeded in cutting the number of people driving to work by 15 per cent. This modest percentage translates into many car miles and congestion avoided. For a firm with 2000 staff mainly travelling to work by car this results in some one million miles less per year. An equivalent cut to the UK's total commuter mileage would be 13 billion less car miles.

National planning guidance now says that all planning applications with significant transport implications should be covered by a travel plan. Businesses looking to expand or relocate will often find that a travel plan is required by the local planning department. Now a guide (*The essential guide to travel planning*, available from www.dft.gov.uk) draws together experience built up by businesses with leading-edge travel plans to explain how facilities managers and others can set up a travel plan for their company.

Objectives

Travel plans are designed to reduce the adverse environmental impacts of transport to and from a specific site or building. Travel plans seek to reduce overdependence on cars, especially when the driver is alone, through encouraging changes in travel behaviour in favour of more environmentally benign modes, such as public transport, walking and cycling. To complement this, particularly where it is more difficult to provide alternatives to the car, travel plans incorporate measures to reduce the impact of vehicles (e.g. through greener fleet management).

Travel plans also seek to reduce the need to travel: both travel undertaken by employees when carrying out their duties, for example, when attending meetings, exhibitions, training courses or visiting clients, and travel to and from the workplace. Alternative methods of communication, such as telephone conference calls, are encouraged, as well as vehicle sharing.

TRAVEL PLANS: A MANAGEMENT TOOL

For facilities managers, the travel plan is a powerful management tool that offers considerable benefits to the company or organisation. Travel plans can help a business to:

- stay operational at a time of expansion
- reduce the costs of car parking provision and/or business travel costs
- improve, or at least maintain, staff journey times to and from work
- contribute to a healthier environment
- enjoy the additional bonus of a healthier workforce.

Guide to effective travel planning

Convincing the CEO

The first step is always to convince the CEO and other senior managers that the travel plan combines resource, facility and site management in one package and that it is essential for the future of the organisation and its site. In other words, the adoption of a travel plan is a strategic issue which needs director-level support.

Establishing the baseline

A survey of staff travel habits needs to be conducted. The questions should be designed to ensure that a comprehensive understanding can be gained of how people are currently travelling and what changes they may be willing to make in the future. This should be complemented by a series of audits to obtain a clear picture of current arrangements regarding the following:

- personal security for cyclists and pedestrians on site and along key routes to the site
- business travel: staff travel expenses budget, mileage undertaken and by what method
- car parking arrangements and charges
- traffic counts by vehicle type, time of day and day of week
- access to site by bus, train, on two wheels and on foot
- shift patterns and their impact on staff travel to work.

A package of complementary measures

Possible measures to take include:

- car sharing
- charges for car parking
- encouraging public transport use
- enhanced bus services
- improving cycling facilities
- improving the pedestrian environment (on and off site)
- reduction in the need to travel.

The emphasis placed on different measures varies enormously, according to local circumstances. If, as is the case at a major Liverpool hospital, over two-thirds of staff already travel to work by public transport, then the emphasis will be on helping them to avoid switching to the car. However, if the site is never going to be well served by public transport and your staff live in rural areas some distance from it, sharing car journeys to work is likely to be a more important measure. The priorities for action will be determined in part by the baseline assessment and in part by what is judged politically and economically feasible.

Consultation about the plan

A transport working group should be established, which should act as a reference point throughout the development of the travel plan and then assist with its implementation. Membership should comprise:

- facilities manager
- human resources manager
- environmental manager
- staff representatives
- staff willing to champion change
- council officers (on planning, transport, car parking, cycling, local agenda 21).

Because a travel plan involves everyone, it is vital that everyone is on board, so the next stage is to consult all staff about the plan and the changes it envisages. Interest groups can be established (e.g. a cycle opportunity group or bus-user group). Also employ the usual methods of raising awareness: an in-house newsletter, team briefings, leaflets accompanying salary slips and notice boards.

Adoption of the travel plan

With the results of the consultation process collated and analysed, the plan now needs to be finalised, costed and approved at the appropriate level. A travel plan cuts across traditional disciplinary boundaries and is of strategic importance. So, in most organisations, it is presented formally to the board for endorsement.

Resources for implementation

It is accepted that someone has to drive the travel plan forward. For organisations with many staff on one site, or where there are many employees in an identifiable geographical area, a travel coordinator should be appointed. The coordinator should be:

- a good communicator at all levels
- able to use information and communications technologies
- able to motivate people
- a pragmatic organiser.

This person will need to be fully resourced and supported and will need access to the decision making process. All of these costs will form part of the business case for the travel plan and are usually offset against savings in staff travel expenses budgets and the reduced need for more car parking, for example.

Project management

As with any strategic plan, the facilities manager's project management skills will be important. By now, the priorities, costings and a timetable have been established and the transport working group should be ensuring coordination across departments and with external bodies. There will be actions that can be taken immediately, for example:

- provision of cycle lockers and better cycle security
- promotion of existing bus services
- telephone conference calls
- improvements to pedestrian routes
- priority parking for car sharers.

Other measures take more time to plan and introduce, such as revising car parking and staff travel policies (severing the link between the pay packet and travel by car), flexible working schemes and improvements to the public transport infrastructure.

Marketing the travel plan

Facilities managers are familiar with energy conservation campaigns. Successful ones have involved all staff in the process and in many instances have produced dramatic efficiency gains. Travel plans also require a constant marketing effort to be made in order to reach every nook and cranny of the organisation and to keep people informed of developments, new incentives for car sharing, walking and cycling, support for public transport users, and so on.

Setting targets

Starting with the figures derived from the staff survey, it will be practical to establish desired targets for changes in the way people travel to your site. In the worked example shown in Table 8.1, the emphasis happens to be on more car sharing and public transport.

The table assumes there are 3000 staff in the work premises on any one working day, and that all methods of transportation are reasonably available to most employees. Decreases or increases in percentages directly relate to more staff coming by bus and fewer staff coming by car, for example. Extrapolating to 2010 will produce fewer staff coming by car, and so on.

Monitoring and evaluating results

Measurable targets need to be established for each element of the travel plan. For example, take-up monitoring will show the number of annual bus passes sold, the number of cycle lockers in use and so forth. Frequent staff surveys will be expensive and are likely to produce diminishing returns. However, conducting a repeat of your survey of staff travel habits every two years will be essential to check whether you are achieving the desired modal shift. Other monitoring methods include checks on key budgets, periodic comparisons of service levels, and recording anecdotal and intuitive evidence.

Table 8.1 Staff travel targets

Mode of transport to work	Actual (%)	Targets (%)		
	2003	2005/06	2007/08	2009/10
Car, as driver alone	60	55	50	45
Car, as passenger	17	18	19	20
Car, dropped off	9	8	8	7
Bus	9	11	12	13
Train	1.5	2	3	3 or 4
Bicycle	1	2	3	4 or 5
Motorcycle	1	1.5	2	3
Walk	1	2.5	3	4

Note: The strategy could include reducing road transport vehicle carbon dioxide emissions (by reducing total business vehicle mileage and improving the average fuel efficiency of vehicles) and having at least 10 per cent of all fleet cars being alternatively fuelled, as well as reducing single car occupancy rates. Homeworking and flexiworking could also be expected to increase, leading to a decrease in car use.

Encouraging behavioural change

Facilities managers will need to employ a range of skills in implementing travel plans. Those more used to building roads and maintaining buildings will find their traditional working methods are needed when it comes to site infrastructure changes, but new skills are required when it comes to encouraging behavioural change. Travel plans have to be handled with sensitivity and care, as reducing overdependence on the car calls for changes in lifestyle, which is distinctly intimate and personal territory. For this reason, words such as 'consultation' and 'partnership' must not only be part of a new facilities management vocabulary, but also be translated into meaningful actions, capable of being verified and evaluated.

Transport best practice with the Energy Saving Trust

The Energy Saving Trust provided a transport best practice service to the road freight transport industry for many years. Covering such topics as energy-efficient fleet management, its impartial and authoritative advice on energy-efficiency techniques and technologies in industry, transport and buildings is widely respected. It now offers an expanded service covering travel plans, which includes free publications on travel planning and access to free expert advice and assistance.

Reducing vehicle emissions is a key transport issue, which has been agreed as a priority across government. To reduce emissions of carbon dioxide, it has been agreed to reduce business vehicle mileage, improve the fuel efficiency of fleets, use alternatively fuelled vehicles wherever feasible and reduce single-occupancy car commuting.

Results from research show the costs of a travel plan will vary greatly depending on the type of organisation, the location, and the input from the local authority and public transport operators. The highest cost of £431 compares with the average of maintaining a car parking space, typically £300–500 per year. For further details see the Information section at the end of this chapter.

CAR PARK MANAGEMENT

Parking restrictions

Overly draconian parking measures were feared at one point in the progress of the government's *PPG13* (see section on Legislation earlier in this chapter), namely that car parking spaces at hospital developments were to be linked to bed numbers. Now, new hospitals and school developments are all to be planned to maximise accessibility by non-car modes of transport while simultaneously allowing good access for emergency vehicles and those who need to use cars. Office developments should have one car parking space per 30 m^2 (originally it had been 35 m^2) for all developments above 2500 m^2. Any threats to new office developments are thus eased.

Parking is, however, going to get tougher. Consider, for example, government plans to raise £2.7 billion a year 'for better transport schemes' by allowing local authorities to charge people who work in towns and cities between £150 and £500 a year to park at their own place of work. These plans will hit the low paid, as employers are likely to pass on charges to the workforce. The answer is to give massive relocation incentives and tax breaks to work-intensive business operations to encourage them to move to accessible locations where people can get to work cheaply and easily or park for free. There is also the issue of congestion charging to consider, something being considered by many local authorities in the wake of the London exercise.

Planning a car park

The first questions to ask when planning a car park are:

- Why do you want a car park?
- How does it fit in with your transport strategy?
- How big does it need to be?
- What other facilities are involved?
- What material will it be built from?
- What buildings will the car park relate to?

The car park may be designed to serve office blocks, in which case it will be predominantly busy for 30 minutes at the start and end of the day, and quiet in between. A shopping centre car park will have more regular patterns of usage, except that demand will double or triple in the weeks before Christmas. A multiplex cinema will be busiest at the end of the evening, as all films are timed to finish within a 30-minute window so staff can go home.

When developing a new car parking facility, it is important to:

- identify both the users and the demand
- perform a traffic impact study
- determine the land availability
- aim to provide the maximum number of spaces while maintaining high standards of circulation
- draw up conceptual layouts.

Design

The car park is (or should be) the gateway to the business, town, office, retail centre or facility: it is the first thing a visitor will see. If it is dirty, scruffy, badly lit and smelly, that is the impression the visitor will have of the facility. If it is clean, well maintained, well lit and of pleasant design, the visitor will take that memory away. Given that 2.4 m is needed for a car space, the concrete supports need to be spaced at least at 2.4 m or multiples of 2.4 m intervals.

Aside from an architectural overhaul and/or repainting and cleaning, ensuring that car parks are well maintained can be achieved through outsourcing.

Floor graphic advertising

In the USA, outdoor floor advertising is worth over £150 million a year. At present in the UK, the concept allows owners or operators of car parks to tap into a large source of revenue that builds year on year. The owner is able to make money from what was previously just a cost centre.

Advertisements of up to 4 m× 4 m can be produced (subject to local space availability) using a combination of flat line colours. Designed to withstand up to three years' constant use, these advertisements are produced using materials that comply with all current health and safety legislation. The materials used have a luminescence some five times greater than highway cats' eyes, making the messages highly visible at night.

Car park owners and operators, as well as supermarkets and National Health Service (NHS) trusts, are all ideal locations as they offer a parking facility for consumers or office workers. Income derived from the advertising can be applied to finance refurbishment costs, or to generate significant incremental revenue.

Lighting and security

Staff working late or hospital visitors, for example, walking to or from their cars in the dark will need good lighting. Under the Chief Police Officers' secure car park scheme, the advice for the design of car parks is to avoid dark corners and corridors. Much emphasis is put on the design and management of the facilities. Now, many staff are involved in continuous on-site surveillance, and closed circuit television (CCTV) facilities are often deployed. (Security issues are covered in detail in Chapter 12.)

In the UK today there are many underground and multistorey car parks. Quick visual improvements can be achieved through improving the lighting and painting using bright colours, but structural difficulties are more awkward to rectify.

FLEET MANAGEMENT

A fleet can be a vital part of the company's operational effectiveness, whether that means salespeople arriving punctually for meetings or equipment arriving on site in plenty of time and in suitable condition. Facilities managers should realise that proactive cost management and policy advice from fleet management companies can be useful.

In some quarters, the company car is no longer seen as a 'perk'. The government has long been pursuing a policy of 'greener' motoring, with the overall aim of getting drivers out of big, petrol-thirsty saloons and into more environmentally friendly vehicles.

Companies may need guidance on aspects of running a fleet, from vehicle acquisition and disposal, to maintenance and accident management; not to mention fuel costs, which in the existing UK climate increasingly form a heavy burden on any business. However, where a fleet management company can be at its most valuable is in providing expertise on the financial, tax-related and environmental issues. If a business intends to buy a normal combustion engine car it will only be able to offset between 10 and 25 per cent of the cost against their tax bill, depending on its emissions. To offset the full capital cost of a company car against their corporation tax bill, companies will have to buy from band B or better (hybrid or electric cars with very low emissions) to qualify for the full tax allowance.

The higher rates of vehicle excise duty (VED) being imposed on buyers of new vehicles were among a string of changes introduced in the 2008 Budget. VED is a huge revenue-earner for the Treasury. The extra cash generated by new vehicle duties on cars in 2009/10 will be £465 million, and in 2010/11 the extra VED raised will be £735 million.

From 2009/10 there will be six new VED bands including a top band (band M) for cars emitting over 255 g of carbon dioxide per kilometre. These cars will pay an increased VED rate of £425. But cars emitting 150 g or less per kilometre will pay less. From 2010/11, the most polluting new cars will pay a first year VED rate of £950, while those new cars with a 130 g/km or less emission level will pay nothing.

The 2008 Budget replaced the capital allowance treatment for business cars with an emissions-based approach. Cars are placed in one of two capital allowance pools according to their carbon dioxide emissions, with cars with emissions above 160 g/km receiving a lower allowance. First year allowances of 100 per cent will continue to be available for cars with emissions not exceeding 110 g/km. The Budget increased company car tax rates on all but the cleanest cars emitting less than 135 g carbon dioxide per kilometre in 2010/11, and enhanced incentives to drive fewer miles through changes to the fuel benefit charge from April 2009.

A lower VED rate is offered for diesel vans that comply with European Union (EU) air quality emission standards. It was confirmed that main road fuel duty rates will rise by 1.84 pence per litre on 1 April 2009, and that rates will then also increase by 0.5 pence per litre above indexation on 1 April 2010. The Vehicle Certification Agency (VCA) has details on new car fuel consumption and exhaust emissions figures at www.vcacarfueldata.org.uk. This site can be used to identify the VED and/or the relevant company car tax percentage bracket, based on carbon dioxide levels.

INFORMATION

Energy Saving Trust transport helpline, Tel.: 0845 602 1425, www.est.org.uk
Summary of 2002 research: www.localtransport.dft.gov.uk/travelplans/index.htm
Department for Transport: www.dft.gov.uk
Essential Guide to Travel Planning: www.publications.dft.gov.uk
Association for Commuter Transport: www.act-uk.com
Carplus, car sharing advice: www.carplus.org.uk

Information Technology and Communications

9

Frank Booty

Changes in the way business is transacted have been occurring owing to the phenomenal growth of the Internet, world wide web and mobile communications. There is still much change and disruption to come, with many traditional mechanisms of doing business – even products and services – set to disappear. For many small and medium-sized enterprises (SMEs) it will be a case of 'adapt or die'. Evidence or an analogy of that scenario may be found in the decline in sales of compact disc (CD) singles alongside the equally rapid rise in the growth of downloading tunes: the music distribution industry was shaken to its core. Now, the next generation internet is being readied, fuelled by video. Until 2013, it is peer-to-peer communication, with video content taking over from 2013 to 2025, and thereafter video communications. Video streams carried over the web have already mushroomed from 2005's nine billion to over 300 billion in 2008. Key applications will include tele-presence and video communications, desktop video and digital signage, video surveillance and internet protocol television (IPTV). The message is that facilities managers must keep abreast of developments and the implications of the ever-changing information and communications technology (ICT) market.

Facilities managers need to be aware of the convergence of the data and telecom markets, a phenomenon which embraces the fields of information technology (IT), telecoms, the Internet and mobile communications. Globally, many telecom carriers are building a web of broadband (high-capacity) Internet protocol (IP)-based networks to offer companies raw capacity and increasingly sophisticated added-value services. What are the factors driving this market?

MOBILE GROWTH

There are at least 2.2 billion mobile phone users globally. Prepaid calls, widespread availability, reduced charges for normal mobile calls, free offers and marked increases in wireless data communications are all factors behind the growth in the mobile phone networks. However, the technologies driving this growth, GPRS and UMTS (general packet radio service and universal mobile telecom system, the name of the third generation or 3G mobile phone standard), hit the headlines earlier this century for the wrong reasons. Sending pictures/video over mobile phone networks has not taken off in a massive way and many of the operators are looking to business data traffic as a

283

means of growth in their businesses. However, there is also competition from wireless local area networks (LANs) and the WiMAX standard (see below). There is a trend towards mobile Internet access.

FIXED NETWORK NEED

The growth curve for voice telephony has levelled, while the already strong growth of data traffic in fixed networks is accelerating, especially in packet-switched networks, which are increasingly based on IP. Most telecom operator revenues still derive from voice traffic, although moves to lower call charges are already noticeable.

The number of Internet users continues to rise, and will at some future date pass the 750 million global fixed telephone connections. Average connect times are increasing, but the unbundling of the local loop is still a key concern.

WHAT WILL CONVERGENCE MEAN?

For facilities managers, convergence will mean:

- the ability to run voice traffic over existing data infrastructure, resulting in reduced support overheads and enabling low-cost add-ons, moves and changes
- the delivery of voice, data, video, and so on, through one service, leading to reduced billing costs
- a growth in specialist data centre providers, tele-hotel companies and communications-related outsourcing in general
- increasingly flexible, wireless communication infrastructures, enabling seamless integration between mobile and fixed work, with implications for office space needs.

BLUETOOTH

Much future communication is expected to be handled between computers. One way for computers to connect to other systems is through a wireless connection. The glue to provide that connection is Bluetooth, an open standard for short-range digital radio: it is not a competing technology for the equally new third generation mobile phone systems, but a supporting one. It operates in the industrial, scientific and medical (ISM) applications band, which is available almost everywhere globally. It is cheaper than competing technologies and provides up to three voice channels for speech applications. While it is not a replacement for a LAN, it can provide the final connection to the network through its ports, enabling wireless access from anywhere within the ports' coverage area.

It has been claimed that the 3G mobile services launched across Europe in 2004 faced lower cost, higher performance fixed wireless competition from the outset. Users realised that the promised unified wireless environment envisaged with 3G was not going to happen. The introduction of fixed WLAN 802.11 (aka Wi-Fi) technology shattered this vision, with 802.16 (WiMAX) broadband fixed wireless access systems services challenging it still further.

IEEE radio technology, known as orthogonal frequency division multiplexing (OFDM), is much more efficient than 3G radio technology wireless code division multiple access (WCDMA). It is not believed that the WiMAX solution will obviate the need for mobile-wireless 3G services, but a low-cost high-speed service with much wider reach than Wi-Fi will introduce new competitive pressures in the mobile wide area network (WAN) market. The task of 4G will be to reconcile the two.

It should also be noted that Bluetooth is already built into twice as many devices as Wi-Fi, mostly cell phones, and so on, but laptop/notebooks and personal digital assistants (PDAs) including both Wi-Fi and Bluetooth could validate potential enterprise applications. Disadvantages of Bluetooth are that it is cellphone-centric and handles a maximum of eight nodes per network.

WIMAX

3G cellular (HSPA and EV-DO) and mobile WiMAX are the technologies that are expected to serve most of the 2012 mobile wireless broadband market, but high-speed OFDM packet access (HSOPA), long-term evolution (LTE) and/or ultra mobile broadband (UMB) will also be available. It is expected that WiMAX, variously as fixed or mobile services, will bring broadband access services to rural and undeveloped regions of the world. Growth in India will be strong and precede 3G rollouts. It is expected to become large in China, but on the heels of 3G TD-SCDMA. Further, if mobile WiMAX can supplement existing cellular networks as an overlay, it will become a mainstream technology.

VOICE BECOMES DATA

The big swing in the future will be the move in the content of traffic carried over networks from voice to data. Currently, more voice than data is carried, but that situation is changing. Note that in digitising voice, the voice patterns become represented as 1s and 0s, which is exactly how data is represented. Networks will be carrying only data; everything (voice, text, video, TV, multimedia, tele-conferences and video-conferences, emails, messages, and so on) will be delivered through one service. The billing will prove a fascinating exercise.

A PROSPECT FOR THE FUTURE

Imagine being always online in a wired and wireless world. A manager in their office receives email via the office LAN. They leave the office for a meeting in the conference room. Email carries on coming via wireless LAN technology, such as Bluetooth.

At lunchtime they switch off the PC and leave the office. Email continues to be delivered, this time to their mobile phone or PDA or a hybrid of the two. Back in the office, when the system is next switched on, Bluetooth updates the company server with the PDA/phone's latest information. While this is going on, a phone call comes in from the manager's superior who is travelling over the Atlantic. Access to a video graphic is required during the call, as well as conferencing with colleagues in Australia and Singapore.

- Synchronisation is handled automatically.
- The capability exists of moving seamlessly between data and telecom bearers, no matter how many devices are being used or where they are located.
- Mobile and fixed devices work seamlessly together.

WI-FI ZONES AND WIRELESS LOCAL AREA NETWORKS

The Wi-Fi Alliance, a non-profit organisation formed in 1999 to certify interoperability of IEEE 802.11 products and to promote them as the global, wireless LAN standard across all market segments, has over 300 members offering more than 4000 products. A global programme under the Wi-Fi Alliance banner promoting the availability of public access terminals in Wi-Fi zones is underway. Hot spots (Wi-Fi zones) are to be found in key European airports, hotels, city centres, coffee shops and increasingly office buildings. Air passengers can now make wireless broadband Internet connections with their corporate intranets from airport lounges or while in the air.

Few, if any, technologies hold more promise for the enterprise today than wireless. But few, if any, technologies also have more potential security risks. Industry analysts disagree over the successful penetration rates for wireless local area networks (WLANs) into the professional mobile PC-installed base: opinions range from 'it's all over-hyped' to 'surpassing 90 per cent'.

Wireless access, as a substitute for digital subscriber line (DSL) and other wireline Internet access, is fast coming into vogue. A popular variant today is Wi-Fi. Soon to be coupled with voice applications, Wi-Fi – as a complement to traditional carrier access arrangements – is emerging as a serious alternative for enterprise communications buyers. As with other emerging technologies, enterprise acceptance will be gradual. But as the technical alternatives prove real, business purchasing is expected to accelerate.

IT INFRASTRUCTURE MANAGEMENT

Intelligent buildings

There is no universal definition for the term 'intelligent building'. One definition is, 'an intelligent building provides a productive and cost-effective environment because its four basic components – structure, systems, services and management – have been fully integrated, creating the optimum design'.

Many of a building's ongoing costs relate to the low-voltage systems that provide and sustain the building's functionality. The reality is that some 75 per cent of a building's lifetime cost will derive from ongoing maintenance and operating expenses. As the average life of a building is 30–40 years, these costs for that lifetime can be staggeringly high. Any innovation that simultaneously reduces cost over time while enhancing building performance must hold substantial appeal. Intelligent buildings offer such innovation.

Intelligent buildings harness technology and link building systems to supply more efficiency, higher productivity and increased comfort. Most buildings feature 10–46 low-voltage systems, each calling for its own control, management and monitoring over the decades of the building's lifetime. Without a common infrastructure that can link them together, these dozens of systems can create a lifetime accumulation of unnecessary cost and commotion. But with a single backbone supporting all of these systems – from security to lighting, heating, ventilation and air conditioning (HVAC) to communications – building operations can become high performance and cost effective.

Future drivers for intelligent buildings will be: IT, robotics, smart materials, sustainable issues and the impacts of social change. Besides technological developments, other factors will include climate change, changes in regulations, and how people work and live within these buildings.

Future-proofing

The cost of installing an inadequate IT infrastructure that does not support the company's future requirements is considerable. Whether facilities managers are responsible for IT or whether they work with an IT department, ultimately facilities and IT management converge. If your business is moving or building a new set of premises, make sure you are involved in planning right from the start.

It would have been nearly impossible to future-proof a business 15 years ago against today's IT demands, and it is still difficult for facilities managers to predict what demand will be in the next few years. But continual changes in the way organisations work are generating demand for bandwidth-hungry applications, more efficient data flows and, all importantly, flexibility.

Any changes to existing systems, or replacement of those systems, should be done with the maximum of forethought. A well-structured infrastructure should build in the flexibility necessary for facilities and IT managers to handle future moves and technological changes.

Resource planning

While infrastructure management or infrastructure resource planning is a recent concept in facilities management, taking the principles behind it into account is common sense. It is good practice to assess what you have and how well it is working before moving on to the next step. Consider the following:

- What are your current IT assets: computers, cabling and telephone systems? How old are they: are they nearing the end of their lifetime? Where are they and who is using them?

- What is the gap between their existing value to the company and what company directors and/or employees would like them to deliver?
- Are there ways of maximising the use of existing assets rather than buying new ones?
- Are there ways that your existing system can work 'smarter, not harder'; through better network set-ups which minimise network congestion, for example?

Assessing the market

The facilities management world is full of companies making claims about what their technology can do for you. Each option has its own advantages and disadvantages, and it can be all too easy to be persuaded down a given route without exploring all the options. Assess the products on offer and look carefully at the experience of the companies offering them. When considering a new IT infrastructure, ask the following:

- Is the system easy to deploy and reliable? Is it likely to be overtaken by future developments?
- Will it fit with existing systems?
- Is it based on open standards, allowing flexibility in future?
- What are the cost implications?

As each technology possesses its own particular strength, businesses should always begin by understanding how they differ and what works best for their own networking needs. Before opting for the latest technology, question what value you are looking for over what you already have. Upgrade networks only when it is necessary to incorporate newer technologies, rather than changing investment unnecessarily simply because an older technology is no longer fashionable.

Cabling categories

To a large extent, category 5 twisted pair cabling has replaced coaxial cable in LAN cabling for horizontal distribution, while multimode optical fibre has become the main media type for intrabuilding backbones. The key factor in choosing the right cabling

Table 9.1 Types of connection protocols available for ISO communications layers

	Past	Present	Future
Management layer	Proprietary	KNX BACnet LonWorks, IP	IP
Comms layer outstation to outstation	Proprietary	KNX BACnet LonWorks, IP (ethernet, FTT Arcnet)	IP (ethernet)
Outstation to field devices	A/D I/O current loop	A/D IOS ID IP	IP (ethernet)
Physical layer	Proprietary coax + optical fibre	Cat 5e/6 + optical fibre	Wireless Cat 5e/6/6A + optical fibre

option is the anticipated demand on network capacity. For office IT, such as word processing and accounts, category 5 offers 100 Mb/s, which is commonly regarded as more than enough. However, for data-intensive applications, enhanced category 5 (5e) or category 6 will be required (Table 9.1).

The Telecommunications Industry Association (TIA) and the International Organisation for Standardisation (ISO) collaborated on the technical work associated with the category 6 standard, with the aim being to have a power sum bandwidth of 200 MHz, twice that of category 5. TIA is developing specifications of connecting hardware to ensure backward compatibility and interoperability with category 5 and 5e cabling. (See Chapter 11 for more on structured cabling.)

VOICE OVER INTERNET PROTOCOL

Facilities managers need to consider the advantages and disadvantages of voice over Internet protocol (VoIP), sometimes referred to as IP telephony. Unlike the conventional phone system, which is a circuit-switched network, VoIP sends calls over the Internet, a packet-switched network. It first converts the voice traffic into data packets. These packets are then routed over the Internet in the same way as other data, reassembled at the point of arrival and then converted back into voice.

While some Internet phone traffic may travel from computer to computer, most VoIP traffic uses the current telephone and fax systems. At the network level, this traffic is managed by a VoIP gateway, which converts the call between the Internet and the public telephone service. This gateway may be located at the company's premises, or at the office of a service provider such as an Internet service provider (ISP).

VoIP can offer new applications which can boost productivity. These include:

- integrated voice, email and fax messaging
- computer–telephone integration for call centre representatives
- enhanced network collaboration.

Features of VoIP

- **Unified messaging**: with a conventional system, business employees may phone, fax, email or share files over the company network or the Internet. VoIP combines these into one channel. For example, technology called 'unified messaging' can be used to funnel each employee's voicemail, email and faxes to a single mailbox, to be retrieved from a single end-station (PC or phone). Calls between offices travel over the existing data network through VoIP. By staying on the company network, these interoffice calls travel free of toll charges.
- **Toll-free telephony features**: IP phones deliver a complete suite of business telephony features, including call waiting, caller ID, transfer and conference calling. These features can be deployed across any number of sites that are connected by a company's data network, without incurring toll charges.
- **Remote access via single interface**: VoIP makes it possible to make voicemail, email, fax and video messages remotely accessible through a single interface

(PC or phone) at each user's office desktop. This set-up increases productivity for mobile sales and service people and eliminates the cost and hassle of managing separate voicemail, email and fax systems. Employees will have faster access to customer requests from any location.

- **Customer service**: integrated call systems can tie phone and email systems to back-office applications where customer account information is stored. When customers call, call centre agents have instant access to their account history and can deliver a more personalised and timely service. In addition, customers visiting a company's web site can use Internet tools to request assistance from a live agent. VoIP enables the agent to deliver voice assistance over the phone while assisting the customer online in real time.
- **Information sharing**: employees can collaborate more easily, sharing documents online in real time, during conversations or during conference calls. VoIP also supports video transmissions over the network, enabling employees to view video-based training modules, product announcements and presentations from their desktops, enhancing productivity, saving time and lowering travel costs.

Integration with the data network

Using VoIP means connecting your voice equipment – phones, fax, machines and private branch exchange (PBX) or key system – to your data network. Users can then be equipped with individual IP handsets which work like traditional phones but plug directly into network LAN switches and deliver enhanced integration with the data network.

Quality

Two main issues affect Internet voice traffic: latency or sound delays and voice quality. Both of these issues are being addressed.

Voice quality is affected by many issues, including how it is compressed and decompressed for delivery over the Internet, and how packets are processed. Often, voice calls sound shaky or robotic, which is obviously unacceptable for many typical business applications.

Voice quality is often driven by packet loss (data packets failing to arrive, not arriving on time or arriving with errors). To ensure steady quality, experts recommend a packet loss of less than 5 per cent. At 10 per cent voices start to quiver; at 20 per cent they start to sound like robots. The latest VoIP products do provide much improved call quality. Increased use of broadband technologies such as DSL and cable modems also help to make VoIP more accessible.

Standards

VoIP standards are not fully established. There is a standard, H.323, from the International Telecommunications Union (ITU) for carrying multimedia traffic over IP networks. This defines how delay-sensitive traffic, such as voice and video, is prioritised on LANs and WANs. But many products are not interoperable with competitors' offerings because of the leeway in implementing H.323. The answer is to install gateways from a single vendor.

Security

Businesses need to take security measures when implementing VoIP. Traffic must be encrypted as eavesdroppers can listen in on conversations wherever packets pass. The issue is being addressed through the H.235 standard, which covers authentication, encryption and other security measures.

Developments in IP telephony

IP traffic is increasing exponentially at a time when overall growth in international traffic is slowing. IP telephony can generally be offered to customers at prices very much below those offered over conventional circuit-switched networks, particularly on long-distance and international calls, because traffic pricing on IP-based networks is largely distance independent. Comments aired in the USA, however, state that VoIP is more about making money than saving money, and users should look beyond a chance to cut long-distance costs. Rather, they should think about a new generation of VoIP-based computer telephony integration (CTI) applications which can produce increased turnover and boost a company's productivity.

CONFERENCING AND PRESENTATION TECHNOLOGY

Technology has made and continues to make enormous strides, but the art of conducting meetings and sharing information has mostly failed to catch up. Now possible are audio-conferences (also known as conference calls or tele-conferences), video-conferencing, data-conferencing and email:

- **Audio-conferencing** extends the two-way telephone call, allowing multiple parties to interact on a single call.
- **Video-conferencing** is a fully interactive two-way video and audio communication using specialist equipment, most of which has data-conferencing facilities included.
- **Data-conferencing** can be used with either conference calls or video-conferencing to allow people to share data files and applications between computers during their meeting.

Email is useful for pre-meeting exchange of information. Business can use portable, desktop or installed projectors; there are some 300 projectors from over 40 manufacturers (see below, Projectors). Room control systems coordinate lighting, curtains/blinds and audiovisual equipment to create different environments, and to integrate different types of audiovisual equipment to make them work as one. Gas plasma display screens of up to 120 cm (50 inches) diagonally, 130 mm in depth and with a 160-degree viewing angle are available, as are wireless handset controls and television displays with widescreen and high-definition television (HDTV) standards.

Conference technology

Today's managers typically spend up to one-third of each business day in meetings. UK executives on average attend six meetings a week, spending 11 hours travelling between them. While new technologies will not replace all face-to-face meetings, they provide less costly, quicker and more productive ways of communicating. Costs of meetings are not only confined to travel. There are also hidden costs, such as administrative time used for meeting preparations, travel cancellations and the cost of time in transit.

Some meetings are particularly suitable for conferencing, for example:

- meetings that need to take place quickly
- meetings involving lots of people
- international meetings
- briefings to a large number of people.

Of course, many sales meetings, relationship-building meetings with customers or colleagues and introductory meetings will still require the personal contact of a face-to-face meeting.

Projectors

Reliability, projection quality, size and portability are all factors that facilities managers should consider when choosing projectors. Ultimately, choosing the right model comes down to deciding on what best satisfies personal or corporate criteria:

- **Fixed projectors**: any company carrying out regular presentations at a central base where quality of presentation is a critical factor will need a fixed solution, with maximum flexibility of input options (PC, VCR and DVD). Typical users will be trainers, directors and other senior members of staff, with applications ranging from internal training and demonstrations to product launches and sales pitches. Room costs will range from £4000 for a basic ceiling-mounted projector to £100,000 for a boardroom incorporating the highest specification products. Features that should be considered are angle, brightness, resolution and maintenance.
- **Desktop projectors**: these are designed for organisations in which different people need access to the system within different areas of the business, in different rooms and even externally, such as conferences and exhibitions. Users could be anyone in the organisation, and ease of set-up and reliability are therefore important. Costs can vary from £2000 to £8000 with weight being a major issue (under 7 kg is ideal, plus a hard carry case).
- **Fully portable projectors**: models cost between £2500 and £5000. Weight needs to be under 4.5 kg. These units are typically aimed at those launching products to potential customers, or visiting colleagues in satellite offices.

DOING BUSINESS OVER THE WEB

Customer service concerns are seen as the main driver of email for business document distribution, while elsewhere moves to an e-business culture are set to embrace the use of Internet technology, providing there is buy-in at board level. Without the right e-business approach, companies risk losing a significant portion of sales volume and profit margin to companies that have more effectively aligned themselves with their customers' supply chains.

By now, most businesses on both sides of the Atlantic will be migrating most of their document distribution to email. A key factor behind increased email deployment is the desire for a more responsive customer service. Email is seen as the key to cutting response times and increasing the quality of customer communications. Furthermore, e-business enables interactive one-to-one marketing and sales but at practically zero incremental cost, whether the company is trying to reach one customer or thousands.

Top–down approach

The approach to e-business must be sold from the top of the company down, and the facilities manager will have a major role to play as the enabler for the organisation. The emerging electronic marketplace is one which demands anytime, anywhere access (the $24 \times 7 \times 365$ model) over a number of devices and converging fixed and mobile networks. Major stumbling blocks affecting take-up of e-business in earnest are security and trust. The former can be a problem of perception while the latter is a tactical issue that can usually be resolved through discussion and agreement.

E-procurement

An inevitable consequence of e-business business-to-business trading is the closer integration of companies' IT systems with the web to allow linking of accounting applications for orders, invoices and payments. It is not a new idea. Electronic data interchange (EDI) systems have been around for over 20 years but have failed to achieve the expected level of take-up because they require extensive customisation to create common information standards between organisations.

More recently, enterprise management systems have been promoted as a more effective means of controlling business finances, offering improved reporting and management information. But these again rely on consistent information coding using agreed cost centres which often vary significantly between clients and suppliers.

The problem is that many of the new systems and processes cannot make use of legacy data, and then perpetuate the problem by creating a new set of bespoke information. In general, (typically SME) suppliers are not yet well geared up for web-based trading.

There is much potential in the technology, not only to assist in bypassing cumbersome manual processes, but also, if handled correctly, to generate valuable management and benchmarking information. Much expectation surrounds universal communications standards such as XML, a preferred choice by many for future e-business applications.

Without common ground on information exchange, Internet trading may never achieve its potential.

For more information relating to the growth and drawbacks of e-procurement, including 'club' e-commerce sites for procurement, see the section on e-procurement in Chapter 4.

Negotiating by email

Billions, if not a trillion, emails are sent each year in the USA, with the UK not far behind. Many routine communications between facilities managers and their user group or supply chain are now based on email exchange. However, research at the Harvard Business School has shown that people are less inclined to share information when communicating via email than through a face-to-face meeting or voice-to-voice communication. To minimise the negative features of email communication, it is important to create a rapport with the other party, by meeting beforehand, engaging in 'virtual small talk', and avoiding terse statements or shorthand which could create tension or misunderstanding. If the negotiation seems to be getting angry or personal then you should stop emailing and make a phone call or arrange a meeting.

Facilities managers should be aware that there are four ways of staff staying in touch, with certain degrees of immediacy: instant messaging (IM), texting (SMS), email and telephone. While all four can do the job of communicating a message, the crucial test of security really means that only email should be doing the job. Telephone, text and email have become well established in the business psyche, while IM is a newcomer. Yet market research shows that IM grew 33 per cent faster then email did when it first broke onto the market, and IM has now passed email in terms of global traffic. Meanwhile, over 1.2 billion text messages are sent every week in the UK (end-2007 figures, according to the Mobile Data Association), a clear demonstration of its popularity; in 1999 that was the number sent in a whole year.

The role of the intranet site

Many facilities managers are now supporting their operations through a web-based interface with their internal customers. A well-designed interface can make a positive contribution to the climate for virtual negotiation. According to research by Cisco Systems, a successful site has three main features: quality of the information architecture, quality of the graphic design, and readability.

Intranet services

Intranet services currently on offer include knowledge database, room bookings, catering, requests, customer satisfaction monitoring and travel portal. Content management is key: technical issues can be handled through the company's existing network structure, but the facilities manager will need to know information such as who entered what information when. Content management software costs upwards of a few hundred pounds.

Action points

When designing a site, consider the following:

- Base the content of the site on a survey of user/customer feedback.
- Make sure information content is regularly updated.
- Make sure navigation through the site is consistent.
- Field test the usability of the site with a focus group.
- Minimise the graphic content to speed downloading.
- Present the material in a format that the reader can glance through and easily retain the main points of information.

In addition, facilities managers should identify the key stakeholder groups and design specific information resources for their individual needs. Most facilities managers will have three key stakeholder groups:

- interactive customers wanting to access services
- senior management who want to understand the organisation and performance of service delivery
- interactive suppliers who want to access the operation of the supply chain.

HELPDESKS

The fundamental purpose of a helpdesk is the speedy restoration of a user to full productivity when that productivity has been impeded by a failure in corporate technology or its use, thus maximising the return on investment in both employee and technology. As well as queries about technology and equipment, helpdesks can be used to respond to calls and requests such as for room bookings or new services and audiovisual aids.

Helpdesks typically adopt call centre practices such as interactive voice response (IVR) and CTI to satisfy customer demands. There are many helpdesk software packages, ranging from enterprise-scale customer relationship management systems to entry-level options which can be downloaded from the web.

DIGITAL VERSUS PAPER DOCUMENT SYSTEMS

Most businesses have long accepted that the much heralded 'paperless' office is a myth. Indeed, the proliferation of PCs in the workplace has actually served to increase the amount of paper printed in most organisations. This is because people find it easier to familiarise themselves with long and/or complicated documents in hard copy format. As a result, paper storage remains a significant facilities issue.

Paper storage

Documents should be kept in the cheapest way possible. If they are no longer current, and unlikely to be needed on a regular basis, they should be transferred to a basement or to off-site storage for whatever length of time is required by statute. The overall cost of off-site storage is typically about one-third of the cost of on-site storage.

Establishing an electronic document system

However information is stored, it makes sense for a document management system to be established. Document management involves the conversion of paper documents into electronic images on a computer, that is, document imaging. The documents are scanned into the system, to be stored on a hard disk drive or an optical disk, when they are indexed. To read any document, the imaging system retrieval tools are used, with the process taking seconds to perform.

An electronic document management (EDM) system enables an organisation to save space and retrieve documents more quickly and cheaply, and makes it easier for operatives to share information. Under British Standards' Code of Practice, PD0008, for Legal Admissibility of Information Stored on EDM systems, electronic documents and scanned images may be accepted as evidence in court.

INFORMATION

4G information: www.4g.co.uk

WiMAX: www.wimaxforum.org

OFDM: www.ofdm-forum.com

MultibandOFDM: www.multibandofdm.org: pro-Intel aficionados who support the future (think 2014) Bluetooth alternative which is also an alternative version of ultra wideband (UWB), a radio technology with higher data rates and lower power consumption

International Telecommunications Union: www.itu.int

Wi-Fi Alliance: www.wi-fi.org

GSI: www.gsigroup.com

essentialfm report, 'Intelligent buildings', 12, November/December 2001

Helpdesk details: www.helpdesks.com

There is a global open standard specifically designed for building automation: Zigbee (based on IEEE 802.15.4) has international support and Building Management System (BMS) vendors regularly attend Zigbee (www.zigbee.org) workshops. BMS control, monitor and optimise such building services as HVAC, lighting and alarm systems (see also computer-aided facilities management, CAFM, in Chapter 11)

The Built Environment

10 Workplace Facilities

Frank Booty

Facilities managers are stewards of the built environment. The day-to-day running of a building is fundamental to facilities professionals. From procuring the most competitive electricity contract to selecting the most appropriate air conditioning system, the facilities manager needs to be aware of the market, costs and benefits of all aspects of workplace facilities and equipment.

The increasing impetus on energy efficiency in business operations, health and safety issues and maximum productivity are now among the business objectives that facilities professionals must address. Ensuring that the appropriate services and facilities meet the needs of their users and are functioning efficiently and cost effectively will bring benefits to all aspects of an organisation's operations.

BUYING GAS, ELECTRICITY AND WATER

It is becoming an accepted view that facilities managers are kept in the dark over energy costs. Lack of transparency in the existing deregulated gas and electricity markets is costing UK businesses valuable time and money. The sheer number of companies selling gas and electricity in the UK is in itself a headache for facilities managers.

Some facilities managers also feel they are taking a gamble when choosing between an established, but potentially more expensive, supplier and one of the many new but equally viable competitors. The popular view is the deregulation of energy industries across Europe, and the consolidation of companies, is intensifying the key issues, as foreign suppliers join the fray.

How to get the best deal

The solutions to the energy-buying conundrum for facilities managers comprise a much simplified procurement process and more transparency in the market. Some facilities managers utilise the (free) services of reverse auction agencies. Any organisation seeking to obtain the best quote on energy enters a few details about its current set-up on an agency web site. All Ofgem-registered gas (or electricity) companies in the UK then 'blind bid' for the organisation's business in a reverse auction to offer

298

the most competitive price. Ofgem (the Office for Gas and Electricity Markets) is the official regulator. Only with a truly open market can those in charge of buying commercial and industrial energy be sure they are getting the best deal.

BETTA

The British Electricity Trading and Transmission Arrangements (BETTA) are meant to create 'a fully-competitive, British-wide wholesale market for the trading of electricity generation' and draw Scotland fully into the scheme. BETTA replaces NETA, the new electricity trading arrangements, which went live in March 2001. BETTA operates the high-voltage electricity transmission system throughout Britain. Hitherto there was a separate Scottish system run by Scottish Power and Scottish & Southern Energy. National Grid looked after transmission in England and Wales, with the two systems joined by an interconnector. However, the Scottish companies were privatised as vertically integrated concerns covering the whole process, from generation through transmission to supply. Ofgem reckons the changes give renewable generators in Scotland better access to the Anglo-French interconnector, making it easier to sell in continental Europe. Before NETA, the previous pool arrangement had been regarded as deeply flawed and seen as 'effectively a means for generators to set a wholesale price which suppliers and large consumers had little choice but to accept'. In contrast, BETTA is a genuine market in which there is real choice.

Large energy users

Suppliers are expected to focus heavily on the small and medium-sized enterprise (SME) market, where better margins and profits can be made compared with the high-volume business. What then can larger users do to improve their position?

When looking at contracts, be prepared to be innovative, look to load-manage, talk to suppliers about the usage profile, communicate regularly regarding any changes in consumption habits, watch the markets and supplier activity and know which suppliers are best suited to match a company's needs, and set budgets in line with an educated view as to where prices might be heading.

Ofgem's advice is to 'shop around to get the best deal'. The role of this government watchdog is to make sure that markets are operating as competitively as possible. What Ofgem has no control over is the Climate Change Levy (CCL) (see section on Climate Change Levy in this chapter). This presents a conundrum. On one hand, electricity prices are lower than they would be, thanks to BETTA, while on the other hand the CCL has the effect of raising prices.

Combined heat and power

Combined heat and power (CHP) is an efficient technology for generating electricity and heat together. A CHP plant is an installation where there is simultaneous generation of usable heat and power (usually electricity) in a single process. The term CHP is synonymous with 'co-generation' and 'total energy', which are terms often used in other European Community member states or in the USA. The basic elements of CHP plant consist of one or more prime movers usually driving electrical generators, where the heat generated in the process is utilised via suitable heat recovery equipment for a variety of purposes, including industrial processes, community heating and space heating.

Efficiency

CHP can provide a secure and efficient method of generating electricity and heat at the point of use. Typically, CHP achieves a 35 per cent reduction in primary energy usage compared with power stations and heat-only boilers. This is because it uses heat from electricity generation and does not suffer any transmission losses, since the electricity is produced on site. So, when there is an optimum balance between heat and power loads, the host company should be able to make economic savings. Today, the existing CHP installed base achieves over 30 per cent reductions in carbon dioxide emissions compared with coal-fired power station equivalents, and over 10 per cent compared with gas-fired combined cycle gas turbines. Further, the latest CHP installations produce over 50 per cent reductions compared with power generated by coal-fired complexes.

District and localised production of heat and electricity avoid massive amounts of efficiency loss from central power generation and lend themselves to considerable savings in energy. The fuel cell also has a key role to play in the overall process for ensuring that buildings reduce their harmful carbon dioxide emissions. Fuel cells convert the energy from a chemical reaction directly into electricity and heat and could provide the future CHP solution for buildings. Although most fuel cells are still under development, they seem certain to play a large part in providing CHP in the hydrogen economy, which promises to eliminate many of the problems the fossil fuel economy creates. It could reduce depletion of natural resources and eliminate pollution (including greenhouse gases) caused by burning fossil fuels.

The CHP Club

The government is eager to achieve its target of 10 GW of CHP capacity by 2010, and has set up the CHP Club to help achieve this. This is particularly targeted at new and potential CHP users, and also aims to help existing users to extend their schemes. It intends to provide members with a one-stop shop: a combination of information, exchange of experience and advice facilities on CHP and related topics, all for free.

The Combined Heat and Power Quality Assurance (CHPQA) programme will enable good quality CHP to earn exemption from the CCL; potential members can visit and register with the CHP Club website. A consultation paper, *CHPQA – A Quality Assurance Programme for Combined Heat and Power*, has been issued.[1] The operation of CHPQA will be reviewed. Any changes made will only apply to subsequent CHP developments and will not be applied retrospectively to existing CHP schemes.

---

[1] CHP quality assurance information, www.chpqa.com.

Totally integrated power

A lack of a uniform approach to energy monitoring, control and delivery has meant that occupants of industrial and commercial buildings have found it practically impossible to integrate energy supply to the same extent as other critical building services, such as HVAC and security. Heating, lighting, access control and fire and security systems all need to be controlled, along with a building's energy requirements.

Totally integrated power, a concept developed by Siemens, offers integrated power distribution from medium-voltage switchgear all the way through to socket outlets on an office wall. The concept covers every aspect of power distribution and electricity management required in an industrial environment. It has been estimated that totally integrated power can bring power use savings of 25 per cent, because it enables electricity users to reduce their total power costs, as well as increasing the efficiency of operation of plant equipment. Operators of public buildings, including museums, hospitals and airports, are said to benefit from totally integrated power just as much as industrial and office markets. The theory is that by expanding the capability of data control and management through an entire building, all aspects of energy delivery and consumption can be integrated.

Open systems in building control

The organisation buildingSMART (formerly the International Alliance for Interoperability, IAI) has the mission 'to provide a universal basis for process improvement and information sharing in the construction and facilities management industries, using industry foundation classes (IFCs)'.[2]

IFCs are a practical tool for information sharing, using object technology; they are information rich, swiftly transmitted and not limited to any one software vendor or system. The IAI's first chapter was set up in the USA in 1995 and the second in the UK in 1996. Other chapters now cover German-speaking, French-speaking and Nordic countries, Japan, Singapore, Korea and Australasia.

buildingSMART has chapters covering facilities management areas such as reactive and preventive maintenance, asset management, move management, change management, helpdesk, energy consumption profiles and performance data, but nothing currently in the signals area. This is the domain of open systems protocols and interoperability, principally involving the use of LonWorks, BACnet and EIB. With these systems, the situation is one where central systems increasingly feature distributed intelligence at all outstations.

The LonMark Interoperability Association is a key driving force in the establishment of interoperable guidelines for building, industrial, transportation and residential/utility automation. LonMark membership is open to any manufacturer, end-user and system integrator committed to the development and use of open, interoperable products using multivendor LonWorks control networks.

BACnet is a data communications protocol for building automation and control networks. Developed by the American Society of Heating, Refrigerating and Air-

---

[2] buildingSMART, www.iaiorg.uk.

conditioning Engineers (ASHRAE), BACnet is a US standard, a European 'prestandard' and a potential global standard.

The EIB bus – EIB stands for European integration bus but is always referred to as EIB – serves the same function as LonWorks and BACnet, offering the sensor-to-sensor and unit-to-unit links of LonWorks and the controller-to-controller-level protocol of BACnet.

As yet, there is no 100 per cent open communication. The Building Research Establishment runs an EIB and BACnet training centre. The Energy Systems Trade Association's Building Controls Group is one possible source of information. Watch out for Zigbee (IEEE802.15.4), a wireless standard aimed at building management systems (BMSs); the intention is to control and monitor buildings entirely from a personal computer (PC) through one converged (ultimately IP-based) building network. This is starting to happen.

Avoiding power failures

Organisations reliant on sensitive electronic and computer-based equipment are vulnerable to even momentary cuts in the electricity supply. Mains failure could bring most workplaces to a standstill. PCs, workstations, network servers, retail point-of-sale systems, medical equipment and telecom systems all depend on a high-quality uninterrupted power supply. A distortion of only a few milliseconds can destroy data, disconnect communication links and damage delicate instrumentation.

The best way to limit or prevent potential damage caused by power irregularities is with battery back-ups known as uninterruptible power supplies (UPSs) and standby generators. The cost of UPS systems ranges from under £100 to some £150,000, and generators start at £25,000 and can cost hundreds of thousands of pounds. When selecting a UPS, there are several important issues to consider:

- What are the potential consequences of power loss in both the short and long term?
- How much load requires protecting and which pieces of equipment are most critical?
- What is the length of time power needs to be sustained?
- What are the power requirements of critical equipment?
- What is the likelihood and possible duration of power losses or fluctuations?

The use and application of UPS systems needs to be part of a company's disaster recovery policy (see Chapter 6). For maintenance of UPS systems, see the section on Maintenance and repair.

Water competition

The government is boosting the opportunities for competition in the water industry in England and Wales in order to bring customers more choice, keener prices and better services. As from 1 December 2005, companies have been able to compete in the water industry under the framework of Water Supply Licensing. Companies are being given clearer rights to enter the water market. Incumbent companies can remain vertically integrated statutory undertakers, retaining their key strategic water resource and environmental duties. The government is continuing to ensure that public health, the environment and the quality of drinking water are safeguarded. Ofwat (the Office of Water Services) continues to review regulatory changes, such as further transparency in incumbents' costing information, to promote competition. Full information can be found on the Defra and Ofwat web sites.

Checking utility invoices

It pays to check all utility bills thoroughly, as costly mistakes in the supplier's favour can and do occur. Anecdotal evidence suggests major users can expect average refunds of 2–4 per cent from rigorous attention to bill validation. Smaller users are not immune; although errors may affect them less regularly, their effect can be quite disproportionate.

Companies may already have some form of bill-checking procedures in place, supported by specialised software. But it is worthwhile reviewing how effective and efficient the systems are by questioning whether the checking process can validate the following items:

- Is it the correct supplier? Ensure, for example, that there is no double billing resulting from a previous incumbent overlapping with a new contract.
- Are the prices correct, including transportation of use-of-system charges if separately itemised?
- Is there continuity, for example, no mismatch of 'previous' meter readings, no double-counted or phantom bulk deliveries?
- Is the arithmetic correct?
- Is there proof of delivery; that is, independent verification of meter readings and bulk fuel drops?
- Is there consistency with earlier billing patterns?

The complexity of some utility invoices can make it a daunting task to apply thorough vetting. Bills for contract electricity supplies and effluent charges, for example, can be quite complex. Worse still, much of the proprietary software sold for the purpose is not up to the job and becomes less relevant as non-tariff contract deals proliferate in the deregulated utility marketplace.

One practical approach is to model the invoice for each account with a spreadsheet. The current supplier's details and contract price structure can be built in as constants. When the variable data from a new bill is entered into the appropriate boxes, the grand total should match that shown on the real bill. Changes of supplier, changes of price and even changes of tariff structure can be accommodated by creating a modified version of the model bill.

Meter-reading errors

Experience shows that arithmetical errors are rare. Meter-reading discrepancies, however, do occur. Of course, an incorrect or estimated reading will usually be corrected on the subsequent read and any resulting overcharging will be balanced by a compensating undercharge, but there are important exceptions to this general rule. One is when the estimated reading occurs on the date of a change of supplier or end of contract year, when the price changes. Too much consumption may then be charged at the higher price and too little at the lower price. More significant is the situation where the meter is changed and the final reading on the outgoing meter is estimated or wrong.

CLIMATE CHANGE LEVY

Energy consumption contributes significantly to the UK's carbon dioxide emissions. Business accounts for nearly 70 per cent of these emissions, and industry and commerce alone are responsible for nearly 42 per cent. The UK government has set an ambitious target for a 20 per cent cut in the country's carbon dioxide emissions by 2010, in line with the Kyoto Protocol. Initiatives have been launched by the government to assist companies in reducing their energy consumption. However, it is the CCL which has provided the impetus in getting businesses to think about how they can cut their energy consumption and costs.

The levy is basically a tax on coal, gas and electricity affecting both private and public sectors. Its existence has increased the viability of renewable energy options and CHP, as these supplies will be exempt. British Gas estimates that the levy will add 10 per cent to electricity bills and 15 per cent to gas bills. See the section on CCL in Chapter 1 for details of levy rates and exemptions.

Objectives

The aim of the levy, which was introduced in April 2001, has been to encourage greater energy efficiency and use of greener power supplies. Reduced levy rates will be accorded to those who demonstrate that they have invested in energy efficiency. Any company pursuing an efficiency path will benefit from the levy; the government estimates that most businesses could save 10 per cent of their energy at little cost. Such a reduction would amount to more than just a 10 per cent reduction on the levy, it would also cut the overall fuel bill by a significant sum.

The levy is not aimed at the SME sector directly, but this group does have a key role to play. According to the Institute of Directors, if every small business in the UK were to cut its energy requirements by 20 per cent, the nation's energy bill would be reduced by £1 billion, and 3 million tonnes of carbon dioxide emissions would be eliminated. Whatever the size of the organisation, every company needs to have an energy management policy. The backing, understanding and support of everyone, from directors to staff, is mandatory.

REDUCING EMISSIONS AND COSTS: LOGBOOKS AND THE ENERGY PERFORMANCE IN BUILDINGS DIRECTIVE

The government is seeking to reduce carbon dioxide emissions by 80 per cent by 2050, and 15–25 million tonnes of carbon (MtC) by 2020. Energy efficiency is identified as delivering half of the necessary improvements. Carbon Trust is a government-backed organisation set up to provide UK businesses and industry with advice on reducing carbon emissions, and administers the enhanced capital allowance scheme. More at www.eca.gov.uk.

Since April 2006, a statutory requirement (Part L) through Building Regulations is being enforced for all new non-residential buildings in England and Wales to be handed over with a logbook containing details of the building services installed, enabling owner/occupiers to monitor and control energy consumption. The Chartered Institution of Building Services Engineers (CIBSE) has published 'CIBSE TM31 – Building logbooks – a guide and standard template for preparing logbooks for non-domestic buildings' as a key tool to enable design teams and contractors to meet this requirement. The log books underpin much of the certification process required under the European Union's (EU's) Energy Performance in Buildings Directive (EPBD). See CIBSE briefings (www.cibse.org/briefings) for an overview of the EPBD. Also see www.diag.org.uk. The Directive Implementation Advisory Group (DIAG) was established to advise the government on EPBD's effective and timely implementation. CIBSE has established an Energy Performance Group to promote compliance with, and cost-effective application of, energy efficiency, and anyone can join at www.cibse.org/energyperformance.

The EPBD directive was introduced because the 160 million buildings in the EU use over 40 per cent of Europe's energy and create over 40 per cent of its carbon dioxide emissions, and that proportion is increasing. Under the Kyoto Protocol, Europe is committed to reducing emissions and the Directive is intended to contribute to achieving this. Heating fuel is the most important component (57 per cent of domestic consumption, 52 per cent of non-residential building consumption). Water heating accounts for 25 per cent of domestic consumption and 9 per cent of non-residential use. Lighting accounts for up to 25 per cent of emissions due to commercial buildings. According to the EU, a cost-effective savings potential of some 22 per cent of present consumption in buildings can be realised by 2010. The EU is now reviewing the EPBD.

Certificates for new and existing buildings should be available when they are constructed, sold or rented out. The certificates should be not more than 10 years old, and be accompanied with advice on how to improve the energy performance of the property to which they are related. Public buildings (initially those over 1000 m^2 in area) that can be visited by the general public need Display Energy Certificates, which show the actual energy usage and must be displayed constantly.

Energy Performance Certificates (EPCs) have been required by all commercial buildings since October 2008. The aim of the EPC scheme is to increase awareness of energy use in buildings, and to promote increased investment in energy efficiency by establishing agreed measurements of relative energy performance and introducing regular inspections and re-evaluations. EPCs are not required on construction, sale or rent, for places of worship; temporary buildings with a planned time of use less than two years; standalone buildings with a total useful floor area of less than 50 m^2 that are not dwellings; and industrial sites, workshops and non-residential agricultural buildings with low energy demand. EPCs are not required on sale or rent for buildings due to be demolished. A guide is downloadable free from www.communities.gov.uk/publications. All new commercial buildings must meet zero carbon targets by 2019, and the public sector by 2018.

Estates of existing buildings outnumber the amount of new build properties. Thus refurbishing and upgrades figure prominently. The information on Part L and EPBD is based on public domain information on consultation exercise and feedback (www.odpm.gov.uk). Building Regulations include Approved Document L1A – new dwellings, ADL2A – new buildings that are not dwellings, and ADL2B – work in existing buildings that are not dwellings. Part L 2005 is the legal framework to implement some of the EPBD (Energy Performance in Buildings Directive) in England and Wales. ADF1 covers means of ventilation.

But why are buildings not energy efficient? 'Because no one cares', is the pundits' reply. Buildings are set up for the provision of adequate services and not for energy efficiency. These two objectives are not mutually exclusive. It is estimated 90 per cent of heating and ventilating control systems in buildings are inadequate. This costs industry and commerce £500 million per annum in additional energy.

In an individual organisation, the person responsible for managing energy could be the managing director/partner, office manager, facilities manager or other staff member. A motivated champion is crucial for implementing and maintaining energy efficiency, typically driven by ethical concerns, motivation to improve economic performance, and technical and professional pride. The characteristics of successful energy champions include self-motivation, clearly defined responsibilities (supported from the top), access to information, ability to motivate and influence other staff, ability to push positive decisions, and ability to demonstrate economic payback and performance benefits. Energy-efficient behaviour should be part of ongoing business improvement and target setting.

Average cumulative costs for an office over 20 years are 68 per cent for people, 25 per cent for rent and rates (London figures), 4 per cent for maintenance and energy, and 3 per cent for construction. So, capital costs of construction are small compared with other cumulative costs. A process called 'continuous commissioning' is advocated for refurbishment as it reduces energy cost and environmental impact, reduces maintenance cost and improves maintenance strategy, improves occupant well-being and life safety, offers a corresponding improvement in productivity, provides an enhanced value of the building/estate, and gives an improved image from knowledge that the building is professionally managed using environmentally sound principles.

Meanwhile, this checklist contains practical advice from a utility firm:

- Include energy efficiency in buying specifications and for all equipment.
- Get office equipment switched onto standby mode: this can help to make energy savings from 25 per cent to over 50 per cent.
- Make staff aware which equipment can be switched off when not in use.

- Install motor controllers to reduce fuel consumption of refrigerators and freezers.
- Reduce building heat losses by up to 90 per cent by insulation (while maintaining adequate ventilation).
- Consider reducing the volume of air to be heated by installing suspended ceilings.
- Fit temperature controls, check thermostat settings, draught-proof windows and doors and raise staff awareness: the maximum recommended office heating level is 19 °C and for each one degree of overheating, costs rise by about 8 per cent.
- Reduce thermostat settings for storerooms, corridors and places where there is a high level of activity: workshops should typically be 16 °C and stores 10–12 °C.
- Install seven-day time-switches; check that settings are correct and correspond to occupancy patterns.
- Remember that radiators can be switched off before the end of occupancy.
- Check that heating and ventilation operate at the minimum recommended settings when the building is unoccupied.
- Check which rooms regularly overheat and correct.
- Fit thermostatic radiator valves, set them correctly and lock them.
- Have heaters and burners cleaned and serviced at least yearly by a qualified contractor.
- Turn off lights whenever possible, especially in toilets, corridors, storerooms and canteens; make staff aware that they can save 15 per cent in this way.

Measuring energy efficiency

By far the most common yardstick for measuring, and managing, energy is by its cost. For a large number of users, interest does not extend beyond the one line on the expenditure budget for heat, light and power. However, if cost is the only yardstick, interest in energy efficiency depends on whether unit costs of fuel are rising or falling. Until 2003/04, prices had been steadily falling, with a corresponding decline in interest in improving efficiency. The CCL was intended to help to correct this. Energy costs have since been rising, and are not expected to come down for the foreseeable future.

Where energy is used in buildings, a step forward in analysis is cost per unit area. Much information is available on cost standards for offices, schools, hospitals and other types of building, which enable users to compare their performance with best practice. Gas and electricity are sold by the kilowatt-hour (kWh), and this is the most commonly used unit for the analysis. Again, a range of information is available, notably through the Carbon Trust, which gives benchmark standards for energy use in various applications.

Energy-efficiency improvement programmes

Programmes of improvement must inevitably consider ways of actually increasing energy efficiency. Sometimes this is done by making major changes to the way requirements are met, for instance by installing a CHP unit, or replacing a centralised steam boiler by smaller units close to the point of use. These are expensive capital projects which have to be carefully assessed and evaluated, with options such as whether to carry through the project in-house or use a contract energy management company.

Far more often, the improvement programme is achieved gradually. Such a programme might typically aim to improve energy efficiency by 2 per cent per year. This is likely to be measured in terms of a performance indicator, which may be in terms of kWh, kWh/m^2, carbon dioxide emissions, and so forth. If it is measured in cost terms, an adjustment for any changes in fuel prices must be made for the results to have any meaning.

The importance of measuring where energy is used should also be emphasised. For instance, many organisations cannot separate their use of electricity for power and for lighting, or distinguish where steam and hot water are used after they leave the boilerhouse. Measurement enables energy use to be assessed against standards and its costs to be allocated to users. Adequate metering is essential for good energy management.

In practice, there can be all sorts of variations from year to year: the weather may have been different, there may be more people and more computers in the office, and some parts of the building may have been enlarged or refurbished. It is, however, possible to apply normalisation adjustments to certain factors, such as heating degree days, and this helps to allow more accurate comparisons. Another option may be to join an accreditation scheme which relies on more than one performance indicator, enabling a better judgement to be made about good management and progressive improvement.

The value of metering data

All organisations with a maximum electrical demand of more than 100 kW will already have metering systems that record consumption every half hour, every day of the week. This Code 5 metering is required under the rules covering the operation of the electricity supply market. The data gathered from these meters is primarily intended for use in billing and in balancing supply and demand requirements between generators and suppliers, but has also proved useful in combating the extra liability associated with the CCL.

The metering information that is despatched for the purposes of the electricity pool is, however, edited. While this makes balancing supply and demand and billing procedures run more easily, this edited data is of little use for energy management, and it is important to get hold of the raw data. Your data collector should be able to provide this. The supplier's permission should be sought (as, technically, the data belongs to them), but it is now standard practice to allow this. The figures for consumption will not provide a great deal of information in themselves. Their value lies in comparing them with, for example, trend data, degree day information and normalised performance indicators. So, the data stream needs to go through some form of analysis.

The results should highlight where energy consumption patterns are exceptional, for example, at night or weekends, or over shutdown periods. They will also indicate any drift from optimum performance and can provide an early warning of serious equipment faults which bring about noticeable drops or increases in consumption over short periods.

This information can be made available within a matter of days, rather than the weeks it normally takes the supplier to send a bill. For this reason, it is important to ensure that the contract you make with the data collector provides for the speedy release of this information to you.

WASTE WATER MANAGEMENT

Sewerage is an essential service, the exact costs of which are difficult to calculate. There are ways in which businesses can keep track of the amount they pay for waste water. Water charges are tightly regulated and in a constant state of review, with the industry regulator Ofwat protecting customers' interests by setting the overall price limits that govern charges. The 10 sewerage companies in England and Wales are given strict guidelines, but accurately pricing the amount of water the businesses send back into the sewerage system is problematic. Firms can sometimes pay more than they need to.

Sewerage charges

Water industry regulations allow water companies to levy a single charge for sewerage, although Ofwat recommends that costs are broken down into three constituent elements:

- foul drainage (including trade effluent)
- highway drainage (runoff from roads and pavements)
- surface drainage (runoff from properties).

The aim is to ensure that charges are related to the services provided. This makes it easier for customers to assess whether they are being correctly billed and to check they are getting the services for which they have paid.

Foul drainage

Foul drainage refers to the dirty water discharged from lavatories, sinks, washing machines, and so on. Charges are based on the volumes of water supplied to the property with adjustments made for non-return to sewer. It is not possible to measure the exact amount of waste returned and it is the responsibility of each water company to decide how it calculates its own system for payment. However, this does not prevent customers from challenging charges. If a query is made, a representative from the water company visits the site and evaluates the situation according to the evidence supplied regarding the ways in which water is used there. The object is to agree allowances.[3]

---

[3] Environmental reporting: see www.defra.gov.uk/environment/water/index.htm.

Frozen food businesses, which use large amounts of water in production processes, hospitals (where liquids are often disposed of elsewhere) and pharmaceutical companies are examples of organisations which have a relatively low return to sewer and may be eligible for reductions. Even swimming pools can claim a discount for evaporation of water. No foul drainage charges should apply where a property is not connected to the sewerage system. The onus is generally on the consumer to come forward to claim rebates.

Highway drainage

The way in which water companies charge for highway and surface drainage tends to vary. Some operate a flat fee, others refer to the rateable value of the property or the surface area drained to a public sewer. According to Ofwat's 1998/99 *Report on Tariff Structure and Charges*,[4] Ofwat's Director-General believes that as highway drainage benefits all those using roads, directly or indirectly, there is a case for recovering the costs of the service from highway authorities or from users of the highway. Currently, the law prevents this option and costs are recovered from sewerage customers. This situation is unlikely to change in the near future, so it is not currently possible for customers to claim a rebate for highway drainage.

Surface drainage

A charge based on runoff from properties forms a significant part of the total sewerage bill and businesses not using the service, or making only limited use of it, may be able to save money. Again, water companies differ in the way they calculate charges and not all of them differentiate between customers who are connected to the facility and those who are not. Ofwat's Director-General wants water companies to provide a clear explanation of the charges on bills and is encouraging them to reduce those charges if the property concerned does not benefit from surface drainage.

Trade effluent

Trade effluent refers to industrial waste, including that discharged at the end of or during manufacturing processes. This covers such things as waste from food factories, dairy product manufacturers, chemical plants and abattoirs. It is also classed as sewerage, but billed separately. Customers pay according to the strength and volume of the trade effluent discharged, both of which have a bearing on the level of treatment needed for the waste. Public swimming pools, for example, may benefit from lower tariffs because of the lower strength of effluent discharged.

Large-user tariffs have been introduced by most of the water companies. These work primarily through the imposition of a high fixed charge and a lower volumetric rate for water supplied, or alternatively one where the standing charge remains the same, but there is a lower volumetric rate for all consumption over an agreed level. The aim is to provide attractive water charges for high users without presenting an incentive to waste water.

---

[4] Ofwat, www.ofwat.gov.uk and list of reports at www.waterview.org.uk/aptrix/ofwat.

Allowances

Water companies may make allowances for customers who can prove that rain water, for example, is drained and taken away by means other than the public sewer. The water may go to a soakaway or watercourse using the customer's own arrangements, or even a trade effluent meter. If significant volumes of water are not returned to the sewer, or if only a small volume of surface water is discharged, it may be worth diverting all the water so that none returns to the public sewer, which is one way to cut charges. Ofwat considers that some businesses may be underpaying for this particular service.

Metering

Some 20 per cent of non-household users, normally small businesses, remain unmetered, and in some areas this figure rises to nearly 30 per cent. Since most sewerage charges are based on the volume of water supplied to a property, meters are the best way of ensuring that costs relate to usage. The regulator is keen to encourage the use of meters and believes it is the sensible option where economically viable.

Remote reading

Remote reading of meters is gaining favour. Accessing each water meter to take a direct reading is not always easy, and manual readings have to be keyed into the energy management system by hand. To combat such problems, manufacturers have developed remote-reading systems, the main technologies of interest being electronic meter reading and radio reading. For electronic approaches, the meter requires an in-built encoder linked to a touchpad which passes data to a handheld interrogator. The data is subsequently downloaded to the energy management system. No access to premises is required to take readings. With radio, the meter reader does not have to touch the meter at all. Most remote reading systems can be retrofitted if required.

Individual water companies will have advisers on commercial and industrial metering issues. Accurate, timely information on consumption patterns can help managers to keep water costs under control. If there are cost savings to be made, the water meter could be the key.

European Community directives

Meeting European Community (EC) directives is a requirement for water quality and compliance adds to the cost of treating sewerage and effluent. So how can reduced tariffs be justified? Ofwat considers that reduced charges are based on customers making less use of the reception and conveyance part of the service, consistent with the way tariffs for large users have developed on the water side. The way sewerage companies calculate charges is far from straightforward and structures differ from company to company. General advice from sewerage companies is to contact them if you believe you are not making full use of their services or can prove you are being overcharged.

Preventing water pollution

Point-source pollution, from sewage and industrial effluents, for example, is easily regulated. However, pollution from diffuse sources such as surface water runoff from urban areas is not. Diffuse pollution, from silt and solids in particular, is the most significant pollution being faced today. Surface water can be contaminated by oil, silt, leaves, dog or cat mess, carbon and a mixture of pollutants from the air, roads, industrial yards, car parks and other hard surfaces, such as drives and pavements. Discharges from surface water outfalls have been found with average concentrations of suspended solids over 200 mg/litre, almost the same as raw sewage. Discharges have also been found contaminated with sewage debris and high levels of dangerous bacteria.

Main causes of contamination

There are five main causes of contamination:

- contaminants deposited on drained surfaces, such as oil, rubber, chemicals, pesticides and mud
- wrong connections of foul water to surface water drains, by accident or ignorance
- public ignorance of where drains ultimately lead
- spills and deliberate disposal, particularly oil, flushed into surface water drainage systems
- sudden flushing of contaminated water into drains, leading to flooding and subsequent groundwater pollution.

Heavy rainfall accelerates runoff, producing a mixture of pollutants flushing rapidly into drains, then into rivers, causing contamination and flooding. Because the natural settling-out process is bypassed, the common result is widespread contamination of natural watercourses and the public water supply.

The Environment Agency is promoting sustainable drainage systems (SUDS) which are designed to overcome some of the disadvantages of conventional piped drainage systems in built-up areas by reducing flooding risks, improving water quality and providing a better environment for people and wildlife (www.environment-agency.gov.uk/SUDS).

Best management practices

The introduction of best management practices (BMPs) means that non-point-source pollution and flooding can be effectively and economically prevented and/or controlled. BMPs are ways of minimising diffuse pollution. Two techniques have been developed, procedural and structural, which aim to:

- slow the speed of runoff to allow settlement, filtering and infiltration
- reduce the quantity of runoff collected
- provide natural ways of treating collected surface water before it is either discharged to a watercourse or infiltrated into land.

Best management practice options

A range of BMP options exists from which designers, developers, planners, drainage specialists and civil engineers may choose. These include grass swales and filter strips, infiltration basins, extended detention ponds, retention ponds, wetlands, porous surfaces and procedural BMPs (aimed particularly at agricultural areas).

- **Swales**: these can utilise the common green space alongside roads or other open areas. Basically a much improved ditch with a broad bottom and gently sloping sides, a swale gives a low sheet flow, slowing the water and enabling pollutants to settle out. Swales obviate the need for expensive roadside kerbs and gullies. Maintenance costs are much lower and drainage of the road surface is guaranteed.
- **Detention ponds**: these are designed to collect storm runoff, holding it for a few hours to let sediment settle out. Outside of storm periods, most ponds will be dry. The main function is to remove solids; removal rates of 80 per cent are possible. Rates for nutrient and trace metal removal are more modest, however, and a retention pond or wetland will improve this performance.
- **Retention ponds**: these retain a significant volume of water all the time. The design can allow for substantial variation in the retained water level and the pond can become an attractive local amenity, as well as an effective filter for nutrients, trace metals, bacteria and organic matter.
- **Storm water wetlands**: these are enhanced wet ponds with shallow areas incorporating a variety of marsh and wetland plants covered in up to 0.15–0.3 m of water. The algae and plant material filter and remove nutrients to a much greater degree than ponds alone. Storm water wetlands must always be purpose built. Leading surface water into an existing natural wetland can harm aquatic life and is not an acceptable practice.
- **Porous pavements**: these are an alternative to conventional paving and allow water to permeate through, rather than run off, the paving. Rainwater can filter directly into the subsoil or can drain into a reservoir (about 1 m deep) before soaking slowly away, discharging to the watercourse or being stored for landscape watering. Porous surfaces are appropriate where runoff is lightly contaminated and close to source. They have been shown to remove up to 80 per cent of sediment, 60 per cent of phosphorus, 80 per cent of nitrogen, and substantial levels of trace metals and organic matter.
- **Infiltration trenches**: these are shallow, excavated trenches, backfilled with stone to create an underground reservoir. From this, the water filters into the subsoil, and can help to replenish groundwater resources.
- **French drains**: these are below-ground systems comprising a trench filled with gravel wrapped in a geotextile membrane. Runoff water is led to them directly from the surface or through a system of pipes. French drains are less costly than kerbs and gullies, and are useful where only small watercourses are available to receive runoff water.

FACTS ABOUT BMPS

- BMPs are mostly always cheaper than conventional systems, usually by up to 50 per cent.
- Maintenance requirements are generally less than those for conventional drainage systems. BMPs trap pollutants at one point, sparing managers the task of having to clean out many small structures.
- Research conducted in the USA shows that pollutants cannot be considered to be hazardous or toxic material.
- Storm water ponds can be designed to be safer for children by adjusting the geometry; slopes to ponds should be gentle, minimising the risk of a child falling into one.
- With sensitive planting and landscaping, BMPs such as swales, ponds and wetlands can be attractive features, as can porous surfaces for car parks and pedestrian areas.
- Benefits include:
  - increases in property values
  - increased wildlife and conservation value
  - sustainable development
  - a cleaner water environment.

Best management practices in action

Examples of BMP sites in the UK include porous car parks at Nottingham Trent University; infiltration trenches at Shire Hall, Reading; storm water retention ponds at Lexmark's site, Rosyth; and grass swales at Freeport Leisure, West Calder.

HEATING, VENTILATION AND AIR CONDITIONING

Indoor air quality

It is estimated that the average person spends up to 90 per cent of their time inside a building. Maintaining air quality is vital to ensure health and well-being, as well as maximum productivity in the workplace. Yet, despite recent air quality initiatives relating to external air pollution and the reduction of emissions, there is currently no legislation and there are no EU directives related specifically to indoor air quality.

When defining air quality, it is important to consider everything from temperature and humidity to air flow and cleanliness, as well as the maintenance of any air conditioning or ventilation equipment used. The Heating, Ventilating and Air Conditioning Manufacturers Association (HEVAC) has set up an initiative specifically on indoor air quality. It is evaluating research on the subject and is keen to create a greater understanding about the problems that may occur if air quality is poor (see section on Promoting cleaner indoor air, later in this chapter).

Health problems (such as asthma, eye irritations and nausea) are known symptoms of poor air. There have even been cases of legionnaire's disease spread by bacteria in HVAC systems. Poor air quality also impacts on productivity in the workplace.

The effect of equipment on air quality

Factors influencing indoor air quality include furniture, carpets, fixtures and fittings, process plant and equipment and general office apparatus such as computers, photocopiers and fax machines, as well as HVAC equipment.

Temperatures increase with the amount of electrical equipment used, small quantities of toxic ozone can be produced by printers and fax machines, for example, and emissions of volatile organic compounds (VOCs) are possible by-products of soft furnishings, carpets and some furniture. Assessing the equipment and furnishings in a space and choosing equipment with minimum impact on indoor air quality is advisable and can mean that smaller and less expensive HVAC equipment is needed as a result.

Saving energy and improving quality

To save energy and improve indoor air quality, HEVAC suggests:

- choosing electrical equipment with a sleep mode where possible to reduce power consumption during inactive periods
- fitting ozone filters to equipment where appropriate
- fitting carpets and furnishings with low emissions of VOCs
- having dedicated extraction equipment to control fumes and emissions
- using solvent-free inks where possible
- checking how easily equipment can be cleaned and following manufacturers' recommendations on cleaning and maintenance.

Indoor air quality can be improved simply by removing or limiting as many pollutants as possible, providing a good quality environmental control system (which includes the removal of pollutants) and ensuring adequate ventilation (natural, where possible).

CIBSE recommends[5]:

- eliminating contaminants at source
- substituting with sources that produce non-toxic or less malodorous contaminants
- reducing the emission rate of substances
- segregating occupants from potential sources of toxic or malodorous substances
- improving ventilation (by local exhaust, displacement or dilution, for example)
- providing personal protection.

---

[5] CIBSE, Tel.: 020 8675 5211, www.cibse.org.

Choosing equipment

Choosing the correct HVAC equipment is essential. It needs to be appropriate for the space in which it is installed and maintained properly thereafter, otherwise it could have the effect of reducing air quality. Plant today is smaller and more efficient than it used to be and likewise (new) buildings are becoming more energy efficient. The Federation of Environmental Trade Associations (FETA) emphasises the importance of maintaining equipment and stresses that using top-of-the-range equipment in the wrong place can be just as damaging as older equipment, especially if it is not maintained and cleaned regularly. FETA has the responsibility for testing air conditioning systems under Part L of Building Regulations.

Air conditioning which brings fresh air inside may be an excellent idea, but if the unit is sited in the wrong place it could actually be counter-productive, dragging more pollution indoors in the form of traffic fumes, especially if filters are not changed regularly. According to CIBSE, the grade of filtration needed varies according to several factors, including the level of external pollution and the exposure limits set to protect occupants, and the amount of protection required for the internal surfaces of the building, air handling plant and air distribution system.

A voluntary system run by Eurovent, the European testing body for HVAC equipment, certifies products. Eurovent tests the equipment's performance and power rating, which ensures that manufacturers' claims about the amount of energy the product uses are correct.

BACTERIAL ATTACK ON HVAC SYSTEMS

Experts first brought bacterial attacks on HVAC systems to London property owners' attention in 1998. In 2001, an increasing number of cases were discovered throughout the UK. The bacteria in question, Pseudomonas, attack the inside of systems and ultimately turn them into sprinkler systems. Pseudomonas attack anything that will corrode. London-based systems appear to be well covered but increasing numbers are being discovered elsewhere. It is crucial to assess risk at the testing stage of HVAC systems as problems start as soon as the pipes get wet and are then left with stagnant water. Pseudomonas, which are harmless to humans, can be spread through using dirty hoses for filling systems and unscrupulous companies selling products claiming to eradicate the problem, or giving incorrect or inappropriate advice.

A nine-point plan has been drawn up and facilities managers should check that consultants and installers are complying:[6]

1. The quality of water for filling systems should be monitored and controlled.
2. Biocide and dispersant should be considered for use with initial fill.
3. Dose with an appropriate biocide prior to the clean.
4. Always test water by sampling in the cleaning process.
5. Ensure that all temporary hoses and equipment are chlorinated before use.

---

[6] For more information visit www.fhp.uk.com.

6. Consider the type of filtration to be used carefully.
7. Consider the type of biocide carefully, its longevity, for example.
8. Consider the type of inhibitor carefully (nitrite-based units provide bacteria nutrients).
9. Regularly monitor the microbiological condition of the water.

Air conditioning

The view of air conditioning as a luxury product is generally being abandoned. Indeed, in a range of locations encompassing offices, shops, restaurants, bars and cars, it has become essential to have an air conditioned environment. There are situations, however, where natural ventilation is preferred on energy efficiency grounds.

The need for air conditioning

Trends towards open-plan offices have led to increased demands for air conditioning systems, as the space allocated per employee is more restrictive in an open-plan area than in a conventional office. Add to this the massive amounts of heat generated by the many electronic systems in today's offices and the need for cooling becomes obvious. Opening windows is not always the answer, as in many cases the windows will be sealed units. Where the windows do open wide, the outside environment may be a busy, noisy and dirt-polluted thoroughfare.

A small office with two people and a typical range of office equipment can generate 3000 W of heat per hour, which is more than a two-bar electric fire. Poor air quality is known to reduce or inhibit productivity. The increasing drive towards air quality is expected to spur growth in the air conditioning market. Market research shows that the office segment accounts for 59 per cent of sales.

Types of system

Fixed

The packaged air conditioning market consists of packaged split systems and packaged multisplit systems, while the central-based air conditioning market includes the fan coil, air handling unit and chiller segments. There are over 250 manufacturers capable of supplying air conditioning systems. Many companies will sell systems through distributors, specialist dealers or facilities management companies. Central-based systems tend to be distributed directly by the manufacturer because they are fitted to a building's requirements.

Portable

It is possible to hire portable air conditioners with solutions ranging from small mobile temporary air conditioning units for short-term or emergency use to permanently installed heat pump units, supplying both cool and warm air, suitable for offices of any size which need cooling in the summer and heating in the winter.

Many customers for portable units are facilities managers in end-user companies or facilities management companies or heating and ventilation personnel. The market exists all year round. If a computer complex's systems go down, the facilities manager can make one call to bring in suitable equipment. The ranges on offer handle offices up to server rooms, with rental hires from about £50 per week for the smallest system to over £250 per week for an 8 kW unit.

Business is typically split 60 per cent for cooling in office environments to 40 per cent for emergency failures of fixed systems, and '80 per cent of the market is London'. Typical scenarios have been setting up new offices in buildings which are not air conditioned and relocations within an existing building. Listed buildings figure prominently too, as planning consent to install fixed systems is difficult to obtain.

Close control

The close control air conditioning market comprises data centres, web server farms, Internet service provider (ISP) computer rooms and centres for telecom switch gear and computer network management apparatus, where key requirements are for temperature control, humidity and filtration.

The telecom market generally needs electrical energy and air conditioning (cooling) in equal measures. Combined cool and power (CCP) claims to eliminate most of the air conditioning power needs, reducing overall energy demand of a 'telehouse', typically by 22–25 per cent. The ratio of cooling and power produced by CCP closely matches the needs of a typical telehouse. Carbon dioxide emissions are reduced by over 20 per cent, securing a number of tax advantages including CCL exemption and enhanced capital allowances (see the section earlier in this chapter, Reducing emissions and costs).

Temperature control

Whatever type of air conditioning system is used, the key objective is temperature control, ideally, for office workers, a pleasantly cool, clean atmosphere with adequate levels of humidity (see later in this chapter, Maintaining adequate humidity). With web server farms, however, there has to be a positive pressure in the room so that when a door is opened there is no ingress of dirt or other contaminants. Temperature has to be maintained in a range typically 22–24 °C, while the relative humidity needs to be maintained at 50 per cent plus or minus 5 per cent.

The machines in these environments (computers, disk drives, network switching equipment, telecom systems, etc.) are particularly sensitive to these two drivers and cannot tolerate any rate of change. While staff in an office might tolerate variations of 10 per cent fluctuations, the machines in data centres, server farms and network switching centres cannot.

The emergence of a new generation of processing platforms, collectively referred to as blade servers, is changing the way banks, financial operations and other intensive users of computer technology run their businesses. Blade servers are designed to address one of the thorniest problems faced by managers today: the proliferation of application servers over the past decade as organisations have scrambled to deploy new web sites and corporate applications. Consequently, many administrators find that they now have more servers than their teams can manage. Worse still, the processing power offered by individual servers is frequently underutilised by the applications they host. Simultaneously, the footprint of conventional servers means they take up more physical space than the organisation can afford to allocate. Many difficulties are caused for building managers and service engineers attempting to control server room temperatures.

By reducing conventional servers to their component parts and repackaging them in rack-mounted chassis that pack the maximum number of processors into the minimum available space, blade servers effectively tackle this crisis. A fully loaded 42-unit blade chassis produces 12 kW of heat, about 50 per cent more than a typical household oven. Experts estimate that air-cooling systems must move 140 cubic feet of air for every 1 kW generated, to keep temperatures below the 25–27 °C point, at which system performance dips and component lifespan is reduced. Intense research work is proceeding in these areas.

Refrigerant policy

Since usage of the R22 refrigerant has been banned in all new systems sold, most systems marketed today will be equipped with R407C refrigerants. Advice on chloro-fluorocarbon (CFC) and hydrochlorofluorocarbon (HCFC) phase-out is available which reflects EC Regulation 2037/2000 on ozone-depleting substances.[7] No one will be permitted to use virgin HCFCs from 1 January 2010 and all HCFCs, including recycled materials, will be banned from 1 January 2015. Recovered CFCs must now be destroyed by an environmentally acceptable technology. Recovered HCFCs can be either destroyed or reused until 2015. There are some rogue companies still marketing equipment with the wrong refrigerant.

Promoting cleaner indoor air

Improving outdoor air quality in terms of lower emissions is a key objective, but what is being done to promote cleaner air indoors?

CIBSE guidelines

While legislation and guidelines on the subject are lacking in the UK, there are standards for indoor air quality set by CIBSE. These are mainly based on comfort levels. In its *Guide A on Environmental Design*,[8] CIBSE states that indoor air quality 'may be said to be acceptable' if:

- not more than 50 per cent of the occupants can detect any odour
- not more than 20 per cent experience discomfort
- not more than 10 per cent suffer from mucosal irritation
- not more than 5 per cent experience annoyance, for less than 2 per cent of the time.

CIBSE also points out this does not take into account contaminants such as radon gas, which may have serious health effects, but is nevertheless odourless and difficult to detect. Workplaces should therefore be comfortable and ventilated by a sufficient

---

[7] EC regulations, The Stationery Office, www.tso.co.uk.
[8] CIBSE, Tel.: 020 8675 5211, www.cibse.org.

quantity of fresh or purified air. Getting the balance right is more complex than it sounds. The Building Research Establishment (BRE) recommends natural ventilation is used where possible, but accepts that there is a tradeoff for many managers. Keeping windows tightly shut, for example, prevents energy from escaping and thus cuts down on energy wastage. It also stops outdoor pollutants such as traffic fumes from coming in. However, with windows sealed, the risk is that temperatures will rise and indoor pollutants increase as a result.

Main pollutants

Air can become polluted in a number of ways, some of which are more obvious than others. HEVAC's indoor air quality (IAQ) initiative cites the following as the main sources of pollution:

- industrial processes producing fumes and airborne contaminants
- bacteria and dust spread by inadequately filtered and poorly maintained HVAC systems
- emissions from building materials, furnishings and equipment
- carbon monoxide produced from gas and paraffin heaters
- tobacco smoke
- pesticides sprayed on potted plants
- mould in damp areas and rotting food, both sources of bacteria
- an excessively humid atmosphere which can help the growth of bacteria.

Maintaining adequate humidity

The Health and Safety (Display Screen Equipment) Regulations 1992 provide that employers are to ensure that 'an adequate level of humidity shall be established and maintained' in offices where there are visual display unit (VDU) users.[9] Exact levels of 'adequate humidity' are not specified, but there is a duty of care to ensure that reasonable conditions are maintained. For staff health and comfort, the optimum adequate level of relative humidity is normally considered to be about 50 per cent relative humidity (RH) and certainly not less than 40 per cent. Several optometric organisations (Association of Optometrists, College of Optometrists and EyeCare Information Service) agree with this level, as do professional bodies such as CIBSE, British Institute of Facilities Management (BIFM), Building Services Research and Information Association (BSRIA), BRE and HEVAC's Humidity Group. The Health and Safety Executive (HSE) says that it is generally considered to be in the interests of worker efficiency that humidity be maintained in the range of 40–70 per cent RH and that in warm offices the relative humidity should ebb towards the lower end of this range. British Standard 29241 (from 7179) recommends 40–60 per cent RH for office terminals.[10] (For more information, see Chapter 1.)

---

[9] Copies of HSE guidance on the DSE Regulations 1992, ISBN 0 11 886331 2, are available from the Stationery Office, see footnote 7.

[10] British Standards BS 29241, see www.bsi-global.com.

Employers' duties

Employers are under a duty to comply with this legislation. This means monitoring humidity constantly to ascertain whether adequate levels are being provided. It is also important for employers to recognise any dry heat symptoms being experienced by staff, in order to identify whether any problems need to be addressed to ensure staff health and comfort. If such problems are not addressed by the facilities manager, the normal course of action would be for employees to complain to their employers, who would take steps to remedy the problem. If the problem continued, the next step would be to involve Environmental Health Officers and, in some cases, the unions. Eventually it could lead to prosecution, as with cases of repetitive strain injury (RSI).

Recognising symptoms

Employers are now under an obligation to identify symptoms of low humidity and recognise that they have a duty to take corrective measures for staff health and comfort. VDUs produce dry heat and this can cause discomfort to staff who work with computer monitors. Dry air acts like a sponge and absorbs moisture from all surrounding surfaces, such as operators' eyes, nose, throat and skin. Typical resultant dry air symptoms include headaches, dry eyes (particularly for contact lens wearers), dry throats, dry skin, frequent cold and flu-type conditions, and tiredness and lethargy. People with dry skin conditions such as eczema or psoriasis and respiratory or breathing problems such as asthma are often worst affected by the dryness. Static electricity is a conclusive indicator that humidity is too low.

Stress

Stress levels can be affected too. A nationwide research study of over 100 workplaces investigated the links between overly dry workplace air and employee stress. Readings showed that the environments of one in five subjects had only 25 per cent RH, making them as dry as the Sahara Desert. One in 10 had only 23 per cent RH, as dry as California's Death Valley. In these excessively dry surroundings, employees suffer a rapid loss of water, affecting their physical and mental well-being while increasing vulnerability to workplace stress. Nationally, 83 per cent of workplaces with 35 per cent RH or less were rated as high-stress environments by their employees.

Staff discomfort

Although essential for comfort of workers, central heating and air conditioning also dry the air considerably. In addition, lighting and solar gain contribute to the problem.

The World Health Organisation (WHO) has long recognised that air conditioning, computers, photocopiers and dust are the scourge of office life. Statistics also show that 6 per cent of staff sickness and absenteeism is due to sick building syndrome (SBS), costing the UK economy between £300 million and £650 million a year (estimated by WHO and a House of Commons environment committee). See Chapter 1, section on Improving well-being: Sick building syndrome, for details of buildings at risk of SBS and WHO-recommended solutions.

However, it is possible to monitor and maintain humidity at comfort levels, resulting in a healthier, happier workforce with less sickness absence and increased productivity. This has to be good news for long-suffering employees and employers alike. It makes economic sense too, as the cost of a free-standing retrofit humidifier which plugs into a 13 A socket is less than two to three days' staff absence due to sickness.

Measuring humidity levels

When workstation risk assessments are carried out, humidity levels should be measured and recorded to ascertain whether there is an adequate level of humidity. A humidity recording kit should comprise a digital thermohygrometer with built-in memory, a supply of charts showing the adequate humidity band, plus information on the legislation and what practical steps to take if the humidity is too low.

It is recommended readings are taken in the same location at the same time each day, for example, beside the workstation in mid-afternoon when computer screens have been running for several hours and the worst (driest) conditions are being experienced. On day one, simply read off the relative humidity, record the readings and repeat daily. If readings are below 40 per cent, the air is too dry. In computerised offices, it is most unlikely that humidity will be too high. By mid-afternoon humidity can frequently fall to as low as 20–25 per cent RH, well below the adequate band.

Managing humidity levels

If humidity is too low, causing staff discomfort, humidity needs to be raised to adequate levels. If the air is too dry, the simple solution is to put just the right amount of moisture back into the air to restore a more healthy and comfortable atmosphere. This usually means aiming for about 50 per cent RH for a normal indoor temperature of about 20 °C. It is important to seek specialist impartial advice on the right type of humidifier, and its output rating, for the size of area. Aspects such as hygiene, noise levels and running costs must also be taken into consideration. Some units also combine air cleaning functions and will help to filter out dust, traffic exhaust and tobacco smoke, as well as ozone from copiers and printers. The result is a healthier, and standards-compliant, workplace.

WASTE MANAGEMENT

Waste management is an ethical problem that no one can bury. Defined as the unwanted residue of an organisation's activities, waste can include anything from toxic liquids and solids, pallets and packaging, expired light bulbs and printer cartridges, to the contents of the waste paper basket.

Disposal

Three options exist for waste disposal:

- the strict reduction of waste generation, through adjustment, redesign of processes and cooperation with suppliers

- the internal, and responsible, recycling of waste materials to provide new or different products
- getting rid of the waste to someone else and making it their problem.

The third option is the most commonly deployed for business; it is an attractive option, particularly as specialist companies move into the waste management field. In the past companies had to pay the local authority or a dedicated contractor to take away unwanted rubbish (to the nearest landfill site or incinerator). Now someone may collect waste for free and use the output as their raw material.

Responsibility over profitability

Of the many millions of tonnes of solid materials, and many more tonnes of consumed water, only 25 per cent is converted into products physically consumed by customers. The balance is the national waste problem. The waste handling industry is understandably booming, worth something like 0.5 per cent of gross domestic product (GDP) and growing, according to research by the Chartered Institution of Wastes Management.[11]

The public wants to see that companies are behaving responsibly. Awareness of the impact of waste is at an all-time high. Legislation is following public concern too. Company and product liability is extending into every aspect of the production process from inception to final disposal. The issue embraces a growing emotional draw to environmentalism and the pragmatism of economics. Landfill sites have become a scarce resource and there is not enough space to bury rubbish as has been the practice in the past.

Companies have to implement a coherent waste management strategy, argues the Department for Environment, Food and Rural Affairs (Defra). The Environmental Protection Agency obliges directors to ensure that the waste generated by their businesses is stored, moved and disposed of by approved means (or they face fines of up to £20,000 or six months in prison per offence).

As of autumn 2007, the Waste Electronic and Electrical Equipment (WEEE) EU Directive requires businesses to recycle their old information technology and telecoms equipment, and other electronic and electrical waste.

A sustainable resource

The concept of waste as a sustainable resource has been around for a long time. The idea has been given a boost and recognition through the introduction of European and UK legislation. A recent guide produced by the Environment Council, *A stakeholder's guide to sustainable waste management*, is most useful. The Environment Council firmly believes 'environmental problems can be solved by involving, listening to and working with people' and indeed applied that principle in developing the guide.[12]

---

[11] Chartered Institution of Wastes Management, Tel.: 01604 620426, www.iwm.co.uk.

[12] Environment Council, www.wasteguide.org.uk.

How do companies turn waste disposal into waste management? The first step is to understand the problem, and the starting point for this is a waste audit. This identifies the materials existing as waste within the company and provides data for subsequent comparison to monitor improvement. A further benefit could be the introduction to a specialist company with which to work in implementing improvements. Kick off with the relevant local authority and/or existing waste contractor.

Responsible environmental management is logically consistent with low costs. Savings can accrue through careful housekeeping, but longer term savings follow investment in research technology or process design. Waste is set to be the next frontier of competitive advantage. By looking in their dustbins, companies can find new ways to improve internal business performance.

NOISE AND VIBRATION

According to the Association of British Insurers, some 80 per cent of claims for occupational disease against employers' liability insurance relate to deafness. This compares with 6 per cent for lung diseases and 4 per cent for upper limb disorders. Hazards of high noise levels at work include:

- incurable hearing damage
- disturbance to work
- interference with communication
- stress.

Allowances paid by the Department for Work and Pensions (DWP) to people who suffer at least 50 decibels (dB) of noise-induced hearing loss in both ears (which is like trying to listen to the television through a brick wall). Any claimant must have been employed for at least 10 years in a specified noisy occupation.

Statutory requirements relating to noise at work generally are contained in the Noise at Work Regulations 2005 (see Noise at Work Regulations 2005, in Chapter 1). Employers are required to take reasonably practicable measures on a long-term ongoing basis to reduce employees' exposure to noise at work to the lowest possible level, and to lower noise exposure where employees are exposed to levels of 90 dB or above. For HSE guidance on controlling noise exposure, see Reducing noise, in Chapter 1.

There are currently no specific provisions relating to vibration other than those contained in the Social Security (Industrial Injuries) (Prescribed Diseases) Regulations 1985 and the Reporting of Injuries, Diseases and Dangerous Occurrences Regulations 1995.

Damage to hearing

It is natural to lose hearing acuity with age. Loud noises cause permanent damage to the nerve cells of the inner ear in such a way that a hearing aid is ineffective. With prolonged exposure, the region of damage moves to both higher and lower frequencies. Damage begins to extend into the speech range, making it difficult to distinguish consonants, so words start to sound the same. Eventually, speech becomes a muffled jumble of sounds.

Some people are much more susceptible to hearing damage than others. There is the condition of tinnitus (ringing or whistling in the ears and a temporary dullness of hearing), experienced when a person leaves a noisy location. Care should be taken to avoid further exposure (symptoms disappearing with continued exposure actually indicate that the ear is losing the ability to respond to the noise). Medical advice should be sought.

The risk of hearing loss depends on the noise dose received and on the cumulative effect over time (except from catastrophic circumstances, as with explosions). A procedure for estimating the risk of handicap due to noise exposure is provided in British Standard BS 5330. Here, a hearing loss of 30 dB is defined as a handicap (understanding of conversations and appreciation of music impaired); but to be entitled for DSS disability benefit, a person must suffer a loss of 50 dB.[13]

Sound perception

Typically, the quietest sound that can be heard (in other words, the threshold of hearing) is zero decibels (0 dB) and the sound becomes painful at 120 dB. It should be noted that zero decibels is not zero sound.

Depending on the method of presentation of two sounds, the human ear may detect differences as small as 0.5 dB. But for general environmental noise the detectable difference is usually taken to be between 1 and 3 dB, depending on how quickly the change occurs. A 10 dB change in sound pressure level (SPL) corresponds, subjectively, to an approximate doubling/halving in loudness.

The SPL of industrial and environmental sound fluctuates continuously. It is possible to measure the physical characteristics of sound with much accuracy and to predict the physical human response to characteristics such as loudness, pitch and audibility. But it is not possible to predict subjective characteristics, such as annoyance, with certainty. A meter can only measure sound and not noise, which is defined as sound unwanted by the recipient. But in practice the terms are often interchangeable.

Sound power level

Sound output of an item of plant or equipment is often specified in terms of its sound power level, measured in decibels relative to a reference power of 1 pW (picowatt). It must not be confused with SPL, which at a particular position can be calculated from knowledge of the sound power level of the source, provided the acoustic characteristics of the surrounding and intervening space are known. Thus,

---

[13] British Standard BS 5330, see www.bsi-global.com.

to provide a picture of the SPL–sound power level relationship, for a noise source which is emitting sound uniformly in all directions close above a hard surface in an open space, the SPL 10 m from the source would be 28 dB below its sound power level.

Minimising noise through design

Sound travels via either transmission or reflection. Noise transmission is the main problem in cellular offices, especially conversational noise leaking from adjoining cells. This can be especially distracting since there is less background sound in the enclosure to mask it. In modern offices most cellular space is created by combinations of demountable full-height partitions rather than brick or block walls. The acoustic performance and fit of these partitions is crucial to good sound insulation. Manufacturers' specifications for the noise reduction coefficient of demountable partitions will be for optimum performance.

It is up to facilities managers to get as close as possible to this by ensuring that the partitions are well fitted (special mastics and packing materials can help to seal gaps where panels meet uneven floorplates). Where there is a suspended ceiling and the panels do not pass through and stop at the upper slab, then sound will travel through the ceiling void and compromise the insulation offered by even the best partitioning. Air vents between cells and holes for piping and electrical cabling will also channel noise. Manufacturers of suspended ceilings should be able to provide a statement of the material's noise reduction coefficient and speech frequency sound absorption.

In open-plan space, noise reflection is more of an issue than transmission. All office surfaces, that is, flooring, wall coverings, ceilings and furnishings, can combine to reduce or increase noise reflection. Hard, flat surfaces will bounce noise around a space. Soft carpeting with in-built cushioning or combined with underlay will contribute strongly to deadening ambient noise reflection as well as footfalls.

Noise reduction through screening in open-plan space is problematic. The recommended screen height to deaden sound transmission between workstations is 1.65 m, but this creates the kind of cellular warren in open space that has become increasingly unfashionable over the past 15 years. Even the most absorbent low-profile screens have little effect on sound transmission.

HSE guidance warns that barriers and screens can provide a limited solution on their own. Near the noise source, most sound is received by direct transmission. A brick or steel sheet barrier can stop this, but even when it is lined with sound absorbing materials, problems with reflected sound further away will remain.

Controlling noise at source

One of the best ways to cut out noise is to control it at source. Modifying existing equipment to make its mechanisms quieter, for example, will cost only one-fifth of building a box around the piece of equipment.

An even better solution to noisy equipment is not to buy it in the first place. According to HSE guidance *Reducing noise at work*[14]: 'Often the single most cost-effective, long-term measure you can take to reduce noise at work is to introduce a purchasing policy for choosing quieter machinery'. This avoids the need for 'expensive retrofitting of noise control measures'.

Another example of controlling noise at source is those call centres which have abandoned the traditional rows of booths in favour of low-screened workstations. Large numbers of employees work in close proximity, avoiding excessive noise by moderating their voices (with the help of sensitive telephone headsets).

Sources of advice

HSE figures have suggested that 1.3 million employees in the UK are exposed to noise levels above 85 dB. Both short bursts of loud noise and long-term exposure to high noise levels can gradually, but relentlessly, damage the inner ear, according to the Royal National Institute for the Deaf (RNID). Employees may only realise their hearing has been damaged years later.

The BRE says that noise is one of the most emotive issues in today's built environment, with the potential to impair health and productivity in the workplace. Once any facilities manager has acknowledged noise reduction is an issue in their workplace, BRE may represent a valuable source of help. Services include:

- accredited measurement of field and laboratory sound transmission
- product development
- advice on best construction practice
- innovative measurement of building element performance with sound intensity
- environmental noise measurement and assessment.

FURNITURE

There is significant investment in new desks and office furniture. On average, new furniture is acquired every 10 years at a cost of some £1000 per workstation. It is sobering to realise that office workers constitute nearly 50 per cent of the working population and many of their waking hours are spent at their desks. The functionality, appearance and general cleanliness of the workspace will influence how efficiently staff work, how they interrelate with colleagues and within their teams, and how they view the office environment and its ambience.

Selecting furniture

Facilities managers may need to replace existing stock: desks, chairs, storage and office desk accessories. The process of selecting the right products, supplier and

---

[14] ISBN 0 7176 1511 1.

manufacturer need not be complex. Most suppliers will agree to lend samples of their furniture for trials. There are many consultants and facilities management contractors willing to take on any aspect of space planning. Some furniture suppliers offer free or low-cost space planning services. Whatever the decision, always seek references or testimonials.

There is also an emerging market for rental furniture. Points in its favour include:

- no capital expenditure
- no cash flow disruptions
- instant flexibility to changing work patterns
- no storage costs relating to redundant furniture
- 'free' space planning services
- immediate disaster recovery.

Workspace design

Furniture and storage options integral to a more cost-efficient and flexible workspace design include:

- **Standard workstation**: as with the principle of the universal footprint, the closer you can get to a single desk size and type for all staff, the less likely you are to have to move any desk. Where individuals need larger work surfaces or more storage than the average or are allocated meeting tables, try to provide these through modular extensions to the standard workstations, so that only these extras have to be relocated, leaving the desk undisturbed.
- **Mobile furniture**: if you are likely to have to move complete workstations, choose a system that is robust enough to stand relocation (not all are) and is ready to move after a minimum of disassembly. Most manufacturers offer desking and filing pedestals with built-in lockable castors to facilitate movement. Check that the wheels are up to the job (again, not all are).
- **Centralised storage**: the more localised the bulk storage, the greater the chance it will have to follow staff in intradepartmental moves. Centralised filing areas on each floor mean that moves around the floor should not necessitate moving anything but a small amount of local filing. There may be some tradeoff with individual efficiency, though, if staff have to walk any distance to refer to documents.
- **Shared worksettings**: the idea that most office space was taken up by desks, each allocated to one employee who would work there all day, went unchallenged until the 1990s. Although this pattern is still dominant, an increasing number of organisations have experimented with alternative models over the past decade, such as reducing personal workspace or removing it altogether and replacing it with a variety of shared worksettings.

The thinking behind shared worksettings was that many white-collar employees were in jobs that took them away from the office for a majority of the time.

These workers still claimed permanent desk space which was seldom used. Even those tied to the office were often engaged in types of work which were inadequately catered for by the traditional desk. The latter point has been reinforced by the growing emphasis on knowledge management and teamworking among white-collar workers. Traditional, rigid layouts cannot support collaborative group work or the regular forming and disbanding of project teams. These drivers have produced a variety of new office configurations, all of which have elements in common:

- **Teamspace**: in which groups of employees retain a small 'home' desk each and the rest of their space is recycled for shared worksettings such as meeting space and breakout areas.
- **Hotdesking**: where employees who spend a large minority of their time out of the office have no allocated space but use non-allocated workstations, either on a first come, first served basis or by booking in advance.
- **Hotelling**: where all workspace is bookable by the hour.

The most common of these configurations, hotdesking, is covered in more detail in Chapter 11.

Workstations

Existing workstations can be easily recycled as shared desking. For hotdesking, an 1800 mm wide desktop is unnecessary and takes up space that could be better used. A surface deep enough to hold a computer monitor – for a shared PC or docking facilities for laptops – and wide enough to spread papers is all that is needed.

Height-adjustable desking

Height-adjustable desking is now more common and should be specified where possible to allow users maximum comfort, however short the stay. When choosing workstations for hotdesking, work surfaces and edgings should be checked for durability. Also ask for references from other users to ensure that the products will stand up to more intensive use from multiple users.

Seating

Workstation chairs need to be easily adjustable for height, back support and preferably seat tilt. If more than one person is going to use a chair, the configuration controls need to be easily located and robust. To alert staff to hotdesking chairs, one idea is to colour-code them with vivid fabrics and designs.

Soft seating for breakout areas gives an opportunity to add a splash of colour to the office layout. Sofas and tub chairs should be simply styled and firmly upholstered, designed for comfort but not for curling up on in front of the fire. Fabrics should be hard-wearing and easily removed for cleaning, especially if they are located near catering or vending points.

Aiding correct posture

The type of chair should fit both the task and length of time doing it. Chairs should have a forward tilt mechanism to open the pelvis, similar to the position in standing, and allow the correct concave curve to be maintained in the lumbar spine while sitting. This consequently leads to correct positioning of the other spinal curves. Seating should support the body in a suitable posture, in other words, the 'S' form of the spinal column should be maintained. The purpose of the backrest is to support the whole spinal column effectively, no matter what seating position is adopted.

The user should be able to move and change postures regularly and freely without any constraints from the chair. Cushioning and covering should allow for air circulation over the skin and the surface material must provide sufficient friction to prevent sliding. For details of HSE guidance on seating, see Chapter 1, section on Seating.

The cost of back problems

Back problems are the greatest cause of illness among people of working age. Most people spend over 70 per cent of their waking day sitting, and this is cited as one of the major factors responsible for back pain and associated injuries. Each year there are 30,000 work-related back injuries, 33,000 work-related back accidents and 500,000 work-related back illnesses. The cost to industry has been estimated at 117 million days of certified sick leave, resulting in at least £5.5 billion in lost production each year. Given all these statistics (even allowing for margins of numerical error) it obviously pays to get the seating right.

Buying furniture

The Office Furniture Advisory Service (OFAS) provides full information on standards and what to look for in purchasing furniture. There are about 350 office furniture and seating manufacturers in the UK, and many more suppliers.

CATERING

Why should a company have a restaurant? Many large organisations do not, arguing that nearby local restaurants, cafes, coffee shops, sandwich bars and fast-food outlets provide all the resources required. This philosophy can have an adverse effect: long off-site lunch breaks and stress caused by having to hurry back to the office are two potential pitfalls, and a lack of control over the staff's diet is another; a healthy diet produces a healthy workforce, which means less absenteeism. Furthermore, a company with no restaurant is less likely to be able to attract staff because of the lack of quality food and subsidised prices, and has nowhere to provide hospitality for customers.

Since January 2006, those involved in the food business, including caterers, have had to meet more exacting demands. The EEC Hygiene of Foodstuffs Regulation 852/2004 seeks to pursue a 'high level of protection of human life and health' while 'including the aim of achieving free movement of food within the EC.' Alongside this, the Food Hygiene (England) Regulations 2005 also came into force on 1 January 2006, dealing principally with enforcement and legal provisions as well as temperature control requirements. The 852/2004 Regulation applies to all food business activity from primary production through to sale to the end-user. The significant new measure for caterers is that they must now have written hazard analysis critical control point (HACCP) documentation in place. The Food Standards Agency has launched an abbreviated system for small businesses called '*Safer Food, Better Business*'. Major caterers have developed online manuals to enhance reporting procedures and access to the guidelines for their staff.

Drinking water

Staff have to be well watered and fed to be content. Research has found that most people said they worked more effectively in a smarter workplace. The two items workers considered most essential to a productive day were natural daylight (93 per cent) and chilled water (83 per cent).

Workers do not drink enough water. Indeed, most people say the amount of water they drink is directly related to where the water cooler is kept, and many of these people were frustrated at coolers being too far away. Lack of water leads to poor concentration, fatigue, irritability and headaches. It would thus appear that a key factor in keeping productivity to a maximum is ensuring that staff have access to a water cooler which is integrated into the design of the workplace. Not everyone wants to eat at work, but everyone needs water.

Company restaurant facilities

Sitting down to eat together at lunch or for a snack is one of the best ways to break down barriers, discuss problems, thrash out misunderstandings or share company gossip. To do this, staff typically adjourn to the restaurant, coffee bar or similar facility. Company size will dictate the type of restaurant or café facilities and levels of services offered. But the driver behind all these plans is to get staff into a mood where they not only want to come to work, but enjoy it as well, with eating and drinking central to that enjoyment.

What do the staff want?

Where company catering and restaurants are concerned, facilities managers must try to understand their customers' needs. Easy-to-organise staff surveys of what foods and drinks are preferred, what times of day staff would like them served, reaction to special menus and themed eating days, for example, can produce extra usage of the restaurant facility.

A focal point

Making the restaurant an attractive, comfortable and interesting place to eat and socialise, as well as promoting it as a centre for networking, meeting visitors or prospective clients and holding impromptu and scheduled meetings, will pay dividends for the company and, in so doing, for the facilities manager. To that end, the restaurant should be easily accessible from all parts of the building and as equally visible.

Fast food

While only operating a fast food-type operation is not recognised as common sense, it may be prudent to offer staff the opportunity to have access to a 'fast' service. The staff survey can help here. Making eating a worthy experience within the company will help to boost staff morale and retention.

Vending machines

Many facilities managers eschew vending machines because of the quality of the beverages produced and the image that has built up around them. However, there is still a market for these machines. It is unlikely that the 'tea lady', complete with trolley, buns of dubious freshness and a steaming urn, will ever make a full-scale reappearance. The future of the vending machine would seem assured. But canny facilities managers will approach the area sensitively and provide the staff with what they want. Assuming that the board can be convinced, this will help the staff to work better, aid staff retention and help to percolate money through to the bottom line.

Times

A restaurant should offer a full range of meals and general refreshments, at times when the staff want them. Moves towards more global trading have resulted in offices being open at what were hitherto deemed antisocial hours. The concept of facilities being offered at practically all hours is not as ridiculous as it used to sound. It is a fact that food still being served fresh at 2.30 pm will both encourage a wider spread of diners sitting down for a meal and increase the numbers of staff utilising the restaurant.

Menu

Some people will want to bring in their own food and drink. This should not be discouraged, and space should be made available. Others will increasingly want organic food or gluten-free products; again the level of demand can be gauged through a staff survey. It is a misnomer to label food 'home-cooked' when it is not. But this is an expression that has crept into the vernacular and now sits alongside such culinary delights as seasonal specials, cordon bleu cooking, chicken tikka masala, regional speciality days and low-fat cuisine.

Planning a restaurant

Design

The company must select a design strategy (embracing appearance, space planning, lighting, utilities and acoustics) to reflect its culture and image. The physical appearance of a restaurant is important and care will have to be taken to differentiate it from any offerings in the local surroundings. Decisions will have to be taken on whether to provide Internet access points, so that staff can plug in their laptops while having a break or talking to customers.

Siting and layout

Siting of the restaurant facility is determined by the availability of utility services and the fact that cooking smells wafting into the reception atrium might not be a good idea for customers waiting there. The restaurant needs to be welcoming and designed so that people can flow freely without obstruction. Key to restaurant layout is throughput, trying to avoid or alleviate queuing problems (one suggestion is to provide a flexible paypoint arrangement, to avoid what is typically the biggest bottleneck). Food should always be on display, tables cleaned regularly and dirty dishes removed efficiently. Clear notices and signs should guide customers through the facility.

Space and costs

If a company opts to serve ready-prepared meals with supporting beverages, sandwiches and soup, its kitchen space requirements will be much less than for an operation providing a gastronomic experience starting from raw ingredients. The amount of space required for food production areas is dictated by the type and range of food offered. Specialist help is needed to design a catering kitchen. If the restaurant is to offer a single-sitting hot meal service only, there needs to be a ratio of three restaurant spaces (that is, areas that can earn money) to every kitchen space (the food preparation area, stores and servery). For two or three sittings this ratio reduces to two restaurant space areas. The operating costs per employee per year average at about £400 but will increase if the company provides a subsidy.

Ambience

Food is always associated with a feel-good factor. Conducting business, meeting friends or discussing issues with work colleagues can be carried out more enjoyably over a meal, snack or drink. What adds to the enjoyment is the ambience of the surroundings. This is aided by the décor, sounds, lighting and building infrastructure.

- **Décor**: this concerns colour schemes, furniture, wall and floor coverings, crockery and cutlery design, plants/flowers/foliage and paintings, sculpture and murals, and so on. A good designer is necessary to ensure optimum results. Too garish a colour scheme makes diners feel ill, while minimalist white walls call for sensitive incorporation of pictures and plants. Likewise with soft furnishings and furniture generally: too wild a pattern and the diners will be turned off. The type of soft furnishings also impacts on sound absorption and thermal barriers, as well as being decorative. Restaurant staff should wear uniforms of a suitably restrained colour.
- **Sounds**: piped 'muzak' is not to everyone's taste but designers consider noise a prerequisite. Noise mostly comes from the diners themselves. Noise creates an atmosphere of being busy. Quiet(er) areas need to be supplied, also with

'muzak', such as water features or wind chimes, to mask extraneous noise. What diners do not want to hear is kitchen noise.

- **Lighting**: daylight plays a major part in the restaurant. Lights are used to create atmosphere and good lighting designers are worth employing. The range and type of lamps available mean that most moods can be catered for and created.
- **Building infrastructure**: the restaurant facility needs to be serviced regularly, with food supplies delivery and waste removal, for example, taken care of. It must also be well served by water, electricity and gas supplies. Heating and ventilation are key requirements, as is humidity control and air temperature. Air filtration to remove particles above 5 μm is mandatory.

Management

Some catering facilities will be staffed and managed in-house, but most will be run by a catering contractor under the management of the facilities manager.

Costing

Knowing how to cost a catering facility is difficult. What should be included? All vending machines, for example? Is the service costed per user or per member of staff? Some companies provide a space (a shell) within the building and request a caterer to provide a defined level of service. The caterer pays no rent but must provide the right quality of service to recoup costs. With some of these deals, the company will fit out the kitchens and furnish the restaurant areas, while the caterer provides the staff, meals and beverages. To make a return, it is generally accepted that at least 50 per cent of the staff should utilise the restaurants regularly. Prices should be kept below those of outlets in the local high street and the ambience, as noted, must be acceptable.

Subsidies

Staff restaurants can be run with or without subsidies. A typical form of subsidy occurs where an employer pays the overheads of the facility, premises and staff, say, and the cost of the food and other items is obtained from the restaurant users. There are subsidies which involve a fixed cash payment from the employer.

Choosing caterers

Few companies have the resources to run a catering facility in-house. This means that most outsource. Nevertheless, the company still needs to set the pace and define what is wanted of its catering supplier. Facilities managers have pivotal roles to play. The choice of caterer will involve setting up a project team whose members must cover all the points already discussed here. The process of selection of caterer must involve food tasting, layout, menu content and discussions with caterers, their employees and existing customers (including site visits). Will the two companies be able to work together, or is there likely to be a culture clash?

Practical advice on how to go about selecting a catering contractor on the basis of best value for money is contained in a booklet from the Chartered Institute of

Purchasing and Supply (CIPS), *How to buy catering services*. Each year the British Hospitality Association surveys the contract catering market in the UK, based on responses from the 16 main companies in the food service management industry. Various statistics are produced by this report, such as average food cost per meal, salary expectations, turnover and types of contract.

Future prospects

More change is expected in the catering sector. Food will become a more potent weapon to attract key workers, with free breakfasts (full English or continental) being offered to entice staff into the office early and beat traffic hold-ups, and to make the workforce more productive. More food options will be made available throughout the day.

CLEANING

Cleaning is one of the most outsourced functions. While it is arguably one of the least glamorous, people will always notice if it is not done. It is also a thankless task: a facilities manager might expend an inordinate amount of resources and budget ensuring that the wards and corridors of a hospital are clean, for example, only to be judged on how quickly and efficiently they put up a shelf in a sister's office. Cleaning is also an antisocial occupation, most workers in that sector having to work after offices have closed for the day. Late and nightshift working is normal.

Market dynamics

The market comprises everything to do with industrial and commercial cleaning: interior cleaning of office space, leisure facilities, retail units, factories, non-domestic buildings generally, local authority units, and so forth, and exterior cleaning (such as of windows and metal cladding), as well as vehicles, industrial equipment and pest control services. Of late, however, the sector has seen many companies hitherto totally dedicated to cleaning diversifying into offering facilities management services as well as janitorial and portering services, linen, washroom and workwear hire and equipment hire.

With the introduction of the Private Finance Initiative (PFI) and Public Private Partnership (PPP) contracts it has not been uncommon for contracts of 30 years (and more in some cases) to be agreed. But according to the Cleaning and Support Services Association (CSSA), which acts as a marker for the industry, contracts have traditionally been from a few months to five years, and sometimes seven, in duration.

Selecting contractors

Service level agreements (SLAs) need to be marketed carefully. The main objectives are establishing the project objectives, identifying key players, securing commitment and agreeing the scope of services within review. Cleaning in the office environment will be broken down into general cleaning, janitorial services, window cleaning and deep cleans.

When selecting contractors, useful documentation for the facilities organisation includes customer satisfaction surveys, service utilisation statistics, job descriptions of services delivery staff, number and type of complaints, service specifications, menu of calloff services, budgetary information and contractors' performance information.

Once SLAs have been established, they have to be periodically reviewed to ensure that they continue to deliver the right services to support the activities of customers. Annual budget preparation should include a review of service levels. The initiation of an SLA project often increases the need to demonstrate the consistency of service delivery. Customers expect the facilities manager to market the services offered, track performance of the services over time and report regularly. This expectation will require the facilities manager to develop better documentation about each service and develop systems that will track the volume of calloffs, frequency of service failures and rectifying measures. Creating an audit trail is part of the overall communications process, but key documents that need to be developed include facilities management services policy, service directory, service specifications, operating procedures and customer feedback records. SLAs and managing contractors are covered in greater detail in Chapter 7.

Costs

How do you cost a cleaning contract? There is no magic formula, although generally, labour, National Insurance contributions, holidays, supervision, materials and equipment would need to be included. The national minimum wage is an important cornerstone of government strategy aimed at providing employees with decent minimum standards and fairness in the workplace. It applies to nearly all workers and sets hourly rates below which pay must not be allowed to fall. It helps business by ensuring that companies will be able to compete on the basis of quality of the goods and services they provide and not on low prices based predominantly on low rates of pay. The rates set are based on the recommendations of the independent Low Pay Commission.

The government has responded to the recommendations made in the Low Pay Commission's 2005 Report on the National Minimum Wage. The adult rate of the minimum wage (for workers aged 22 and over) increased from its hourly rate of £5.05 to £5.35 in October 2006, £5.52 in October 2007 and £5.73 in October 2008. The government is intending that National Minimum Wage provisions will come into effect in April 2009. At the time of printing, this is hoped to be £6.08, a 4.8 per cent increase.

When the Working Time Regulations 1998 (see Chapter 2, section on Working time) were introduced, the cleaning sector suffered a double whammy, as to make up income streams to pay the workers at the proper levels, SLAs had to be adjusted and in many cases the frequency with which cleaning was carried out had to be curtailed. There are no known statistics that indicate how the total contract cleaning workforce might have been reduced as a result of these initiatives, although diversification by some of the key players in the industry could have helped to alleviate any difficulties.

LIGHTING

Energy efficiency

Lighting accounts for some 20 per cent of total electricity consumption in the UK. The CCL levy on electricity used affects all sectors, including industry, utilities, transport and local government. Introducing energy-efficiency lighting schemes is one way of mitigating the effects of the levy, as well as following the Action Energy initiative of cutting carbon dioxide emissions. Replacing obsolete installations with more efficient light sources and introducing high-frequency control gear and lighting controls are also steps that can be taken.

Market analysts have reported how the surge in demand for lighting controls across Europe was attributable to end-users' rising awareness of energy-saving issues and the benefits associated with high-quality lighting and effects. The market is changing with the wider use of electronic gear, especially involving digital addressable lighting interface (DALI). This interface effectively merges all the superior aspects of the light management systems and control gear markets, allowing complete lighting control and improved light output. How DALI integrates with open systems protocols BACnet, EIB or LonWorks remains to be seen (see section on Open systems in building control, earlier in this chapter).

Objectives of good lighting

Good lighting should vouchsafe employee safety, acceptable job performance, and good workplace atmosphere, comfort and appearance, a task that is not limited to maintaining correct lighting levels. Of ergonomic relevance are horizontal illuminance; uniformity of illuminance over the job area; colour appearance; colour rendering, glare and discomfort; ceiling, wall and floor reflectances; job to environment illuminance ratios; job and environment reflectances; and vertical illuminance.

Sources of light

Natural lighting

The cheapest form of lighting is the most natural: daylight. There is only limited application where production is needed beyond the hours of daylight, at all seasons, and where daytime visibility is restricted by climatic conditions. Windows alone, however, cannot always provide adequate lighting for the interior of large floor areas. The most common source of daylight is through side windows. But large areas of glass can cause uncomfortable thermal conditions in summer within the building.

To maximise the benefits of natural light, low screens and glazed partitions can help to allow clear sightlines to windows. Also, 'owned' desks should be closest to the windows, with areas of occasional use, such as meeting spaces, placed further away. Proximity to daylight does bring with it the problem of glare, especially for display screen equipment (DSE) users. The easiest solution is to introduce blinds, preferably of a venetian or similarly adjustable type, to allow local control. The value of antiglare filters placed in front of computer screens is debatable and these should only be used if all else fails.

Natural lighting has to be supplemented, most of the time, with artificial lighting, for which the most common source is electric lighting. The recommended minimum ratio of artificial light to natural light at any long-term worksetting is 1:5.

Electric lighting

The selection of the sources of electric lighting for particular applications is most often driven by factors such as capital costs, running costs and replacement costs. These costs are as important as the size, heat and colour effects called for from the lighting units. Any lamp's efficiency is measured as light output in lumens per watt of electricity. A lumen is the unit of luminous flux, describing the amount of light received by a surface or emitted by a source of light. Typical values for various lamp types are shown in Table 10.1.

Common incandescent lamps (that is, coiled filament lamps where the temperature is raised to white heat by the passage of current, so emitting light) are generally inexpensive to install but suffer from relatively expensive running costs. A discharge or fluorescent lighting scheme (where electric current passes through certain gases and in so doing produces an emission of light) has higher capital costs but higher running efficiency, lower running costs and longer lamp life. In larger workplaces, there is often a choice between discharge and fluorescent lamps. The normal mercury discharge lamp and the low-pressure sodium discharge lamp have restricted colour performance, although high-pressure sodium discharge lamps and colour-corrected mercury lamps do not suffer from this disadvantage.

LEDs, sometimes called solid-state lighting, are different from most light sources in that they do not have a fragile filament and glow with electricity being passed through them. They are not a gas discharge or arc lamp like fluorescent or high-intensity discharge that emits photons, or light, through the electric arc plasma. LEDs are microscopic layers of materials on a substrate, usually silicon carbide or sapphire, which emit photons when a small electric current passes through the layers. The composition of the layers determines the colour of light emitted.

Table 10.1 Lamp output

Type of lamp	Lumens per watt
Incandescent lamps	10–18
Tungsten halogen	22
High-pressure mercury	25–55
Tubular fluorescent	30–80 (depending on colour)
Light-emitting diode (LED)	40–60 (with ongoing development)
Mercury halide	60–80
High-pressure sodium	100

LEDs generate heat as current passes through the chip, but they produce very little heat in the light they emit. The heat emitted must be dissipated through proper thermal management design, the array of LEDs and the enclosure or fixture. LEDs currently generate about 40 lumens per watt for the newest high-brightness white, with near-term performance targets of 60 lumens per watt. LED life varies depending on the type and thermal effects. The current range is about 20,000 hours for white to 100,000 hours for green and amber. The aim is for LEDs to be in the range of 100 lumens per watt. Until this happens, the consensus is that LEDs will not compete much for general illumination. Trends among manufacturers are expected to be to develop efficiencies and reduce costs. A related technology, organic light-emitting diodes (OLEDs), could reach the office market first.

High-intensity discharge (HID) sources operate in the 60–100 lumens per watt range with a variety of applications, colours and wattages. HID and LEDs overlap little, as HID sources tend to come in large lumen packages. New ceramic materials with metal halide sources at lower wattages have increased applications for HID display, accent and task-illumination applications. The future could be with ceramic metal halide.

Standards of lighting

The amount of light (standard of illuminance) needed for a given location or activity depends on many variables, such as general comfort considerations and the visual efficiency called for. The unit of illuminance is the lux (lx), which equals one lumen per square metre. The unit of foot candle is not used; this unit referred to the number of lumens per square foot. To measure the degree of illuminance at a specific work-place, a reliable measuring instrument is called for, such as a pocket light meter, which incorporates a photoelectric cell.

HSE Guidance Note *Lighting at work* (1997)[15] relates illuminance levels to the degree or extent of detail which needs to be seen in a specific task or situation. For example, for work requiring perception of detail in an office, the average illuminance should be 200 lx and the minimum measured illuminance should be 100 lx.

See Chapter 1, section on Lighting, for more on the statutory requirements and health and safety issues relating to lighting.

Lighting output

The lighting output of a given lamp will reduce gradually in the course of its life, but improvements can be obtained by regular cleaning and maintenance, not only of the lamp itself but also of the reflectors, diffusers and other parts of the luminaire. A good, economical lamp replacement policy could include, for example, changing batches of lamps, rather than dealing with them singly as they wear out.

---

[15] HS(G)38.

Lighting quality

While the quantity (or amount) of lighting assigned to a location or task in terms of standard service illuminance is a key feature of lighting design, it is equally necessary to take in the qualitative aspects of lighting. These have direct and indirect effects on the manner in which people view their work activities and any dangers that could be posed. Quality of lighting is affected by:

- **Glare**: this is the effect of light causing impaired vision or discomfort experienced when parts of the visual field are very bright, compared with the surroundings. It can be experienced in three different forms:
    - *disability glare*: caused by bright lamps directly in the line of vision, this is visually disabling
    - *discomfort glare*: caused by too much contrast of brightness between an object and its background, this is often associated with poor lighting design; discomfort often occurs without a person's ability to see detail necessarily being impaired
    - *reflected glare*: this is the bright light reflected off shiny or wet work surfaces, where detail may be completely hidden.
- **Distribution**: this relates to the way light is spread. A standard to classify light fittings has been drawn up called the British Zonal Method which classifies the luminaires according to the way they distribute light, from BZ1 (light passes downwards in a narrow column) to BZ10 (light spreads in all directions). It is important to note that a fitting with a low BZ number does not mean that it offers less glare. The shape of the room, reflective surfaces and position are all equally relevant.
- **Brightness**: also known as luminosity, this is very subjective and difficult to measure. Nevertheless, a good brightness ratio (the ratio of apparent luminosity between a task object and the surroundings) can be computed by making sure that the reflectance of all surfaces in the working area is well maintained. Interior design should take account of reflectance values. If the recommended illuminance level for a particular task (or task illuminance factor) is 1, the effective reflectance values ought to be 0.6 for ceilings, 0.3–0.8 for walls and 0.2–0.3 for floors.
- **Diffusion**: this is the projection of light in all directions with no particular priority. Diffused lighting is popular, and reduces the amount of glare emanating from bare luminaires. Drawbacks stem from the density of shadows caused, which can affect safety standards or even reduce lighting efficiency.
- **Colour rendition**: this refers to an object's appearance under a particular light source, compared with its colour under natural light (a reference illuminant, for example). The colour-rendering properties of luminaires should not clash with those of natural light. They should also be as effective at night, when there is no natural light contributing to the working area's total illumination.

Fluorescent light strips are the cheapest and most common form of lighting in workplaces, but can cause glare and (often imperceptible) flickering. Indirect light sources such as uplighters (using more powerful halogen or sodium bulbs) provide a warmer, flicker-free light and are popular with building users. Unfortunately, the flat surfaces (on suspended ceilings, for example) required to reflect light to desk level may not be easy to combine with good acoustic absorption.

Computer-controlled lighting systems

An initiative to provide funding to UK SMEs and individuals to adopt energy-saving measures early this century resulted in a number of consultancies and suppliers recommending the application of local automatic switching devices, such as stand-alone presence detectors, as a means of reducing lighting energy costs within the working environment. UK Building Regulations have also recognised the importance of local control in achieving energy savings. It has been stipulated that no switch should be more than 8 m from the light fittings it controls.

Lighting specialists, however, believe that a more holistic approach, involving the centralised control of all lighting within the working environment, provides a much more powerful solution. Such schemes allow total control of each open-plan area, both from a central control point and locally by staff themselves. They offer a number of ways to achieve high energy savings while simultaneously offering flexible control of lighting schemes to reduce other building operational costs. Automatic control interfaces with the BMS can provide energy-saving benefits which go far beyond lighting alone.

Even a simple local management system combining a manual switch-on facility with a 'staff absent' switch-off feature can effectively double energy efficiency. Getting staff to switch on lighting themselves ensures that lighting is only turned on when staff actually need it, and it can then be switched off automatically once staff have left their work area. With a computer-controlled system, the switch-on operation can be achieved in a number of ways, for example, by personal infrared transmitter, desk telephone or an icon on each individual PC screen.

Studies have shown that where staff are able to dim personal lighting, they typically set their lights at a comfort level below traditional guidelines. They thereby accrue additional energy savings while working in their own optimum lighting conditions.

Lighting management systems

Lighting management systems (LMSs) provide flexibility through the independent monitoring and control of every luminaire in a building. The key to effectiveness here is the flexibility of the hardware and graphics software which monitors the installation on screen. The system utilises virtual wiring, where light fittings are 'linked' to switches on other local control devices through software, and where essential and non-essential lighting 'distributions' exist in software only. In a virtual wiring system, changes to light-switching arrangements associated with staff or office layout changes are carried out from the PC.

Such systems combine digital dimming technology with LonWorks interoperability. LonWorks provides a platform for implementing distributed, open, multivendor, interoperable control network systems enabling various systems from different manufacturers to communicate and interoperate. Using LonWorks, the integration of a building's lighting control system within the BMS can be achieved. This ensures that facilities managers are able to control energy-saving routines which are triggered by factors such as occupancy, external temperatures, solar-heating effects and light levels.

Staff entering offices can be detected by LMS presence detectors, which broadcast an 'occupation message' over the LonWorks control network. Messages received by the BMS then boost the speed of the heating/cooling unit in the occupied area. Staff who want their lighting on can set the desired dimming level from their desktop.

Motorised window blinds can be centrally controlled (for reducing solar gain), with local control in occupied areas. Once areas are vacated by staff, lighting will be automatically switched off or dimmed to a reduced level and local environmental controls switched to a non-occupancy mode.

LonWorks and LonTalk protocols can be used throughout systems for transmission of data on both vertical and horizontal buses. Systems claim to be able to receive and transmit data to and from other LonWorks systems from any point on the lighting management network.

Emergency lighting

Emergency lighting provides a lower than normal level of light and illuminates exit signs to assist in the evacuation of a building. It also provides light on safety equipment, such as fire extinguishers, call points and hose reels.

Emergency lighting luminaires

Emergency lighting luminaires fall into three main types:

- **Self-contained**: the luminaire contains the battery, a control circuit for charging the battery and an inverter for operating the lamp in case of supply failure.
- **Conversions**: almost any type of mains luminaire can be converted for emergency use by the addition of either a self-contained conversion kit (conversion module plus battery) or a slave inverter conversion module.
- **Slave**: the luminaire is powered from a remote central battery cabinet and is interconnected via fire-resistant cable, which should ensure that, if fire affects the wiring, the salve luminaire would continue to operate. The emergency supply can be d.c. or a.c.

Categories of emergency lighting

The different categories or modes of emergency lighting can be described as follows:

- **Non-maintained**: the luminaire only operates when the electrical mains supply fails or there is local circuit failure.

- **Maintained**: the luminaire operates when the supply fails but also operates in a mains-healthy condition using a separate switched mains supply.
- **Sustained (combined)**: the luminaire contains two or more lamps, one being for normal mains operation and the other only for emergency use.

Standards

An emergency lighting scheme must comply with the requirements of the statutory authority and the appropriate standard. The duration of an emergency lighting system is a minimum of one hour. However, most applications in the UK stipulate three hours. In the UK, the latest addition has been the publication of BS 5266 Part 7:1999 (hereafter known as Part 7) which is the UK's adoption of the European Standard EN 1838:1999. Part 7 forms the standard for emergency lighting applications and replaces the Code of Practice BS 5266 Part 1:1988. The important changes in the requirements are noted later.[16]

Emergency lighting systems within buildings are required to be verified to the new requirements. The best way to ensure compliance is to seek a product approved by the Industry Committee for Emergency Lighting (ICEL), part of the Lighting Industry Federation (LIF). For visually impaired people, the general advice is not to increase the average maintained illuminance in a space above the recommended levels laid down in the Society for Light and Lighting (SLL) Code for Lighting 2004. Instead the advice is to offer localised or task lighting to assist those finding it useful. SLL suggests that the various forms of electrically powered way guidance systems are much more acceptable to visually impaired people.

Illuminating escape routes

Part 7 states that a minimum level of 1 lx is required on the centre line of the emergency escape route (formerly 0.2 lx). However, a deviation is listed, which is applicable in the UK, for a minimum level of 0.2 lx but is only for 'routes which are permanently unobstructed'. In practice such a criterion would be difficult to guarantee. Points of emphasis, whether on an escape route or in an open area, should be provided with the appropriate illuminance. These points include a change of level, stairs, change of direction, firefighting equipment, call points and first aid posts. The ratio between the minimum and maximum level should be no greater than 40:1. In addition, a clause has been added for the limiting of disability glare within the field of view.

Part 7 states that the minimum level of illuminance at floor level should be 0.5 lx. Originally it had been an average of 1 lx. The effect of this may result in the requirement for additional luminaires. Similar to the escape routes, the ratio between the minimum and maximum level should be no greater than 40:1.

All these minimum levels must be obtained, even at the end of discharges from aged components, while ignoring the effects of reflections from surfaces such as walls. So with a new installation, in practice, the light output measured should be two or three

---

[16] British Standards BS 5266 Part 7:1999 and EN 1838:1999, see www.bsi-global.com, www.icel.co.uk, and www.tso.co.uk.

times the minimum to ensure that the standard is met throughout the full life of the installation.

Illuminating safety signs

Part 7 specifies that the safety signs used to indicate emergency escape routes and first aid areas have to be illuminated and be capable of being seen in all relevant viewing directions. In addition, viewing distances are specified for internally and externally illuminated signs, based on the height of the pictogram on the signs. The European Safety Signs Directorate, enforced in the UK by the Health and Safety (Safety Signs and Signals) Regulations 1996, stipulates a common European pictogram for means of escape from buildings. All worded-only 'exit' signs should have been replaced by December 1998.

Maintenance

Taking care to install reliable emergency lighting is commendable, but the lighting needs to be regularly inspected and maintained. Facilities managers should not fall into the trap of leaving it until there is a fire inspection, as by then it could be too late. Batteries fail, chargers can stop working and lamps can be faulty. For users, that means an expensive refurbishment and premises closure while remedial work is undertaken.

Testing equipment

Manual testing today is viewed as labour intensive and expensive. Self-testing systems complying with relevant standards and European directives are available from reputable manufacturers and, despite the higher capital cost, they are efficient and reliable. A key advantage for users is the ability to produce records to the fire authorities which demonstrate that the lighting has been tested and proven to be functional.

Testing of emergency lighting comprises two areas:

- automatic self-testing emergency lights suitable for smaller installations
- centrally controlled test and report systems for large-scale installations.

Timers within standalone self-testing systems can carry out tests at weekly, monthly, six-monthly and annual intervals with system failure indicated either visually or audibly. For buildings with more complex emergency lighting, testing units exist which can provide detailed information on fault detection. On these, the exact location of faulty luminaires will be shown and tests can be carried out at times to suit the user.

INFORMATION

General

British Institute of Facilities Management (BIFM), Tel.: 0845 058 1356, www.bifm.org.uk
Building Research Establishment (BRE), Tel.: 01923 664000, www.bre.co.uk

Building Services Research and Information Association (BSRIA), Tel.: 01344 426511, www.bsria.co.uk

Carbon Trust: www.thecarbontrust.co.uk

Carbon Trust, energy helpline, Tel.: 0800 085 2005, www.thecarbontrust.co.uk/energy

Centre for Facilities Management (CFM), Tel.: 0161 295 5357, www.cfm.salford.ac.uk

Chartered Institute of Building (CIOB), Tel.: 01344 630 700, www.ciob.org.uk

Chartered Institute of Building Services Engineers (CIBSE), Tel.: 020 8675 5211, www.cibse.org

Facilities Management Association (FMA), Tel.: 020 8897 8521, www.fmassociation.org

HSE, Tel.: 0870 154 5500, www.hse.gov.uk

IPD Occupiers Property Databank, Tel.: 020 7336 9200, www.ipdglobal.com

Royal Institution of Chartered Surveyors, FM Faculty, Tel.: 020 7222 7000, www.rics.org/fm

The Stationery Office, www.tso.co.uk

Gas and electricity

BACnet, www.bacnet.org

buildingSMART, Tel.: 020 8660 1631, www.iaiorg.uk

CHP Association, www.chpa.co.uk

CHP Club website, www.chpclub.com

European Cogeneration Directive,
www.europa.eu.int/eur-lex/pri/en/oj/dat/2004/l_052/l_05220040221en00500060.pdf

The EIB bus, www.eiba.com

The Energy Systems Trade Association's Building Controls Group, www.esta.org.uk/bcg

LON Mark Interoperability Association, www.lonmark.org

Ofgem, Tel.: 020 7901 7000, www.ofgem.gov.uk

Water management

Parliamentary information relating to water management, www.defra.gov.uk/environment/water/index.htm

Ofwat, Tel.: 0121 625 1300, www.ofwat.gov.uk

Water pollution

Environment Agency general enquiry line (for information on groundwater protection), Tel.: 08459 333111

HVAC

Heating, Ventilation and Air Conditioning Manufacturers Association (HEVAC), www.hevac.com

H&V Contractors' Association (HVCA), Tel.: 020 7727 9268 www.hvca.org.uk

Waste management

The Chartered Institution of Wastes Management, Tel.: 01604 620426, www.iwm.co.uk

The Environment Council, Tel.: 020 7836 2626, www.wasteguide.org.uk

Noise

Association of Noise Consultants, Tel.: 01763 852958, www.association-of-noise-consultants.co.uk

Royal National Institute for the Deaf (RNID), Tel.: 0808 8080123, www.rnid.org.uk

Furniture

Chartered Institute of Purchasing and Supply (CIPS), Tel.: 01780 756777, www.cips.org

Office Furniture Advisory Service (OFAS), Tel.: 01344 779438, www.ofas.org.uk

Catering

Automatic Vending Association of Britain (AVAB), Tel.: 020 8661 1112, www.ava-vending.org

British Hospitality Association (BHA), Tel.: 020 7404 7744, www.bha-online.org.uk

Cleaning

Cleaning and Support Services Association (CSSA), Tel.: 020 7481 0881, www.cleaningassoc.org

National Minimum Wage Helpline: 0845 6006 678.

Lighting

Energy Savings Trust, see www.thecarbontrust.co.uk/energy

Industry Committee for Emergency Lighting (ICEL), www.icel.co.uk

Lighting Industry Federation (LIF), www.lif.co.uk

Society of Light and Lighting www.lightandlighting.com

Space Design and Management

11

Louis Wustemann and Frank Booty

While most of a facilities manager's functions are carried out behind the scenes, space planning and management is the most conspicuous part of the job. The way that an organisation's workspace is laid out and maintained is a vital part of its corporate image, giving an important message to visiting customers, potential recruits and existing employees.

More fundamentally, a comfortable, safe and efficient working environment is essential to enable employees to perform at their best, and to ensure that good use is made of the space available so that facilities managers bring added value to an organisation.

An essential consideration for space planners is the effect the layout and work-settings have on communications and productivity. The chance interaction between staff that generates valuable ideas cannot be made to happen, but research shows that poorly thought-out layouts and the wrong adjacencies make it much less likely.

How can facilities managers plan the workspace to serve its users safely and comfortably and to ensure maximum productivity? This chapter gives advice on successful space planning and the effective management of space changes and office moves ('churn'), setting out both practical design considerations as well as the technological support available for planners in the form of computer-aided facilities management (CAFM) systems.

SPACE PLANNING

Measuring space

Before beginning to plan or replan people into any given space, you first need to know the size of that space. Area is usually estimated in one of the following ways:

- **Gross internal area**: the whole internal area of the building measured from wall to wall.
- **Net internal area**: the gross internal area minus service cores, toilets, lift lobbies and stairways.

347

- **Net usable area**: the net internal area minus any areas that cannot be used for the purposes of space planning, such as areas behind doors and narrow gaps between columns and walls.
- **Net lettable area**: the area on which an organisation pays rent in a leased building, usually somewhere between the net internal and the net usable area.

Planning grids

The planning grid is a means of imposing a notional structure to help plan the occupants into the floor space. The most common grid sizes are based on modules of:

- 90 × 90 cm
- 120 × 120 cm
- 135 × 135 cm
- 150 × 150 cm.

The 90 cm module (creating blocks of 0.8 m^2) was the common standard in the office buildings of the 1960s and 1970s, but grids based on 150 cm (2.25 m^2) sections have become more common, as the larger grid square allows more flexibility of planning and is more compatible with common distribution patterns of cabling and other services.

Bricks and mortar

Much of the building infrastructure is likely to be handed to the space planner as a fait accompli in many cases. Nevertheless, an understanding of the properties of the fixed elements is useful to make the best use of the space.

While small offices built before the 1920s are likely to have upper stories floored with the boards over timber joists, most space planners will be faced with a concrete floor slab, probably with a raised floor through which cabling is run.

When replacing a raised floor or commissioning one from scratch it is worth considering whether the services can be fitted under a low-profile floor that sits as little as 10 cm above the structural floor, rather than the traditional systems that may be up to 50 cm above it. The extra height gained will help to reduce claustrophobia caused by low floor to ceiling depth and create more airy offices.

Suspended ceilings feature in most mid to late-twentieth century offices. Hung from the bottom of the floorplate above, they provide a simple distribution system for services such as ventilation, heating and lighting. Modular systems are easiest to match up to planning grids and partitioning systems and best support reconfiguration of the space below. Carefully specified ceiling materials and coatings can also improve the acoustics in any space.

Partitioning

Interior partitioning in offices ranges from blockwork, through plasterboard and stud walls, to aluminium and monobloc partitions. Which is used depends on various factors, including how long the interior configuration is expected to last. Changing the first two means messy small-scale demolition, while the second two are often demountable and offer the opportunity to reconfigure the workplace.

Factors to consider when choosing a partitioning system include:

- compatibility with other structural elements (suspended ceilings, raised floors, etc.)
- fire resistance
- acoustic properties
- load-bearing capacity, if you are likely to want to use it to hang shelving
- ease of relocation: some systems are more demountable than others.

Cabling

No building over 15 years old was built to cope with the volume of cabling now needed to handle information and communications demands at desk level.

In small spaces, cabling is handled adequately by perimeter trunking with wall outlets, but on most larger floorplates it is run through a raised floor. In cases where such floors cannot be used, a suspended ceiling is often used for distribution with wiring run down structural pillars or specially installed power poles.

Structured cabling offers a unified alternative to the mess of different wiring types that grew up to serve telephone and data systems. It provides a single system that can be configured to offer telephone, ethernet, multimedia and even closed circuit television (CCTV) connectivity all at once. The floor void is usually flooded with wiring and outlets provided at every 2–3 m throughout the floor grid. This means that wherever the workstations are located or relocated they can simply be plugged into the nearest outlet.

Structured cabling systems comprise two basic components:

- **Backbone cabling**: this begins where transmissions enter the building from outside and are routed to a series of telecom rooms serving an individual floor or discrete area of the building.
- **Horizontal cabling**: this is pulled from each room to the individual workstations or outlets on that floor.

The choice of cable – fibre optic, copper or wireless – is crucial and will have an impact on the design, installation and performance of the system.

Structured cabling must be considered as early as possible in the process of planning changes to a building's systems or planning a new building. Combining structured cable planning with heating, ventilation, air conditioning, plumbing and electrical systems can prevent problems of cables interfering with ductwork, electrical wires, or ceiling and flooring grids. When cabling projects are completed, all details of cable routes, types, connections and associated hardware and software, along with documentation of standards and testing procedures, must at least be filed with easy access, and at best incorporated into the company's CAFM system (see later this chapter for more on CAFM systems).

When user requirements alter, the change is enabled by changing the patching configuration in the equipment room without having to touch the horizontal cabling to the floor outlets. The most common means of connecting equipment to subfloor cabling is via one of the following:

- **Floor boxes**: these are recesses in the raised floor with standard phone sockets for voice and data plus power sockets, usually covered by hinged lids, which are flush to the floor when closed. These provide easy access for plugging and unplugging equipment, but need effort to relocate when layouts change.
- **Grommets**: these are essentially apertures through which cables are passed to connections in the floor void. They look neater and can be easily relocated, but once in place make it more difficult to access the plug-in points below.

Cable management

If a cabling infrastructure is what you have behind the scenes, cable management is the art of dealing with cabling between outlets and equipment. Bad cable management leads to unsightly runs of wiring across the office and, at worst, trip hazards. The simple rule is that the more outlets you have, the closer any workstation will be to one and the less cable you have to manage.

One way to run cables into open space where no local outlets are available is to use screen-based furniture where the cables are run through the screens that hold up adjacent workstations. This makes reconfiguring the layouts very time consuming.

Where cabling has to be run in open space at floor level for short runs, cable bridges that cover and protect the wires forming a shallow bump on the floor are an option, but these should never be run across circulation space. (See Chapter 9 for more on cabling.)

Space per person

UK health and safety law specifies 11 m^3 minimum workspace per person. This suggests a minimum space allocation of 4.2 m^2, while architects and planners have traditionally used an average of 10 m^2 per person when calculating maximum occupancy of the net internal area. This figure attributes a proportion of all support and ancillary space (such as circulation and catering areas) to the individual, not just their work area. The British Council for Offices suggests a good practice range of $12\text{--}17 \text{ m}^2$.

These averages give you a rule-of-thumb idea of how many people you can fit into a given net internal area. How you actually fit them into the net usable space depends on the shape of the space and your organisation's space standards. A space standard is the number of square metres allotted to each employee in the organisation as workspace, taking in their desk, chair, local storage and immediate access to the workstation. These can be used as a minimum allocation per person, an ideal to aim for where space permits or a rigid model for planning large numbers of people into a given area. Space standards vary enormously between organisations, as do the principles for determining them. The most common strategies are:

- **Single standard**: also known as a universal footprint, where every member of staff receives the same space allocation and/or furniture. This makes for very simple space planning and churn management, but only the most deliberately non-hierarchical organisations carry the principle through to the most senior management.
- **Space by seniority**: the most common form of space planning since the mid-twentieth century. The size, and often type, of the individual's space is dictated by their rank. For example, in an organisation where non-managerial staff receive 5 m^2 in open space, supervisors might be allocated 7 m^2 also in open plan, while managers would have a 10 m^2 cellular office.
- **Space by need**: space is allocated based on an assessment of each individual's job needs. Those who can prove they need a lot of personal storage are allotted more room and those whose work is highly sensitive or confidential are placed in cellular space rather than open plan.
- **Space budgeting**: space is allocated to departments or teams based on a total of their individual notional space, but they are then free to arrange this space as suits them. A team of 10 people with a combined entitlement of 60 m^2 could use small workstations and limited personal space to free up more of their team area for shared meeting and breakout space.

Simple standards

For simplicity of planning and managing churn it is best to stick to the smallest set of space standards possible and to make them multiples of each other. For instance, in a status-based system, if the basic standard for staff in open plan is 4 m^2, junior managers and those with extra space needs might be allocated 6 m^2 and middle managers 8 m^2 (in partitioned cells where necessary).

Except in the smallest footprints, the space standard does not dictate the orientation or the arrangement of the workstation. Combining the space allocation of a team it is possible to produce regimented bays of desks that can be planned straight across an open floor, variations on a single theme or, especially using the newer workstation shapes (such as 120-degree radial desks) seemingly random groupings, all keeping strictly to the relevant space standard.

Workplace options

Occupants of open-plan space frequently complain about lack of privacy and environmental control, but a return to highly cellular accommodation is unlikely to come soon. The cost of churn and the new emphasis on knowledge sharing, face to face as well as electronically, means that segregation is not yet on the cards again.

Apart from support and ancillary areas (computer rooms, catering areas, etc.) the two most common types of cellular space found in offices are 'owned' offices for one or two people and meeting rooms.

Where individual contained office space is still an entitlement (because of rank or the need for a high degree of privacy or security), such offices are usually planned in blocks, making economic use of shared partition walls, easing the provision of power and voice connections and making planning of the surrounding open-plan areas less complicated.

Open plan

The principle of offering people different types of workspace in addition to their standard workstations, to suit different types of work, has spread in recent years. (See Space-saving tips, later in this chapter.) These facilities include:

- **Breakout areas**: providing breakout areas with soft seating and coffee tables, or even formal meeting furniture, beside intercirculation in open layouts or beside vending or catering points both creates these nodes and increases local meeting space. If they are located near to workstations, breakout areas should be at least partially screened to avoid distracting those nearby (see below, Quiet areas, for more on this).
- **Touchdown space**: this started as one of the types of workspace offered in hotdesking offices, but the recognition that all office users need ad hoc work surfaces when they are away from their desks has led to their adoption in more conventional offices. The cheapest touchdown space is simply any spare desk placed in the corner of an office or reception area. The most space-efficient option is a counter fixed to a wall to cater for short stops by employees who need a surface to lean on or spread documents briefly. Wall-mounted touchdown surfaces, which need be no deeper than 450–600 mm in most cases, are best placed at the entrances to office floors to maximise their use by staff and visitors. Touchdown surfaces should not have sharp edges or corners. No seating is needed, although bar stools are provided by some organisations. These must be easy to move in and out and adjustable. Power and data connections will allow people to plug in laptops.
- **Quiet areas**: quiet areas or meeting cells should ideally be placed in dead areas of the floorplate (against windowless walls, for example, as they are not designed for long-term use). Where areas of open desking are designated as quiet areas, it seems common sense to put them as far away from the noisier breakout areas as possible. In practice, as long as the two are not next door to each other and there are enough acoustically absorbent surfaces in the work-space, they can be in reasonable proximity (say 10–15 m apart) without significant sound leakage. Paradoxically, the lower the screening around quiet areas, the less people will make noise around them because they can see others concentrating.
- **Carrels**: also known as study booths, carrels are hotdesks with some form of screening to provide enclosure for concentrated work in open space. They vary from rows of small 800×800 mm library-style booths to free-standing desks surrounded by 1800 mm free-standing screens on three sides, and are a flexible alternative to permanent enclosures.

Meeting rooms

Dedicated meeting rooms are often an underused part of the office space, simply because they are overproportioned. Research by Alexi Marmot Associates found that two-thirds of meetings in a typical organisation involve six people or fewer, and almost half involve four people or fewer. This suggests that organisations need very few large meeting rooms. That said, small-scale cellular space, for two to six people, is likely to be well used, especially in organisations where most staff work in open-plan space and may need enclosed space for interviews, important phone calls and short periods of work requiring heavy concentration. The late 1990s vogue for tiny one-person cells as small as 2 m^2 has now passed after experience showed they were little used, presumably because they induced claustrophobia.

For an administrative department, one small meeting space in the work area for two or three people might be sufficient for every 40 staff. In a consultancy this would rise to one for every 16 staff. Meeting rooms should always be accessible by main circulation routes and, where they are expected to be heavily used by visitors, should be located near lavatories and coat storage if possible.

Using structural glass for doors or walls of cells creates a less blocky appearance and gives passers-by an immediate indication of whether or not the room is occupied. But if more than one wall is fully glazed, etched glass is recommended for some of the expanse to avoid occupants feeling as if they are in a fishbowl.

Circulation

Circulation includes stairs, corridors and routes through open floor space. The first two are usually fixed, but the third needs careful planning. The primary consideration is the legal requirement for 'escape distance' (see Fire safety: Means of escape, in Chapter 1, for details of maximum escape routes). Beyond the necessity of quick access to fire escapes, it is necessary to consider whether each route will simply be used for getting to individual workspace or to encourage ad hoc communication and how much traffic they may need to cope with in peak periods.

Circulation can be divided into two types:

- **primary circulation**, including stairs, lift lobbies and corridors
- **secondary circulation**, providing access within the open floors.

The latter can be subdivided into two types: intercirculation, which takes people from the primary routes through open floors to team areas, vending or storage points; and intracirculation, which gives immediate access to each workstation off the intercirculation and is often accounted for partly in the individual space standard.

The recommended width for intercirculation is 1.5 m, which allows three people, or one person and another in a wheelchair, to pass comfortably. The minimum clearance for one person without turning sideways is around 55 cm. Where lifts open straight into offices the recommended minimum circulation space in front of the doors is 3 m.

Space planners should bear in mind the requirements of the Disability Discrimination Act (DDA) when devising layouts (see Chapter 2, section on Disability discrimination) as this may help to avoid the need for later and more costly adjustments to accommodate employees with disabilities.

Intercirculation is usually planned through the middle of the workspace, so that access to workstations is equal on both sides. In cases where a shallow floorplate (under 10 m) is wrapped round a large service core, circulation can follow the outside of the core.

With the vogue for less regimented layouts, routes that wind through offset groups of workstations have become more common. These give the office a more pleasant appearance but the designer has to make sure they do not take the travel distance to fire escapes over the legal limit. Where possible, workgroups or teams should not be divided by circulation routes.

Adjacencies

Co-locating team members and staff in the same department is an obvious priority, but co-locating different teams and departments is a more subtle business. Unless you are sure that existing arrangements make for maximum interactivity between parts of the organisation whose work is interdependent, this should form part of your research before planning new space (see section elsewhere in this chapter, Relocation and churn).

Research by BT found that employees interact with those in the same work area as themselves for 10 per cent of their working time, but this drops to 0.4 per cent of their time for employees on different floors. Nevertheless, two departments which need constant interaction may be better served if they are placed on the same end of separate floors near a staircase than if they are at remote ends of the same floorplate.

DETAILED PLANNING

Once you have set your space standards and taken into account the size of the space you have to plan and the desired adjacencies, in the detailed planning your decisions will be affected by a mix of the following factors:

- open-plan versus cellular space
- team versus individual space
- incorporating room for expansion
- flexibility and ease of churn
- aesthetics and staff comfort
- the planning grid.

Group planning

When laying out multiples of workstations with similar space standards in open-plan space, it is time consuming and inconvenient to work in single units. Most planners group workstations together in clusters of four or six or more, combining the space standards of the workstations into a larger unit, preferably fitted to the planning grid. These units, often arranged as bays to create some sense of team identity in the open space, can then be replicated across the floor, simplifying the planning. There is no need for this form of cluster planning to produce a monotonous and rigid layout. Once the size and access requirements of a cluster have been set, variations on the theme will break up the appearance of the layout without complicating the plan.

Avoid orienting workstations facing away from neighbouring circulation space. Occupants will find it hard to avoid being distracted by a natural defensive response to being approached from behind, a feeling of being 'crept up on'.

Space records

As the planning process progresses, space planners will need to assemble different types of plan to record information on the proposed layouts, either on paper or on a computer-aided design (CAD) system (see section later this chapter, Computer-aided facilities management: space usage). They include:

- **Block and stack plans**: block plans can be prepared early on to map the desired adjacencies on a single floor. Where more than one floor is involved, stack plans group all the floors together to show vertical adjacencies as well.
- **Sketches, three-dimensional renderings and mock-ups**: impressions of the planned décor and layouts give the opportunity to try out colour schemes and furniture arrangements and to bring flat plans to life for non-specialists.
- **Detailed space plans**: the final plans will be the blueprint for the new layouts, with all services, circulation routes, storage and workstations marked with details of which teams are based where. These must be updated as soon as and as often as there is any churn.

Office aesthetics

Beyond the essential environmental hygiene factors, space planners need to account for other more subtle environmental issues.

Coloured wall and furniture finishes have become popular. Before deciding on a colour scheme for your workspace, consider the impression and atmosphere you want to create. Reds, oranges and yellows encourage activity and energy but can also cause fatigue, while blues and greens promote contemplation, concentration and creativity. Light colours produce the impression of coolness and spaciousness, while dark colours create a closed-in feeling.

Even in an interior where walls must stay monotone or pale, say to maintain required light levels, bright accent colours can be introduced via chair coverings and storage cabinets, now available from most suppliers in a bewildering choice of hues.

Other ways of softening an office environment are by hanging artworks or indoor planting. Plants are high maintenance, particularly in deep office space where natural light levels are low, but are also highly valued by office users. Studies also suggest that the proximity of greenery helps to reduce stress levels and improve air quality.

Space-saving tips

Every facilities manager wishes to reduce property costs. Making better use of existing space is the easiest way to put off property expansion. One of the most obvious means of space saving is by moving staff who spend a high proportion of their time away from the office to some form of hotdesking or desk-sharing arrangement. Another is to take advantage of the switch from cathode ray tube (CRT) monitors to flat screen displays to reduce space standards. Switching to flat panel displays can allow you to cut the depth of a standard workstation by 300 mm with no detriment to the user, which adds up to a substantial space saving on a large office floor. Other ways of reclaiming space include rationalising storage and cutting the number of meeting rooms.

Rationalising storage

Much paper filing is unnecessary or kept in the wrong place. Documents accumulate in piles around workstations through inattention, suggesting the need for more storage cabinets, which are then duly provided. Migrating filing from floor-based units to shelving above desk level on load-bearing walls and partitions where possible will release space.

An increasing number of organisations are now setting a fixed filing allocation (commonly one or two linear metres) for all members of staff except those who can make a strong case for more.

Another option, especially useful for organisations which need to keep a lot of paper records, for regulatory reasons for instance, is off-site storage. All non-essential records can be archived at cheap, secure warehousing space, freeing up as much as 25 per cent of the prime square metrage on office floors.

The ultimate in space-free records storage is through the use of document image processing (DIP) systems, which allow all incoming documentation to be scanned and stored electronically. DIP systems may one day be the norm, but they are currently expensive to install and suffer from the popular lack of confidence in information technology (IT) systems as the sole repository of vital data.

Fewer meeting rooms

Meeting space in most organisations is usually oversized (see earlier section in this chapter, Meeting rooms) and underused. Dedicated meeting rooms can be downsized. There are usually plenty of underused spaces that can be recycled as meeting space. Company restaurants are the obvious example. Chairs and tables in catering areas remain unoccupied for most of the working day and the simple provision of a couple of mobile whiteboards to be wheeled in outside mealtimes will allow a cafeteria to double as non-bookable meeting space. Corners of reception spaces and lobbies are other candidates for touchdown tables.

Third party planners

For those who feel they lack the expertise or time to carry out space planning, consultants and facilities management contractors offer external assistance with anything from individual moves to the whole property management operation.

Independent specialists or the planning services offered by larger architects, surveyors and even removal companies, will almost certainly bring experience and ideas for space efficiency that most facilities managers would not have access to otherwise. More specifically, they may be able to provide benchmarking data to help managers to compare standards with other organisations.

When choosing a contractor, look for someone who understands your organisation, can provide references from other satisfied clients, will keep your plans secure and will commit themselves to a detailed service level agreement (for more detailed information on procuring and managing contractors, see Chapter 7).

RELOCATION AND CHURN

Churn is defined as the number of people who relocate within a workspace in a year, divided by the total number of occupants and multiplied by 100 to give a percentage figure. If 20 per cent of your staff move once, you have a churn rate of 20 per cent for that year, but if 10 per cent of the staff move twice your churn rate is still 20 per cent.

Benefits and costs

In fact, 20 per cent would be an enviable rate for many facilities managers, who have to cope with levels well in excess of 50 per cent, giving the impression of perpetual motion in the workspace. But the rate of churn, which seems to be rising in most industries, is not necessarily a cause for concern. As more organisations adopt project management principles, high churn that reflects the grouping and regrouping of staff to deal with time-limited tasks and projects is a sign of dynamism and a healthy response to constant change in the wider world.

What facilities managers should concentrate on is making churn less disruptive and, above all, less costly. Indicative churn costs for a simple 'briefcase' move with no disturbance to the layout are around £150, rising to over £500 where furniture and storage are shifted, and jumping to £1600 where any alterations (such as moving partitions or reconfiguring wiring) are needed. Cutting the cost of churn is a matter of building as much flexibility as possible into space plans, building services and fixtures, in order to maximise the number of briefcase moves.

Cutting the cost of churn

Where managers have any influence over the fit-out of a space, the specification of some of the following features will add to the initial cost but generate savings in churn expense year after year:

- **Open plan**: the fewer fixed partitions there are in the space, the easier it is to move people round and the lower the risk of structural alterations when people do move. Where partitioning is used, take care to choose a system that is truly demountable.
- **Structured cabling**: the flood distribution of wiring through a floor (see earlier in this chapter, Cabling) and the 'plug and play' facility offered by structured cabling systems mean that information and communications technology (ICT) needs a minimum of reconfiguration.
- **User-configurable phones**: a telecoms set-up (available for all but the oldest systems) that allows users to relocate their extension number to any handset makes moves easy, especially short-term relocations of days or weeks. Voice over Internet protocol (VoIP) systems which route phone calls over IT networks offer the simplest plug and play telecoms.
- **Space standards**: the closer you can get to a single standard or universal footprint (see earlier section in this chapter, Space per person), with the same space allocation for most employees, the less often you will have to rearrange a layout or move furniture to accommodate individual or group moves.

All of the above will cut churn costs, but the single most effective way to manage down the price of moves is to plan ahead. Strategic facilities planning, carried out two or three years ahead in conjunction with property and human resource functions, will give you a much better idea of the moves that will definitely have to be scheduled and a chance to avoid the wasted effort that comes from moving the same employees twice in a matter of months.

Relocation

Moving from one set of premises to another is churn on a grand scale. Space planning for a work group due to relocate should start at the earliest stage while the new premises are being chosen. Space planners should work with property managers or agents to check that any proposed new location has the right type of space as well as simply the right square metrage.

Relocation often brings a chance for a more fundamental change to space use and/or working practices. This sort of change needs serious planning and sound project management. Space audits of existing offices will offer up metrics that should help facilities managers to decide whether changes to space standards or new patterns such as hotdesking are possible. You may have to work out adjacencies and basic layouts many months in advance of a move to allow for internal partitioning of the new space, so the earlier a rough idea of new arrangements is available the better.

As one of the main measures of success of a move will be the cooperation of the relocating staff, the months leading up to the move should be spent communicating the planned changes and carrying out trials of new furniture, layouts or workstyles in the old location to iron out any obvious problems.

At the move-in date, make sure there is support on hand for any features of the new space that are unfamiliar to its occupants as this will reduce the 'downtime' associated with the move.

Hotdesking

In a hotdesking office workspace is not parcelled out to individual employees but shared between a department or team. Employees sit wherever they can find convenient space. The arguments in its favour are compelling on paper. Although most organisations accept the costs of maintaining and servicing buildings that are only open around one-third of the available hours in the year (allowing for nights and weekends), occupancy rates can be reduced to as little as 10 per cent of the available time owing to the needs of many nominally office-based staff to leave the office on research, client or supplier visits, and with homeworking on the increase.

Making it work

But simply taking away 'owned' workstations and replacing them with fewer, shared desks is not likely to be greeted with any enthusiasm by those affected by the change; they simply lose the status of fixed workspace and gain a new set of responsibilities, including tidying up after themselves. Where hotdesking has succeeded it has been accompanied by a wider change to working practices that brings benefits to the hotdeskers themselves.

The most common quid pro quo for employees giving up their desks is to offer them more pleasant workspace for the times that they are in the office. Some of the space saved from removing 'owned' desks is commonly recycled for new types of worksettings (see earlier in this chapter, Workspace options), such as touchdown points, soft-seating areas and study booths, as well as standard desks. Even with these provisos, hotdesking schemes commonly generate space savings of 30 per cent or more.

How many hotdesks?

There is no rule of thumb for the correct ratio of desks to employees in hotdesking arrangements. Ratios in existing schemes vary from 3:4 (hotdesks: employees) to 1:10. The level of provision will depend on several factors, including:

- the proportion of an average week that staff spend in the office
- whether they are equipped to carry out desk work elsewhere (at home, for instance)
- whether it will be appropriate for employees to use empty desks in other departments if hotdesks are all full.

The best pointer to a correct ratio is an occupancy study, carried out over several days, probably spread over a fortnight or more, to gather data on how many of the existing desks are used by peripatetic staff on an average day and at peak usage.

The perennial question about hotdesking is how to cope with the eventuality of all the employees wanting to come in on the same day. Although this overload almost never occurs in practice, providing a range of space-efficient work-settings (touchdown counters, etc.; see earlier section in this chapter, Workspace options) can make it possible to cope with unexpected spikes in demand for workspace.

Efficient hotdesking

The aim of introducing hotdesking is to save space, but not at the expense of staff productivity. If employees have to spend much more time finding a desk, configuring their phones or connecting to the company network, any gains will be wiped out. The following points will help to ensure a smooth operation, although some may be unnecessary for small hotdesking schemes.

Layout

Place the worksettings designed for the shortest stops nearest the entrance to the hotdesking space, with desks for longer, concentrated periods of work nearer the back. By the same token, any 'owned' workstations for staff who are not hotdesking should be farthest from the entrance, away from the comings and goings of the hotdeskers.

Managing shared space

Although there should be adequate provision for employees dropping in and out without notice, hotdesks and meeting rooms should be bookable by telephone, in person and, ideally, over an intranet. Staff should be discouraged from booking the same space for weeks at a time and other territorial habits, such as leaving material at workstations. A clean desk policy is essential for success. A dedicated concierge or receptionist may be useful to take bookings, maintain facilities (replenish stationery, log IT faults, etc.) and, located near the entrance to the hotdesking area, can provide added security in a space with a transient population.

Communications and IT

Telephones must be configured so that hotdeskers can receive calls wherever they are sitting. Most private automatic branch exchanges (PABXs) already allow temporary reassignment of extension numbers. The more expensive addition of a 'follow-me' phone facility is the best solution. Access to sockets for voice, power and data must be easy for laptop users. Desktop power pods will avoid making visitors scrabble under workstations.

The ultimate easy access IT solution for hotdesking offices is to install a wireless local area network (WLAN) to allow employees to log on from laptops – most recent models are Wi-Fi enabled – anywhere in the workspace, even breakout areas. Busy areas must be equipped with sufficient wireless access points to avoid slow network access from too many users sharing bandwidth.

Desks earmarked for longer stays should be equipped with screens and keyboards that users can hook up to their laptops, since portable computers are unsuitable for prolonged use. (See also Chapter 9.)

Furniture and storage

Standard workstations, mobile furniture and centralised storage are all-important features of an adaptable workspace and are covered in more detail in Chapter 10, section on Furniture.

Although permanent storage at hotdesks is inappropriate, as built-in storage at desk level encourages people to territorialise them by leaving things in the drawers, employees must have some means of moving files between bulk storage and their workspace. Most furniture systems now include wheeled storage pedestals which can be pushed or pulled around.

Chairs may have multiple users every day and must be quickly and easily adjustable.

Stationery

Stationery supplies, which should be easily accessible and sited somewhere near the common areas, are best located near communal printers and photocopiers and pigeon-holes or postboxes; this is another feature that facilities managers need to allow for where employees do not have post delivered to a desk.

COMPUTER-AIDED FACILITIES MANAGEMENT

The rise of the networked personal computer has coincided with the establishment of facilities management as a recognised profession and has led to a rush by software companies to provide tools designed to make paper plans redundant and to help managers keep track of an organisation's assets and model changes with ease. Computer-aided facilities management (CAFM) is sometimes known as computer integrated facilities management (CIFM). The future facilities scenario may also contain the terminologies of facilities management cybrarians, facilities management cyberspace and total infrastructure and facilities management (TIFM). CAFM is the starting point.

The programmes available range from the simplest space-planning software, available free from many office furniture suppliers, which can be used for planning small spaces, to full-blown CAFM systems that support any or all aspects of the facilities manager's job. For the purposes of this chapter, computer-aided systems have been divided into five basic types. However, any one package may contain some of the functions and tools listed under each category.

Infrastructure: buildings, space and asset management

Property-based CAFM holds the data needed to ensure that property and its contents are properly catalogued and monitored. Property databases hold details of all topics associated with construction, ownership, lease and occupancy, and can aid routine tasks by, for example, flagging up lease renewal and refurbishment dates or calculating multiple occupancy costs so that they can be recharged to groups of occupants.

Space usage

Space usage is a prime function of this type of CAFM system. Object-oriented databases link to CAD space-plan drawings to provide a complete picture of space usage down to workstation level. Detailed information, both textual and graphic, can be produced on current space usage, and space needs can be planned for. Churn decisions regarding structural alterations and moving equipment and people can be made online, resulting in work orders being issued and CAD drawings being updated automatically.

Networks and cabling

CAFM systems allow you to manage telecoms and cable infrastructure effectively. Network connectivity can be tracked, an electronic inventory created of the physical cabling and telecom network connections, and network plans drawn up.

Health, safety and environmental management

Health and safety CAFM

Health and safety CAFM helps the facilities manager to ensure a safe working environment and demonstrate compliance with legislation and best practice. Systems come ready loaded with health and safety legislation guidance on topics such as the European 'six pack', the Reporting of Injuries, Diseases and Dangerous Occurrences Regulations (RIDDOR) and the Control of Substances Hazardous to Health Regulations 1999. These can link to the asset register or a human resource database, and suggest and diary appropriate actions such as checking first aid provision or organising manual handling training. Hazard registers can be compiled and maintained on topics such as asbestos, actions scheduled and worksheets created. Relevant health and safety data can be appended to work orders, and the resulting health and safety-related actions recorded. Many health and safety CAFM systems include display screen equipment (DSE) workstation self-assessment software.

Environmental CAFM

Environmental CAFM allows you to monitor, compare and manage the consumption of energy, and to monitor emissions and flag up hazard levels by integrating all energy-related data from existing building control systems and meters across sites and existing energy-related databases, plus all entered data. Energy usage is recorded and compared with pre-entered industry norms. The facilities manager can then identify areas for improvement, and once energy-saving measures are in place, continue monitoring. Reports can then be produced to demonstrate energy saving.

Pollution levels can be monitored across sites and over time, and staff can be alerted automatically if levels exceed preset values. Similarly, indoor air quality can be monitored for inadequate ventilation, improper temperature and humidity or excessive carbon monoxide.

Maintenance, repairs and contract management

With a maintenance CAFM system you can keep a full maintenance history of every piece of equipment or plant held on all sites, and the entire fabric of every building. Users can establish total downtimes and ongoing cost of ownership, which subsequently helps in assessing performance and planning replacements and refurbishments. Maintenance CAFM also enables a preventive maintenance schedule to be planned and implemented, so that essential services are maintained and equipment is ready when needed. Equipment such as heating systems can be interfaced so they can flag up their own faults, or trigger maintenance works orders after an optimum number of operating hours, and day-to-day maintenance repairs can be requested and recorded.

Online purchase requisitions, material requests and work requests can be submitted directly to internal departments or external partners. Scheduling, work assignments, shop assignments, work weeks, personnel histories, department assignments and relocation histories can all be stored and interrogated.

Helpdesk, service desk and security management

Help and service desk CAFM

Helpdesk CAFM helps the facilities management team to provide a responsive fault reporting and resolution service to building occupiers, and to fulfil occupiers' requests for routine services. Data can be collated to show which equipment or parts of a building generate the most faults or complaints. CAFM allows for automation of some of the helpdesk functions within service desk CAFM by allowing occupiers to access services such as room bookings, vehicle servicing or catering online.

Security CAFM

Security CAFM enables the facilities manager to monitor all the physical, personnel and electronic aspects of security – access control, CCTV, movement detectors, intruder systems, security staff routines – and manage them as an integrated system from a single control room. Access control systems can be interfaced, enabling the online issuing and cancellation of tokens and photo-ID visitor passes, and alterations to access times and levels, while if necessary occupants can be located within a site on screen. Alarms can be monitored remotely and fire safety can also be integrated, with heat and smoke detection and break-glass call points automatically alerting staff to the location of any suspected fire. These in turn can activate fire alarms and automatic voice evacuation messages.

Financial, budget and inventory management

Financial and inventory CAFM allows you to track and manage costs associated with capital assets, operations, consumables, work orders, training, that is, anything that impinges on facilities management work. At a strategic level, these systems enable accurate measurement and forecasting of the cost of all activities undertaken by the facilities management function, and the control of its budgets.

Facilities staff can requisition spares from inventory or order them directly from suppliers according to preset agreements and price structures. This enables the close management of facilities management-specific inventory of equipment, spares and consumables, such as cleaning materials.

Specifying CAFM

It is essential to specify the 'owner' of the CAFM project. Who has ultimate responsibility for ensuring that the system meets the objectives set? And who therefore has the authority to make sure that necessary resources and support are available when required? Primary contacts within individual departments or divisions need to be pinpointed, as they will be the people who you will ultimately rely on to feed through information on any changes, whether it be a layout modification or change of occupancy.

Choosing a CAFM system

With so many CAFM products on the market, how do you ensure that you are choosing the best system for your needs? Be rigorous in studying a CAFM system portfolio and take up references from existing users.

In planning a CAFM system, the best place to start is probably the space management aspect, since space is the second largest business cost after salaries. As most CAFM systems take a modular approach, it therefore makes sense to plan the first phases of the project to incorporate the space management requirements, and bolt on other applications later.

Ask the right questions

To select the right CAFM product, you will need to ask the right questions. Some examples of questions, and the reasoning behind them, are listed below.

Is the system standalone or are there any extra software packages to buy?

Does the system need additional database software, or is this integral to the CAFM system and not, therefore, an extra expense? Recommended add-on databases are typically Oracle, Sybase or SQL, but these require licences. Check compatibility as well; many systems have to be mounted on a CAD system. Is the CAFM system available as Software as a Service (SaaS)?

What is the scope of the CAFM system?

Check that the CAFM portfolio will deliver on all requirements. For example:

- For **management information** there should be information on publishing on the web, planning, benchmarking and performance measurement.

- For **business support** there should be data on finance and contract management.
- For **property management** there needs to be information on property portfolio control, estate diary and real estate development.
- For **space and asset management** there needs to be information available on design, costs and inventory, recharge, moves and changes and visualisation of space usage. Can the system be used to place assets, and then list them and data points by space ID, floor, building and so forth? Can the CAFM system also contain scanned or typed contract information regarding buildings, floors or equipment? Can it be used to create zones for cleaning, security, occupancy, etc.?
- For **call handling** the information must relate to the helpdesk, vendor tracking, maintenance procedures, purchasing and vendor invoicing.
- For **building management**, information needs to be available on environmental monitoring, HVAC, plant management, energy inventory, and reports and actions.
- For **security** there needs to be information on access control, alarm monitoring, fire monitoring, and incident reporting and tracking.

Will it be difficult to integrate existing data into the new system?

Any package must interface with existing systems to avoid expensive data re-entry. Check that the system has an open architecture that can complement, integrate and/or support existing systems. For example:

- Can you import and export between databases?
- Does the system support open database connectivity (ODBC) with, for example, dBASE, Microsoft Access and Paradox?

How can I access the information?

Depending on your needs, on-site access may not be sufficient. You may wish to access the data remotely, via an Internet browser or 'thin-client' scenario, whereby you can run small programmes running on your personal digital assistant (PDA) or laptop to access the CAFM system remotely.

Will an off-the-shelf package do?

Off-the-shelf products will be cheaper than customised ones. Check, however, that the product really will be able to meet your needs, and be aware that project-scope creep and sales pressure can change the boundaries.

Is it future proof?

There is always a worry that existing technology will soon become obsolete. Be clear about your current and future specifications from the outset, to ensure that the product will be able to grow with your business needs. Ensure also that the system has developing procedures and continual development in place so that databases can be upgraded easily as technology changes.

How are graphic images handled?

Facilities management automation will often involve linking some kind of graphic image with a database. Check what types of image files can be accommodated and whether the system can scan and digitise existing paper images.

How is the information presented?

- Is the database information available in both tabular and graphical formats?
- Can you change between view formats easily and are changes entered into one format automatically updated in another, or do you have to go back to the master database?
- Can reports integrate data and graphics for professional-looking reports?

Are there any user groups?

Hopefully you will have talked to other purchasers of the CAFM system when going through the selection process. But contact should not end here: keep in touch with them. If your provider has a user group, join it.

Choosing a CAFM provider

Once the system has been shortlisted, careful attention should be paid to choosing the right provider to implement it.

Suppliers should be able to offer all of the following services:

- project management consultancy
- software installation and customisation
- data management and capture
- software support and training.

Consultants worth their salt should offer to perform an information audit and needs analysis before any implementation. They should agree with you exactly what is required to be input into the system, what outputs are required, and what, how many and how often reports are required.

Data entry and protection

Accuracy is crucial in the setting up of original databases and where banks of data are to be transferred across from current legacy systems. If you engage temporary clerical staff to enter raw data, take the time to explain the context of what they are doing, and check their work thoroughly, especially in the early stages.

Thought should be given to exactly what needs to be captured in order to achieve the desired results. What will the resultant data analysis be used for and how much depth is required? Who is the target audience and what kind of reports will they expect to see? It is important to identify your exact needs for the initial phase, as it is complicated to insert or delete columns into the database design once data entry has begun.

Where data already exists, do not transfer it to the CAFM database until you have cleaned it, otherwise errors and miscounts will be replicated. For instance, a desktop copier which has been moved to three different offices may end up appearing three times on a paper-based asset register.

Accuracy is also needed when entering ongoing data. Free-form text is useful for informative notes shared by users, but if data is to be sorted and collated, events must be described in the same words every time. This means having only relevant fields on screen, and pick-lists for which the choice of entry is unambiguous.

Early in the project explore with your provider the levels of security offered by the CAFM system. Firewalls and encryption are essential where data is available to third parties. Internally you need to set up appropriate user access levels, good password practice and clean screen policies. Be aware also of your responsibilities under the Data Protection Act 1998. If your system holds details of identifiable individuals, such as staff, within a space plan or access control system, you need to register with the Data Protection Registrar.

Planning the transition

In the run-up to switching to CAFM, the following will help to ensure a successful transition:

- Identify exactly what needs to be achieved with CAFM to justify the investment. Take time to develop a series of measurables, and also a process for tracking them to ensure that the project is on target. These could range from the logistics of a particular process to quality of data. The project plan should also have a time-frame with target dates to ensure that everyone involved has clear objectives to focus on.
- Plan and make provision for the ongoing management and maintenance of the system. From the moment CAFM is up and running, you will need to keep the system updated with every change that takes place on the ground.
- Check that you have people available in-house who are willing to train on the computer system. Who will be in overall charge of maintaining the database?
- Do you have the appropriate hardware and software to run the system? Check whether current terminals are powerful enough and what, if any, modifications to other programmes or the network will be needed to share data.

Implementing the system

By the time it comes to implementation, the project plan should be fairly well advanced. You will have defined your data capture requirements, and the processes developed for ongoing maintenance should include a change management methodology to capture changes during the initial implementation period. The next steps are:

- Obtain existing CAD plans and ensure that their format is compatible for linkage to the database. Take some time to verify that the floor plans are accurate and up to date before commencement of the data entry. Although established processes should deal with changes as they occur, it is worthwhile kicking off the project with pristine data.

- Identify common space and allocate ambiguous space to the relevant departments or cost centres.
- Commence data entry and CAD updates, while adhering to the agreed standards.
- Submit individual building reports and floor plans to the relevant contact for verification.
- Produce your first set of reports for your target audience, using the agreed data requirements and reporting format.

Review your measurables and the original cost–benefit analysis at the end of the initial implementation phase. Can you prove the CAFM system's financial viability by showing that actual savings have reached what was originally estimated? Is it providing the data that was expected? Before making it your default system, are there processes that can be tightened up or improved?

The first implementation of CAFM may be complete, but new uses for this powerful resource will quickly become apparent. It is most likely that a continual stream of requests will require the addition of several more modules in the system.

Key sources

A snapshot of CAFM/CIFM products includes:

Aperture, www.aperture.com
Archibus, www.archibus.com
CAFM Explorer, www.cafmexplorer.com
Concept, www.fsi.co.uk
EFMS, www.mcs.be
Facility Center, www.tririga.com
Integrated FM, www.integratedfm.com
Manhattan, www.manhattansoftware.com
Maximo, www-306.ibm.com/software/tivoli/
Planet FM, www.qubeglobal.com
QFM, www.swg.com
Real Products, www.wms-plc.com

Note: CAFM is also increasingly available through the SaaS web-based delivery method. Check with product manufacturers.

Access, Safety and Security

12

Frank Booty

Statistics from the Home Office and Metropolitan Police have shown that the incidences of armed raids have increased over time, with perpetrators more willing than previously to threaten or use violence for limited gain. Further, organisations are becoming increasingly aware of the need to act to protect themselves against terrorism.

What options are available to facilities managers seeking to maximise security? Closed circuit television (CCTV), access control and staff training are among the practical measures that can be taken.

CLOSED CIRCUIT TELEVISION

The purpose of a CCTV system is to deter and detect crime and unauthorised actions involving threats to property and people, and to aid with the identification of aggressors before, during and at the end of an incident. There are also other applications for CCTV that many people are not aware of, such as monitoring a premises' environment to meet health and safety requirements for public and staff, staff training and site traffic control.

It is a legal requirement for operators of public space surveillance CCTV in England and Wales to hold a licence (as of 20 March 2006). This is part of the Private Security Industry Act 2001 (which also requires security guards to hold a licence from January 2006; see below). A licence will not be required if the cameras are used solely to identify intruders or trespassers on a site, or for the protection of vehicles or buildings against theft or damage. However, a security guarding licence may be required for these activities.

A key move recently has been the swing to digital cameras and systems from analogue (but there are still many analogue CCTV systems in the UK). Digital recordings can be searched by time and date, incidents can be tagged for later more detailed viewing, and images can be downloaded to computer or automatically burned onto compact disc (CD). Digital systems also enable images to be sent over the Internet for remote viewing.

369

Threats

Threats can affect both premises and their contents and those who work in them. Threats to premises, particularly external areas, are reasonably easy to identify (exits, entrances, perimeters, etc.). Internal threats are more subtle and are often influenced by the function of the space; shoplifting and staff pilfering are two examples.

Identifying threats is a prerequisite when planning a CCTV project. Past experience is seen as the best guide, with the best people to consult being staff, as they are most likely to have spotted potential security breaches or suffered threats. Areas that will need surveillance are those where valuable items are stored, including cashiers' departments where money is kept or handled; areas where staff may face dangerous situations; and all public areas.

The guiding principle is to achieve a balance between crime prevention and protecting personal privacy. It is therefore important to ascertain not only that CCTV is necessary, but also exactly what its purpose and key objectives are.

Security audits

For most CCTV projects it is advisable to conduct a security audit. Such an audit should assess who, what and where needs monitoring or protecting; procedures for cash movement and tills; the location of sensitive or valuable equipment; types of cameras and other security systems, such as alarms, access control and facial recognition; siting the cameras to cover designated areas and avoid blind spots; on-site and remote monitoring; and evacuation procedures.

Security consultants can play an important role in planning a CCTV project. Aside from being independent, they should have experience and knowledge of security issues and situations. Indeed, one of the many major factors to consider when choosing a security consultant is the background and expertise of their team. Another criterion should be membership of industry or professional associations, such as the British Security Industry Association (BSIA), although not all reputable consultants and suppliers are necessarily members (it is not mandatory to join such a body).

Overall security measures and systems should be audited once or twice a year, paying close attention to reported incidents and the response. All employees must be made aware that, whatever their job or status, they have collective and individual responsibility for security in their workplace.

Choosing CCTV cameras

The choice of cameras and accessories depends largely on the application and location for which they are intended. External cameras may need supporting lighting for night-time use, and weatherproof housing. The choice of lens is vital to cater for such factors as focal length, depth of view and variable lighting levels. For internal locations, less obtrusive cameras may be required, to avoid affecting the interior décor. This type of demand has led to the development of cameras such as dome models.

Every site and building is unique and facilities managers will need to translate the purpose and key objectives of the project into a technical specification to arrive at the best CCTV installation. The following factors should be borne in mind when identifying objectives and technical needs:

- Colour is appropriate for indoor surveillance, since colour aids identification, but externally the level of lighting present will dictate whether black and white or colour cameras are appropriate.
- If the budget is limited, a smaller number of higher resolution cameras will be more effective than a large number of low-specification devices.
- Various levels of control over cameras are possible: pan, tilt and zoom can cover a far larger area. Dome cameras conceal the direction in which the camera is pointing, making it impossible for individuals to tell whether actions are being surveyed.
- Criminals will eventually find out whether a camera is dummy or operational and the cost differential is usually minimal, making dummies an unwise purchase.
- External cameras need weatherproof, robust housings and may need demisters, washers and wipers.
- When specifying equipment, consider how future demand may change and try to make expandability and upgrade possible.

Cameras should be sited carefully and mounted low enough to give a clear picture, but high enough to prevent vandalism. CCTV can be integrated with other elements of company security; cameras can be triggered by alarms from intruder or access control systems, for example.

CCTV operation

For layout and operation of the CCTV control room, refer to British Standard (BS) 7499 and BS 5979 (a security consultant will be able to advise the level of control and sophistication necessary). Do consider the following:

- Do not overspend on unnecessarily sophisticated equipment; it should be easily operable by the minimum number of staff in the control room.
- Multiplexers enable continuous images to be viewed and recorded from up to 16 cameras.
- Images from several cameras can be shown sequentially, or on a split screen simultaneously.
- Restrict access to the control room to authorised personnel.

If live surveillance is considered necessary, be aware that a person's typical concentration span when viewing monitors is only 20 minutes. Sufficient staff will be necessary to build in breaks and constant rotation. Cameras can be triggered by movement detectors in normally quiet areas, with an alarm to prompt a guard to monitor the screen. Alternatively, remote monitoring is available by an alarm-receiving centre.

The integrity and success of any scheme are only as good as the staff who operate it. Staff should understand the purpose and objectives of the system, be familiar with the code of practice, procedures and incident response protocol, and be thoroughly trained. The operation of the scheme should be monitored regularly and a longer term independent audit put in place.

Remote monitoring

One CCTV security company specialising in the remote monitoring of commercial property claims savings of up to 75 per cent over traditional manned guarding security. From a central remote monitoring facility, the company controls barriers to allow a vehicle through a site to a chosen building, monitors loading and unloading of the vehicle, and allows the vehicle to exit the site.

Such an arrangement also enables remote control of lighting, air conditioning and pumps, especially in emergency situations. The software provides visual perimeter protection using complex programmable dome cameras combined with live audio facilities. This enables verbal communication between monitoring-centre operators and unauthorised people picked up on camera, or to verify visitors' identification.

Other companies offer remote desktop video surveillance over any distance to enable monitoring of remote or unmanned premises with a personal computer or notebook computer. Costs of such systems are usually affordable, even to small businesses.

Low-light CCTV cameras can be used to identify vehicle number plates near remote facilities where there can be problems from vehicle headlights dazzling conventional cameras, or where there are poor light levels.

SECURITY REVIEW

Any CCTV installation should be treated as part of an overall security review. Apart from any security systems, this should also include:

- integration of CCTV with any manned security
- establishment of correct security and safety procedures
- promotion of an organisation-wide security culture
- the training to achieve these objectives.

Recordings

Recordings can be stored on videotape, disk (e.g. CD) or other electronic media to be viewed offline or in case of an incident. Information recorded must be adequate and relevant, and must not exceed what is necessary to fulfil the agreed objectives. If recordings are to be used as evidence in the case of criminal action, they must have been handled in accordance with Home Office guidelines.[1]

---

[1] See the Home Office booklet on using and storing tapes as evidence: *CCTV – Looking out for you*, available from the Stationery Office, www.tso.co.uk.

Guidelines for handling recorded media include the following:

- Recordings must be indelibly marked with date, time and camera, and be labelled.
- All record-keeping must be accurate and complete.
- Tapes should be kept for 31 days before being wiped clean and reused.
- Tapes should not be reused more than six times.
- Storage of tapes and disks must be secure, especially where they contain incident data, both for protection of potential evidence, and for the protection of individuals' privacy.
- Access to storage must be controlled.
- Ensure good practice in the use of tapes and disks, and release to third parties. Destruction of tapes and disks should be secure and controlled.

Data protection

The Data Protection Act 1998 covers CCTV recordings of individuals who can be identified and where the system can be programmed to search for a specific part of a recording if it is known that an image of a suspected individual is recorded. If your system is capable of this you are obliged by law to register with the Data Protection Registrar. Do this as soon as you decide on such a system; to stay within the law you will have to take into account the principles behind the Act when designing the scheme.

VIOLENCE AND AGGRESSION IN THE WORKPLACE

The objective of an aggressive act is typically the attainment of some material benefit, such as cash or high-value goods. Aggression, expressed as violence, is essentially an issue between the people involved, comprising the aggressor, the aggressed, colleagues of the aggressed and bystanders who happen to be in the area.

At-risk areas

Aggression often takes place in areas accessible to the public. Typically these include shops, post offices and petrol stations; retail finance services; Department of Social Security (DSS) and local government enquiry offices; transport ticketing areas; and hospitals and schools.

The common thread is the interface between the public and members of staff. All these environments exist for the purpose of serving the public and accessibility is therefore a necessary part of that service. The thrust of management effort has been to provide a better service to customers. The dilemma is that unrestricted access is the antithesis of good security.

Studies by criminologists have shown that crimes, particularly those involving violence, take place in environments where the aggressors believe that:

- they are unlikely to be seen by the public, police or security officers
- if seen, there is little likelihood of being identified

- there are few people to report the crime in time for there to be an effective reaction
- there are many good escape routes by foot and vehicle.

The people involved

- **Staff**: staff at transaction counters are mainly under 30 years old and most are female. Their objectives and reactions are predictable in that their objective is minimum injury and their reaction can be trained and managed.
- **Raiders**: raiders are also mainly under 30 years old, most are male and are aware of the risks of being apprehended and the various penalties that will be suffered if caught following a raid, with or without injury. A raider, operating alone or as part of an organised group, relies on the ability to dominate the situation by surprise, threats and violence. To stay and take a hostage will incur delay: to delay is to invite capture and to injure will intensify the police hunt and increase the penalty.
- **Public**: customers or members of the public who are present when a raid occurs are wholly unpredictable. However, in most cases people are initially shocked and take no action.

Personal safety: training and counselling

Courses in personal safety in the workplace seek to train staff how to recognise and remedy confrontational behaviour in others, examine their own attitudes and eliminate areas that can spark aggression. Typical course content would include planning safety in travel to, from and during work; controlling stress and tension; and non-verbal communication. Importance should be placed on taking control in the event of an attack, getting away and the consequences of fighting back. Assertiveness and the law on self-defence need to be discussed and risks should be put into perspective, backed up by statistics.

A crisis management team needs to be put in place to provide a proactive response to any incident. In addition, management must identify external sources of assistance, and build a model of intervention. With adequate crisis intervention, there should be fewer individual long-term needs.

Principal strategies for defence

These are:

- **Hardening the target**: cash transaction counters can be equipped with armoured doors, walls and windows, akin to a bank vault. However, this can be unwelcoming to the customer and there are weak points of visibility/access at the windows and doors.
- **Removing the target**: removing cash into an armoured enclosure as above is one solution, but is not available to most transaction counters where there is a tradition of open access across the counter.

- **Removing the means**: raiders rely on the tactic of surprise to achieve immediate and total dominance of the area by fear and aggression. Conventional transaction counters have no active defence. This advantage should be removed. Some counters are equipped with bullet-resistant glazing and offer a degree of passive protection, but still no active defence.
- **Reducing the payoff**: good cash management by the use of teller-assist units and time-locked minisafes can reduce the amount of money lost to a raider.
- **Surveillance**: this serves no purpose unless there is a practical action that can be taken as a result of the surveillance revealing a threatening situation. CCTV is only a passive deterrence against violence which can also help to identify offenders.

These strategies for effective defence against crime can be adopted on their own or in any combination. Adopting all strategies in different facets of the overall security is usually seen to offer the best strategy.

ACCESS CONTROL, ASSET PROTECTION AND GUARDING

Access to the workplace should achieve a balance between convenience for employees and bona fide visitors, and the denial of access to areas where people do not have a need or right to be. Access control must not interfere with escape in case of emergency or fire. Advice is available from security consultants, approved security retailers, the police and the local fire service.

Access risk analysis

Access must be a basic consideration at the risks analysis stage; every workplace is different, with variations in the business activity and risks. Risks will range from opportunistic or organised crime to the ignorance and carelessness of staff. When assessing access risk:

- Start at the perimeter of the property and work in to the main door(s) of the building, other external doors and windows, the reception area, and on to all internal doors.
- Decide who is to be allowed free or controlled access at these various points and at what times.
- Consider how to secure and allow entry and egress through access points.
- Consider how to identify people who are authorised to be on-site.

Security measures

Access restriction

Access restriction serves to deter attempted entry and prohibit actual entry. The restriction of access to an area that any member of the public needs to enter will inevitably be a compromise between security needs and the business needs of the organisation. With staff-controlled door systems, staff can be actively involved in deciding who they allow to enter their premises and exclude any suspicious characters.

Perimeter boundary

Consider the following:

- Invisible barriers can be created by infrared beam systems.
- Fences or walls have to fit design criteria and be aesthetically appropriate; but check planning permission.
- Power fencing conforming to health and safety standards can be used to deliver a low-voltage electric shock to intruders.
- Tunnels, culverts and manholes giving access to the premises should be secured.
- Perimeter entry points need to be kept to a minimum, but should allow for smooth flow of traffic at peak times.
- Gates should be high and strong enough to deter entry when locked.

Building exteriors

Consider the following:

- How vulnerable is the building exterior at quiet times or at night when the premises are empty?
- Keep entrance doors to a minimum, and locked when not in use.
- All windows should have locks.
- How easy is access from roads, footpaths, flat roofs, roof lights and fire escapes?
- Secure ground-floor windows with steel bars or wire mesh.
- Ensure that goods in and out are always manned during opening hours.
- Good external lighting is a deterrent to intruders in areas with good natural surveillance.

For external doors and doors leading to vulnerable areas, fit as a minimum thief-resistant mortise deadlocks conforming to BS 3621. Specialised advice on locks should always be sought. Always buy from a member of the Master Locksmiths Association.

Rising screens

Rising screens are designed to protect personnel against actual and threatened violence (both physical and psychological). They have a track record of successfully defeating violent raids in a number of countries, environments and cultures. Rising screens are installed into the service counter where staff and public have direct contact, and provide a tangible barrier to projectiles, syringes and physical contact at the touch of a button. Aggression is restricted to the fraction of a second between realisation of a raid and the action of raising the screen. Staff are able to remove the threat the moment the presence or threat of a raider is perceived or suspected.

Screens are housed in a steel body shell which is built into and supports the counter. The screen is a telescopic arrangement of two sheets powered pneumatically to rise from the counter to the ceiling; body and lower screen panels can resist bullet penetration from powerful handguns (.44 Magnum and sawn-off shotgun with solid ball shot). Upper panels, not generally required to be bullet resistant, are made of sheet aluminium and are impenetrable to sight, sound and thrown objects. Special situations can have panels resistant to 7.62 mm armour-piercing ammunition.

Authorised entrants

Identity (ID) badges can form part of an access control strategy; they need to be unique to a facility and not easily copied. Electronic access control systems can combine security with convenience. They allow doors to remain locked, giving continuous security, but automatically unlock them when authorised employees or visitors are identified by their unique codes. Individuals can be identified by voice entry, PIN pads into which numbers are keyed, contact systems such as magnetic swipe systems, non-contact systems where cards or tags pass within range of readers, or new technologies such as biometric fingerprint or eye retinal reading systems.

Intruder alarms

Intruder alarms are a good visible and audible deterrent. They should conform to BS 4737. Take advice from insurers first, and only use installation companies who are members of, or approved by, an appropriate body. There are two basic types of intruder alarm: audible-only and monitored, where the system is connected to a 24-hour alarm-receiving centre (ARC) by telephone or radio link.

Over 90 per cent of alarms are triggered by events other than genuine break-ins. The Association of Chief Police Officers has reached an agreement with the National Security Inspectorate (NSI) [formerly the National Approval Council for Security Systems (NACOSS)], BSIA, the Association of British Insurers (ABI) and the British Retail Consortium to reduce the number of false alarms by 10 per cent each year.

The police have identified three levels of response:

- level 1: immediate
- level 2: response may be delayed due to higher priorities
- level 3: no police response, key holders only.

Newly installed and properly monitored systems, or where the presence of an intruder is suspected, receive response level 1. If four false alarm calls are reported within 12 months, response drops to level 2. If seven false calls are reported in the next 12 months, response drops to level 3. Restoration to a higher level will only be achieved when an organisation can prove that no false alarms have been made for three months.

Personal attack alarms

Personal attack alarms serve to summon assistance in the event of an incident. They can be interfaced to other systems to attract a security officer's attention at a CCTV console to start a recording sequence, sound a siren or start a strobe light. The alarms, of course, cannot stop the aggression, but may deter the raider.

Asset protection

Cash, stock and equipment are prime targets for thieves, both internal and external. Large items of electrical equipment and any easily portable equipment, such as tools, laptops and mobile phones, are vulnerable to theft and burglary, on and off the business premises. The following good practice tips should help to minimise the risk.

Company cash and valuables

- Consult insurers and train staff in cash handling.
- Payment of wages by cheque or credit transfer reduces the need to carry cash.
- Department petty cash should be monitored and managed by accounts staff.
- Cash should be banked frequently.

Stock

- Do not carry high levels of inventory.
- Employ trustworthy people.
- Access control should admit only authorised personnel.
- Check goods issued and receipts.
- Issue standard quantities for specific use.
- Perform regular spot stock checks.

Equipment

- Encourage employees to carry laptops inside a bag; business people carrying laptops are frequently mugged.
- Clear security marking of assets is a simple way to deter thieves and identify items if stolen.
- Security guards can check parcels or equipment being taken off the premises.
- Consider having spot checks of staff leaving the premises to deter internal thieves; a code of practice should be drawn up and clauses written into staff contracts which authorise the procedure (a search is only possible with the agreement of the individual).

Manned guarding

If manned guarding of the premises is to be part of an overall security system, a decision must be made to employ staff directly or to outsource the service to an external supplier. This will be determined by the size and nature of the organisation and other measures, such as CCTV. Outsourced guarding can be either static (on the premises continuously as in-house staff would be) or mobile (a patrol which tours a number of premises out of hours, calling in to check each one at regular intervals).

The Private Security Industry Act

The Private Security Industry Act 2001 became law in May 2001. This means that the security industry in the private sector in England and Wales is now subject to a regulatory system, as in other European Union member states. A Security Industry

Authority has been created to license and approve security firms and the individuals (employees, directors and partners) who work in them. The moves to require security guards to hold a licence became effective from January 2006. Licences will last for three years, providing conditions are not broken. In-house security departments are exempt from the legislation.

Security personnel

Security personnel need detailed job descriptions, called 'assignment instructions'. These should cover all duties and responsibilities, which may include access control, issue and control of ID badges, record-keeping, CCTV monitoring, and incident and emergency response. Guards may also need to make regular premises patrols.

Existing staff taking on security responsibilities must receive appropriate training, which is available through the Security Industry Training Organisation. Security staff need an understanding of the limits of their powers of search and arrest, as specified in the Police and Criminal Evidence Act 1984. Security staff should also be trained and certified as first aid providers and in fire safety. All training should be regularly updated.

Information protection

The continuity and survival of a business depend increasingly on its information. Analysis of the potential risks to an organisation of the loss of information should take into account production processes and design drawings, proprietary software programmes, customer lists, marketing plans and tender information, financial data, legal records, and personnel records. In addition, business documents need to be protected against disclosure, copying, theft, fire and vandalism.

Electronic information is vulnerable in specific ways: networks can be hacked into, emails read, information copied or corrupted, and virus infections introduced. An information security policy should be formulated following an assessment of the risks and with reference to business objectives.

All companies should be aware of the provisions of the Data Protection Act 1998, which sets out rules for handling personal data about any living, identifiable individual, held or processed by computer or other technology.

FIRE RISK

Fire poses a devastating risk to any business. Minimising the risk of fires is fundamental to any safety strategy. Industry analysts point out that no one has died in an office fire since 1990, but that is no reason to avoid minimising risks. Of the 500 people who die in fires each year, some 450 die in their homes.

Causes of fire

As the following list of the commonest causes of fires shows, most fires are caused by people doing something they should not or by not doing something that they should: fire raising and arson; careless disposal of cigarettes or matches; combustible material left near a source of heat; accumulation of easily ignitable rubbish or paper; carelessness on the part of contractors and maintenance workers; electrical equipment left switched on when not in use; misuse of portable heaters; and obstructing ventilation of heaters, machinery or office equipment.

Fire is a chemical reaction, requiring three fundamental elements to be in place: oxygen (always present), fuel (the combustible substance which can be a solid, liquid or gas) and ignition (source of heat energy). As all three points of the fire triangle must be in place to cause a fire, it follows that managers should attempt to prevent this from happening. Both managers and employees should strive to prevent sources of fuel and ignition from coming together.

MINIMISING FIRE RISK

Poor housekeeping is the greatest single cause of fire. This checklist of best practice guidelines is recommended to minimise risk:

- Where smoking is permitted, suitable deep metal ashtrays should be provided. Ashtrays should not be emptied into combustible waste unless the waste is to be removed immediately.
- Combustible waste and contaminated rags should be kept in separate metal bins with close-fitting metal lids.
- Cleaners should preferably be employed in the evenings when work ceases. This will ensure that combustible rubbish is removed from the building to a place of safety before the premises are left unoccupied.
- Rubbish should not be kept in the building overnight or stored nearby.
- No-smoking areas should be strictly enforced, especially in places that are infrequently used. Suitable 'No smoking' notices should be displayed prominently.
- Where no smoking is enforced owing to strict legal requirements (e.g. areas where flammable liquids are used or stored) or in areas of high risk or high loss effect, it is recommended that the notice reads 'Smoking prohibited – dismissal offence'.
- Materials should not be stored on cupboard tops, and all filing cabinets should be properly closed and locked at the end of each day.

Anything that burns can be regarded as fuel. A premises' structure may be made of materials which will readily burn when exposed to fire or flame. Some potential sources of fuel commonly found in premises are upholstered furniture, carpets and curtains; office storage, filing and other storage cabinets; wood and paper; packaging materials; paints and adhesives; petrol and paraffin; and liquefied petroleum gas. A good housekeeping approach assists in reducing risk to an acceptable level.

Statutory requirements

Note that in the wealth of fire legislation, premises are not policed by enforcing authorities. Employers are required to find out for themselves how to comply with the legislation. Consequently, many companies buy in fire safety advice as and when needed from an independent fire safety consultant. A list of consultants is available from the Institution of Fire Engineers. See Chapter 1, section on Complying with health and safety law, on fire safety for statutory requirements relating to fire risk assessment. Radical changes in fire safety laws came into force in April 2006. These have implications for all UK businesses: the new law consolidates existing fire legislation and repeals the Fire Precautions Act 1971 (until 2006 fire safety laws were scattered across over 100 pieces of legislation). The Regulatory Reform Fire Safety Order improves fire safety by placing the responsibility for fire safety on the employer or 'responsible person' for that building or premises. He or she will be required to assess the risks of fire and take steps to reduce or remove them. It is essential that facilities managers understand their responsibilities under this legislation.

INFORMATION

Further information may be obtained from the following sources:

Association of British Insurers, Tel.: 020 7600 3333, www.abi.org.uk

Association of Security Consultants, www.securityconsultants.org.uk

British Computer Society, Tel.: 01793 417417, www.bcs.org.uk

British Security Industry Association, Tel.: 01905 21464, www.bsia.co.uk

CCTV User Group, for model code of practice and procedural manual, www.cctvusergroup.com

Centre for Crisis Psychology, Tel.: 01756 796383, www.ccpdirect.co.uk

Data Protection Registrar, Tel.: 01625 545745, www.dataprotection.gov.uk/summary.htm

Institution of Fire Engineers, Tel.: 0116 2553654, www.ife.org.uk

International Institute of Security, Tel.: 01803 663275, www.iisec.co.uk

International Professional Security Association, Tel.: 01495 757153, www.ipsa.org.uk

Master Locksmiths Association, Tel.: 01327 262255, www.locksmiths.co.uk

National Security Inspectorate, Tel.: 01628 637512, www.nsi.org.uk

Security Industry Authority, www.the-sia.org.uk

Security Systems and Alarms Inspection Board, Tel.: 0191 2963242, www.ssaib.org

13   Maintenance and Repair

Frank Booty

All properties and facilities need maintaining and, for many, maintenance is at the core of the facilities management role. Organisations are beginning to realise that planned preventive maintenance (PM) is more economical than ad hoc replacement of parts when they fail. As well as ensuring that a planned maintenance schedule is effective, managers must be careful to comply with statutory requirements relating to both electrical systems' testing and workers' safety.

Facilities managers also need a strategy in place for maintaining critical information technology (IT) equipment. Whereas vendor-specific service agreements offer excellent service and assurance, such provision is compromised when a number of products from different vendors exist within one workplace. What solutions are available to the facilities manager, concerned with keeping the administrative burden of dealing with contracts to a minimum and ensuring that downtime does not affect the bottom line?

PLANT AND PROPERTY MAINTENANCE

Statutory requirements

Construction (Design and Management) Regulations

A growing number of larger companies, local authorities and other public bodies have been pursuing increasingly formalised procedures for checking the health and safety standards of contractors wishing to work for them. This process has been accelerated by the introduction of regulations which require organisations to satisfy themselves that potential principal contractors are capable of dealing with the health and safety issues associated with projects relating to their premises (see Chapter 1, section on Construction work and building management).

Statutory maintenance tests

Do note the important issue of statutory maintenance tasks. Failure to check your equipment, system or process and record the results at the prescribed frequencies is serious. There are also certain facilities, such as lifts or elevators, where insurance

FACILITIES MANAGEMENT HANDBOOK

implications mean that a relevant official has to inspect the system during the maintenance process.

Costs

Costs for maintaining different types of building range from thousands of pounds per 100 m^2 per year for air conditioned offices down to hundreds of pounds per 100 m^2 per year for warehouses. Annual maintenance expenditure typically covers areas such as decorations, fabric, services, cleaning, utilities, administrative costs, overheads and external works. Sources of information on costs include the Building Cost Information Service (BCIS) and Occupiers' Property Databank (OPD). Other information is available from the Building Research Establishment (BRE), the Building Services Research and Information Association (BSRIA), the British Institute of Facilities Management (BIFM) and the Chartered Institute of Building Services Engineers (CIBSE).

When maintaining properties, fabric costs relate to the floor to wall ratio, the density of partitioning, the standards of fittings and frequency of use. Any services maintenance programme will depend on the quality, age and condition of the heating, ventilation and air conditioning systems installed. The amount of churn within a company will also affect maintenance costs.

Types of maintenance

According to British Standards, maintenance is a combination of all technical and associated administrative actions needed to retain an item or system in (or restore it to) a state in which it can perform its required function efficiently and as expected. Essentially, there are two types of maintenance:

- **planned** (programmed, preventive and cyclical)
- **unplanned** (reactive, normal response and emergency response).

Planned maintenance is a maintenance organised and executed with forethought, control and application of records. It encompasses condition-based maintenance, which is progressed following information received about a system or structure's condition from routine or continuous monitoring processes.

PM concerns the care and servicing by personnel for the purpose of maintaining equipment and facilities in satisfactory operating condition by providing for systematic inspection, detection and correction of incipient failures before they occur or before they develop into major defects. It includes tests, measurements, adjustments and parts replacement, performed specifically to prevent faults from occurring. While PM is generally considered to be worthwhile, it is important to note that there are risks such as equipment failure or human error involved when performing PM, just as in any maintenance operation. PM is also sometimes augmented by reliability-centred maintenance, which attempts to determine the best PM tasks, and by predictive maintenance, which models past behaviour to predict failures.

To simplify matters, PM is conducted to keep equipment working and/or extend the life of the equipment, while corrective maintenance, sometimes called 'repair', is conducted to get equipment working again: it concerns actions carried out to restore a defective item to a specified condition, or tests, measurements and adjustments made to remove or correct a fault.

Unplanned maintenance includes breakdown, corrective and emergency maintenance. A repair is the restoration of an item or system to an acceptable state through renewing, replacing or mending worn-out or damaged parts. It is typically considered that the ups and downs of unplanned maintenance work should be run alongside planned work schedules. The reaction to unknown and unplanned events can thus be offset against meeting statutory obligations and service needs.

Condition assessments

Sound property management involves regularly checking a building's health. Condition assessments are now recognised as key tools for both strategic capital planning and tactical project prioritisation. Such systems integrate life-cycle data and condition assessment information with other facilities management technology systems, such as computerised monitoring and management systems and project management software.

Facilities managers can identify problems at their earliest stages and evaluate a building's future maintenance and repair needs through a systematic approach to assessing the condition of a variety of building components and systems. These may include building structure, building envelope, mechanical systems, electrical systems, interior finishes and lift safety. Building assessment protocols are applied to each component or system defining the scope of the audit for that category, the procedure to be followed and items that should typically be measured.

Checksheets highlighting potential problem areas should be provided to assist facilities managers in conducting the assessments. From there, facilities managers will be able to judge how much work would be involved in repairing areas that require attention.

Refurbishment costs

Building owners and developers considering the relative merits of refurbishment compared with new build may be surprised to discover that renovation projects can exceed 80 per cent of the cost of building a similar facility from scratch and are often about 66 per cent, according to Building Maintenance Information (BMI), which collects and analyses data on property occupancy cost from subscribers, and is a service of RICS Building Cost Information Service Ltd.

Serviceability

Serviceability is defined as the ease with which corrective maintenance or PM can be performed on a system. Higher serviceability improves availability and reduces service cost. Independent cost analysis studies indicate that a large proportion of overall life-cycle costs of most systems and products is given over to maintenance and support. Indeed, for some systems, maintenance and support have been demonstrated to account for up to 75 per cent of overall life-cycle costs. As long-term cost-effectiveness is a key competitive advantage in the current global marketplace, more and more companies are making plans to reduce system maintenance and support costs.

Managing maintenance contracts

All critical systems and response times should be detailed in specific service level agreements (SLAs) (see Chapter 7, section on Service level agreements). Facilities managers can measure the performance of a particular service against the target given in the SLA through a helpdesk or computer-aided facilities management (CAFM) system (see Chapter 11, section on Computer-aided facilities management).

Voucher systems

Maintenance systems, run as a part of a helpdesk or CAFM system, typically utilise job tickets or vouchers. These are issued to relevant tradespeople or contractors, providing them with information such as the location of the fault, description, agreed response time and estimated time for the work to be completed. Planned maintenance vouchers often include service routine details to assist personnel to process the required work. Cost codes track labour costs and all spare parts and consumables used during the work, so that these can be recharged efficiently and appropriately.

Software features

Many modern maintenance software packages use barcodes to collect data and costings on particular jobs, which saves a lot of time and effort otherwise expended on manual data entry. Such software packages can also be used to chart the history of items of plant, systems and schedules, as well as to make records of maintenance frequencies and issuing of spares. They can also assist in the processes necessary to perform statutory maintenance tests required on certain systems (electrical installations, for example).

Number and frequency of maintenance visits

There will be occasions when a planned maintenance service will have to be rescheduled. Companies will always want to know that the precise number of planned maintenance calls agreed in the SLA or contract has been actioned or delivered, so recording an event that has been pushed forward or back must take place.

The frequency of maintenance visits should be agreed, usually according to manufacturers' service requirements or recommendations. Maintenance software packages often feature the ability to plan workloads over defined periods to enable the best possible fit with the resources available to perform tasks. Unplanned activities can also be allowed for.

MANAGING THE MAINTENANCE SCHEDULE

- Check to see whether any statutory obligations have to be complied with.
- Find out exactly what plant and equipment there is within the estate and produce a detailed asset register (this will probably be computerised), which should be checked and updated regularly.
- Ensure that all heath and safety regulations are being followed.
- Conduct a risk assessment on all systems and tasks.
- Make sure that all systems and components essential to the company's business continuity plan are known and that they are properly included in maintenance priorities.
- Develop a planned preventive maintenance schedule for all essential items of equipment, systems and fittings.
- Make sure that an accessible maintenance and test–result recording and documentation system is operational.
- Set up a helpdesk (or at least a work request system) that links to all other procedures and systems within a company.
- Anticipate unplanned maintenance work and agree relevant charging structures.
- Make friends with the finance director, as capital investment budgets will need to be drawn up and life-cycle costing exercises undertaken.
- Keep your team informed.
- Talk to your customers. Ask them what they think of your services (by conducting regular surveys) and find out what they want.

Employee safety

Effective maintenance is key to ensuring a healthy and safe workplace. Managers who control premises have a legal duty to ensure the safety of employees (see Chapter 1, section on Maintenance and repair). Facilities managers also have to be satisfied that any contractors working on site are complying with health and safety guidance and regulations, and that they have conducted risk assessments where necessary. Compliance with health and safety should typically be embedded in specifications or SLAs from the outset, and regular two-way communication on health and safety matters is recommended.

Risk assessments (see Chapter 1)

Workplace injuries (or worse) are substantially reduced through the adoption of rigorous risk assessment practices, which highlight areas where maintenance is necessary. Risk assessments can be conducted using software systems or on paper and should be undertaken for workplace tasks or issues such as air quality, humidity and temperature; manual handling; display screen equipment-related issues – checking closeness to screens, rest breaks, etc.; first aid facilities; static on carpets; cleaning; adequate lighting; cable runs and any 'rogue' trailing leads; control of substances hazardous to health; and noise.

Facilities managers should compile a risk register, where known risks can be ranked from low to high. As insurance companies will give reduced premiums against companies that can demonstrate where risks are low rather than high, the facilities manager who acts proactively here will be warmly welcomed by the finance department.

Preventive maintenance programmes

A PM programme allocates specific maintenance tasks to particular periods in a timetable. PM programmes save money and eliminate downtime. Maintenance programmes range from basic essential checks to comprehensive tests run according to manufacturers' specifications. In general, full maintenance programmes are carried out at the end of a season when systems are shut down after several months of operation. This ensures readiness for the next period of operation. Ideally, a system should also be checked out immediately before start-up to make sure that it is ready when needed. This applies particularly to heating systems in hotels, but is equally valid in industrial premises. Providing site owners authorise a full inspection and maintenance programme, with replacement of parts where required, they should be able to rely on the findings of such a programme for the forthcoming season of operation.

Reading the signs

Maintenance engineers should be trained to spot the tell-tale signals from ionisation probes and ultraviolet cells which indicate the condition of parts of a system. They should also take full-load current and bearing temperature readings from system motors, which verify their state of health.

Companies new to this process are recommended to have a full inventory check or technical audit of the heating, plumbing and ventilating plant at a site. A planned PM schedule can then be worked out to cover all the equipment.

This may, for example, require inspection of oil burners every three months. If the planned maintenance schedule complies with the manufacturer's specifications, the system is unlikely to malfunction.

However, any deviation from these specifications – any cutting corners or cost savings – will inevitably result in unplanned maintenance problems and increased costs. System operating costs will increase with the age of the system, as will maintenance costs and system downtime.

Faults with new installations

Of course, installing a brand new system is no guarantee of trouble-free operation and new systems also require planned maintenance. The frequency of major failures due to poor installation or inadequate design is surprisingly high. This is where a site survey procedure often highlights system faults and design defects, which can be rectified economically early in the life of the system. Before the end of the first year of operation, it makes sense to have all the building systems checked for defects, while the installers remain liable for rectifying any installation faults. It is common to find access control panels that are not wired correctly, and missing thermostats.

Spare parts holding

To support the maintenance programme, a minimum spares holding is recommended. This would include belts for motor drives, inexpensive items but nevertheless essential for the system to operate correctly.

Air conditioning

PM is equally valid for air conditioning systems. Filter elements need replacing regularly or the consequences can be costly.

One company had a unit with a badly blocked filter element, which had been pulled out of its track and was resting on the drive belt. Friction burn marks were evident on the element. This might have caused a serious fire but was, fortunately, spotted in time. The antidote is to maintain a spare parts holding of filter elements, costing £200. A spare parts holding of under £1000 guarantees minimal maintenance costs, and assures minimal downtime and system reliability, insurance which is well worth considering.

Test instrumentation

Sophisticated monitoring equipment is essential. All engineers should carry flue gas analysers and be able to interpret the results to achieve safety and maximum efficiency. This can often save money over time. Engineers have been known to encounter boilers that are operating at efficiency levels as low as 40 per cent, when they should be operating at about 80 per cent. This increases the fuel consumption of the system and, more importantly, can lead to blocked and sooted-up flues which, in turn, can lead to carbon monoxide emission hazards.

Staff training needs

Frequently, there are basic maintenance procedures which can be carried out by staff without the need for a maintenance engineer. These range from the act of pressing a boiler reset button to restarting an entire system. For example, replacing filter elements at prescribed intervals is a relatively simple task which can realistically be carried out in-house, provided the necessary instruction has been provided. When it comes to complex computer-based building management systems, property managers need to be able to interpret the mass of information they are provided with and training is, again, a key requirement.

Environmental issues

Some of the latest sites have been designed to reduce certain gaseous emissions, which are known to affect the composition of the ozone layer. Awareness of gaseous emissions is something with which everyone in the industry should be increasingly concerned. There may also be future standards for the operation of energy controls, including those applying to building management systems.

TURNKEY SERVICE MANAGEMENT

When buying a new electrical appliance, facilities managers must decide whether to take out a service management agreement. With technology advancing at exponential rates, and appliances now supporting more and more critical applications, any downtime due to appliance failure becomes less and less acceptable. Yet, despite manufacturers' claims, nothing is infallible. There will be occasions (albeit infrequent) when problems arise and repairs are necessary. This is when the service engineer becomes your best friend.

Manufacturers offer numerous after-sales packages to provide users with the assurance that, should a product suffer from a failure of some kind, a service engineer will be on standby to offer assistance. Domestically speaking, such service management agreements (usually defined as extended warranty packages) may initially appear costly, but when weighed against an independent engineer's call-out and hourly charges, they could bring benefits after only a single use.

Within the corporate arena, service management agreements play a more integral role in the initial purchase of a product. When committing to purchase a new piece of equipment, assurance is necessary not only that the organisation is purchasing at the best price available, but also that the equipment is accompanied by a comprehensive service package. In the event of failure, the cost to the organisation in loss of productivity alone could easily outstrip the initial capital outlay for the equipment.

Managing multiple vendors

With so many organisations looking for the best deals each time they buy new equipment, it is unlikely that every appliance will be supplied by the same vendor, which has ramifications for the service manager. Organisations looking to manage multiple electronic devices, for example, should consider a single maintenance contractor.

Each individual vendor may offer a uniquely tailored service solution for its own range of products, which means that an organisation may end up holding multiple agreements with multiple vendors. For example, consider the possibility that 'brand X' operates six different products within a particular company, as do four other vendors, each in turn offering a tailored service agreement for each product [email server, network printer, uninterruptible power supply (UPS), personal computers, etc.]. In the event of one of these products failing and an engineering call-out being required, the host organisation must sift through numerous contracts to find the correct one for the correct piece of equipment. The amount of time taken simply to identify the appropriate service department or contact may have potentially cost the organisation thousands of pounds in lost productivity.

The complete service solution

In essence, the complete service solution is a simple idea: one organisation provides the complete service management for all devices. In practice, it is not quite that simple; it is unlikely that a photocopier engineer could provide technical support for a server fault, for example.

Behind the complete service solution lies the fundamental premise of service management: expert technical assistance. As such, the complete service solution offers expert assistance for a specific set of products, be they photocopiers or UPSs. Consider the UPS as the example in this case. The UPS is a niche product (particularly in the high-end segment where a single unit may involve a five-figure capital outlay), and each unit can be fundamentally different. It is hard to compare directly the UPSs of two manufacturers, as each unit integrates inherently individual technology and specifications around a common design, that of offering power protection and assurance 24/7.

Within the UPS marketplace, an immediate response to any service need is imperative. This is why certain manufacturers in the industry have paid close attention to the traditional service management problems, and have devoted time to evolve this practice into a complete one-stop service solution. The UPS industry has devised two modes of service management: third party and outsourced.

Third party service management

Within this model, a single manufacturer operates as the sole service provider to the customer, thus providing direct service and support to every brand within the host organisation. In the event of a failure, the user contacts the manufacturer, and it will send one of its service engineers to address the technical problem. Despite the apparent advantages of this method (one contract for all servicing needs), the manufacturer must be able to service all brands of UPS to a high level. As mentioned already, all UPSs encompass individual specifications and technologies, so it would be essential for the managing organisation to employ expert engineers in all of the UPSs currently in the marketplace.

Outsourced management ('total support')

Operating along similar lines to the model above, the outsourced management solution is executed through a single manufacturer acting as a service facilitator (or agreement host). In the event of a failure, the user contacts the chosen host manufacturer for technical support. Once the call has been received, it is analysed to discover which particular brand of UPS is at fault and whether the agreement host is equipped to support the device in question. Instead of sending the host manufacturer's own service engineer to service all manner of units, this service model means the host manufacturer will (where necessary) outsource the maintenance of a particular unit to the specific vendor. Through this practice, the user is benefiting from cost-effective, vendor-specific service management while enjoying a centralised support contact process.

Of the two models above, the outsourced management method offers the user the advantages of a focused service management agreement without the aggravation of having to deal with multiple contracts. Within the high-end UPS sector, total support has been recognised as the optimum service management method and has received recognition from many well-known organisations, such as Tesco and Bank of Scotland. But should this practice be restricted to just the high-end electronics industry?

Needs of smaller organisations

Traditionally, smaller units have not been covered by total support service management agreements, but with advancing technology allowing smaller and smaller units to support larger systems, smaller unit users are demanding the same level of service management as their larger counterparts.

As a result, manufacturers have adapted the total support model to cater for this sector. For example, in supermarkets the organisation-wide electronic point of sale (EPoS) cashier system may be supported by a large UPS, but individual tills will still have their own separate UPSs. Larger UPSs may require specialised support (from the host organisation or outsourced vendor), but in most cases involving smaller UPSs, immediate support can be offered by the host organisation as the majority of failures in these units are due to the natural lifespan of the internal batteries. In cases such as these, the host is able to 'hot-swap' the batteries on site no matter what brand of UPS is being operated. In this way, the supermarket can call on technical support for all sizes of UPSs, thus assuring power protection for all of its systems.

Maintenance management

Maintenance management is 90 per cent information management and 10 per cent engineering. To carry out maintenance efficiently, a proper information system is essential. Common sense says that it should be a computerised monitoring and management system (CMMS); not paper; not a spreadsheet. If your environment is equipment intensive, with potential failures that can disrupt operations or jeopardise safety, not having a CMMS has serious implications. It means that the organisation is running with inflated operating costs and risks that are not properly mitigated. Note that deciding which CMMS software to use is only half the story: the support and expertise provided by the software vendor is as crucial as the software itself.

KEY PERFORMANCE INDICATORS

Key performance indicators (KPIs) are an important management tool to measure business performance, and are often used to measure maintenance. Unfortunately, unlike operations, there are few 'hard' measures of maintenance output and the measurements that are used are often easy to manipulate.

Maintenance KPIs have to be integrated with operating KPIs and must be 'balanced', and there are three other key criteria that should be considered when deciding what aspects of maintenance to measure:

- KPIs should encourage the right behaviour.
- They should be difficult to manipulate to 'look good'.
- They should not require a lot of effort to measure.

Some measurements may encourage people to do things that are not wanted. A common measurement is 'adherence to weekly work schedule' for non-shutdown maintenance work. It is easy to achieve a high adherence to schedule by scheduling less work, through overestimating work orders. However, what is really wanted is higher productivity, which can often be achieved by challenging people through scheduling more work. So the wrong measurement may work against the company.

Like 'adherence to schedule', other common measurements are easy to manipulate. Some examples are the percentage of time spent on PM work, and the percentage of rework, emergency work and 'break-in' work.

Look for KPIs that are truly relevant and satisfy the three criteria listed above. A good example comes from a plant trying to improve shutdown planning, where a new target of completing all planning two weeks in advance of a shutdown has been set.

Because all shutdown work orders have a shutdown code, a report from the CMMS listing all purchase requisitions against work orders for a specific shutdown that were originated less than two weeks in advance will provide a useful measure. It supports the right behaviour, is unlikely to be manipulated (and if it is it should be obvious) and is easy to measure (the computerised system does all the work). It will also provide information on where to take action and where to recognise good planning efforts.

Innovative KPIs such as the above example may be of greatest value when measuring the success of efforts to change practices, and may be discontinued when the new practices become a habit.

INFORMATION

For more in-depth details on the topics discussed here, contact the organisations listed below:

British Institute of Facilities Management (BIFM), Tel.: 0845 058 1356, www.bifm. org.uk

Building Maintenance Information (BMI) and Building Cost Information Service (BCIS), Tel.: 020 7695 1500, www.bcis.co.uk

Building Research Establishment (BRE), Tel.: 01923 664000, www.bre.co.uk

Building Services Research and Information Association (BSRIA), Tel.: 01344 465600, www.bsria.co.uk

Chartered Institute of Building (CIOB), Tel.: 01344 630700, www.ciob.org.uk

Chartered Institute of Building Services Engineers (CIBSE), Tel.: 020 8675 5211, www. cibse.org

Council for Registered Gas Installers (CORGI), Tel.: 01256 372200, www.corgigas. com

National Inspection Council for Electrical Installation Contracting (NICEIC), Tel.: 020 7564 2323, www.niceic.org.uk

Occupiers' Property Databank (OPD), Tel.: 020 7336 9200, www.ipdglobal.com

Facilities service cost benchmarking protocol

Category	Subcategories (contract bundle items)	Principal examples/ cost elements	Items included in costs	Items excluded
Services maintenance and repair	Heat and ventilation	Boilers	Handymen	Catering equipment
		Radiators	Supervisors	New installations
		Chillers	Blue-collar staff	Alterations
		Ductwork	Specialist services cleaning	Improvements
		Filters	Tenants' areas service charge element	Grounds lighting
		Fire extinguishers	Water treatment	General management time
		Humidifiers	PPM, emergency, ad hoc, etc.	Security systems
			Smoke tests	Cleaning equipment
			Statutory testing	Computer installations
	Plumbing	Sprinklers	Emergency inspection and testing	Data cabling
		Sanitary fittings	PAT tests	Churn
		Water supplies	Legionnaire's testing	Line management
		Drainage		
	Electrical	Switchgear	Hardwire testing	
		Wiring	Testing and inspection of earth bonding	
		Internal lighting		
		Small power		
		Floor outlets		
		Fire alarms		
		Generators		

393

Category	Subcategories (contract bundle items)	Principal examples/ cost elements	Items included in costs	Items excluded
		Building management systems Energy management systems UPS		
	Lifts	Lifts Escalators		
Building maintenance and repair	Fabric	Structural	Handymen	New installations
		Roofing Partitions Doors	Supervisors Blue-collar staff Maintenance equipment Tools	Alterations Improvements Management time Churn
	Decorations	Finishes	Materials Tenants' areas service charge element	Office equipment Grounds maintenance
	Fixtures and fittings	Fixtures Fittings	Signage Essential spares Plantroom cleaning Gutter clearance	Furniture
Grounds maintenance and repair	Maintenance	Services (including drain clearance)	Gardening	Internal landscaping
		Hard landscaping (fences, pavings) Soft landscaping (planted areas, lawns)	Horticulture Tree surgery Fences	
	Cleaning	Power washing	Service charges	
	Utilities	Lighting	Snow clearance Car park areas	

Category	Subcategories (contract bundle items)	Principal examples/ cost elements	Items included in costs	Items excluded
	Security	Special grounds patrols/guarding Specialist grounds security system maintenance	Road maintenance (including gritting)	
Furniture maintenance and repair	Maintenance	Chairs	Service charge element	Additional purchases
		Desks		Moves
	Repairs			Special aids and adaptations
	Isolated replacements	(Only considered if uneconomic to repair)		
Equipment maintenance and repair	Maintenance	Loose electrical appliances	Service charge element	Additional purchases
	Fitting out			Blinds and curtains
	Repairs	AV equipment Rental costs		IT equipment Photocopiers
	Isolated replacements	(Only considered if uneconomic to repair)		
Alterations and fitting out	Churn	Moves and changes costs	Construction project management Professional fees	Major refurbishment Statutory improvement
	Fitting out			
	Improvement Adaptations		Statutory fees Furniture and fittings	
Cleaning	Windows and cladding	Windows	Deep clean toilets	Deep clean kitchens
		External walls	Carpet shampoos	Catering area cleaning
	Internal areas	Floors	Stain removal	Contract management time
		Internal walls	Cleaning materials	Cleans due to churn

Category	Subcategories (contract bundle items)	Principal examples/ cost elements	Items included in costs	Items excluded
		Ceilings Sanitary fittings	Supervisors Tenants' areas service charge element	Plantroom cleans Grounds cleaning
		Toilet area surfaces	Graffiti removal	
	Supplies Furniture and equipment	Janitorial supplies Furniture, fittings and business equipment	Atria glazing Roofs and integral gutters	
	Special cleans	Exposed services Catering special cleans Computer special cleans Lighting special cleans Ductwork special cleans Lift shaft special cleans	Litter picking Cleaning equipment maintenance Health and safety requirements (e.g. slippery surface signs)	
	Pest control	Pest control	Wires, nets, poisons (eradication contract)	
	Waste disposal	General office waste disposal	Equipment maintenance (e.g. shredders)	
		Waste recycling service Hazardous/ environmentally sensitive waste disposal Secure waste disposal	Hire of containers	
Laundry	Laundry services	Uniforms Pillows Sheets	Consumables Dedicated staff costs Contract costs	Tablecloths Chefs' whites New linen/towels, etc.
		Towels	Linen hire	

Category	Subcategories (contract bundle items)	Principal examples/ cost elements	Items included in costs	Items excluded
Security and reception	Reception	Reception	Staff costs	Fire alarms
		Door guard/ commissionaire	Day guards (external and internal)	New installations
	Guarding	External guarding	Night patrols (external and internal)	Alterations
		Internal guarding	Uniforms	Improvements
	Surveillance	External surveillance	Radios	Management time
		Internal surveillance	Security cards	Changes due to churn
	Duties	Visitor escorting	Camera operators	
		Lost property	Tenants' areas service charge element	
	Security system maintenance	Car parking control	Barriers	
		Card readers	Swipe cards	
		CCTV equipment	CCTV Pass issue Key management	
Utilities	Energy	Electricity	Tenants' areas service charge element	Telephones
		Gas		Management time
		Oil		Environmental testing
	Water and sewerage	Water		
		Sewerage		
Internal décor	Horticultural	Trees	Contract charges	
		Cut flowers	Cleaning costs	
		Planters		
	Works of art	Works of art	Insurance premiums	
Archiving		Physical documents	Contract charges	

Category	Subcategories (contract bundle items)	Principal examples/ cost elements	Items included in costs	Items excluded
		Electronic media	Maintenance contracts	
		Film-based media Document/file disposal Specialist filing systems		
Reprographics	Machines	Convenience copiers Central black and white photocopying	Paper management	Utilities
	Consumables	Central colour photocopying Origination/DTP controller	Maintenance leases	
	Dedicated staff	Off-site printing Laminating equipment Binding equipment and consumables		
Stationery	Departmental	Standard paper products Bespoke paper products (headed paper, etc.)	Envelopes Office equipment	Paper for photocopiers Writing materials
	Personal	Non-paper consumables	Other: cards/diaries/ files Flipcharts/foils/pens	
	Storage materials Management		Line management	
IT communications	Switchboard	Switchboard service operation Switchboard service maintenance	Staff costs Equipment maintenance	
	Equipment maintenance	Telephone/fax equipment maintenance	Line management	

Category	Subcategories (contract bundle items)	Principal examples/ cost elements	Items included in costs	Items excluded
		Radio pagers/DECT mobiles maintenance AV conferencing equipment maintenance Telephone cable maintenance		
	Call charges	Mobile phone charges National line charges Intersite leased line charges International line charges Bureau services charges		
IT computers	PC equipment Software	PC equipment maintenance Software maintenance Line charges Website management	Equipment/ hardware costs lease costs ISDN/other line costs ISP subscriptions Domain subscriptions Consultancy costs Maintenance contracts	New purchases (hardware and software)
Distribution (mail/ postroom)	Mail processing/ delivery	Processing and delivery of incoming mail Collection and preparation of mail for external distribution Processing and delivery of internal mail/messages	Dedicated staff Uniforms Equipment maintenance Dedicated IT	

Category	Subcategories (contract bundle items)	Principal examples/ cost elements	Items included in costs	Items excluded
	Charges	Post office charges	Scanning/X-ray	
	Couriers	Local couriers' charges	Postal charges	
		National couriers' charges	Courier charges	
		International couriers' charges	Packaging (e.g. Jiffy bags, bubble wrap)	
	Equipment	Postroom equipment maintenance		
Text preparation	Typing services operation	Typing services operation	Staff costs	
			Equipment costs	
	Bureau charges	Bureau charges		
Transport/fleet management	Pool cars	Pool cars maintenance	Staff costs	
		Pool cars cleaning	Equipment costs	
		Pool cars operation/ chauffeurs	Uniforms	
	Delivery vans	Pool delivery van maintenance	Overalls	
		Pool delivery van cleaning		
		Pool delivery van operation/drivers		
	Service vehicles	Service vehicle maintenance		
		Service vehicle cleaning		
		Service vehicle operation/drivers		
Catering	Staff dining	Staff dining: breakfast menu	Subsidy costs	Fitting out
		Staff dining: lunch menu	Chef, cooks, supervisors, waiters, servers, till staff	General cleaners

Category	Subcategories (contract bundle items)	Principal examples/ cost elements	Items included in costs	Items excluded
		Staff dining: sandwich bar		Utilities
		Staff dining: snacks and beverages	Specialist cleaners	Purchase
	Vending	Vending: sandwiches	Preparation	Finance
		Vending: confectionery and snack products	Food cost	
		Vending: canned/ bottled drinks	Drink cost	
		Vending: beverages	Alcohol	
			Sandwiches	
	Hospitality	Hospitality: sandwich lunches	Leases, rental	
		Hospitality: beverage service	Crockery	
		Hospitality: cold buffet lunches	Disposables	
		Hospitality: hot buffet lunches	Linen	
	Private dining	Private dining rooms: breakfast menu (silver service)	Laundry	
		Private dining rooms: lunch menu (silver service)	Consumables	
	Special functions	Special functions: seated dinners (silver service)	Deep cleaning/ specialist cleaning	
		Special functions: buffets (silver service)	Income from till receipts	
		Special functions: cocktail parties	Catering contractor management fee	
	Bar	Bar: coffee lounge	Consultants	
		Bar: alcoholic and soft drinks		

Category	Subcategories (contract bundle items)	Principal examples/ cost elements	Items included in costs	Items excluded
	Maintenance and repair	Catering equipment maintenance		
Porterage		Workplace removals General services	Supervision Staff Uniforms Trolleys	Waste disposal Moves
Helpdesk/MIS		Helpdesk operation Booking services Equipment/software maintenance	Supervision Staff Uniforms Specialist equipment/software maintenance	
Travel		Booking service operation Travel charges		
Fitness centre		Fitness centre management Fitness centre equipment maintenance		
Nursery/crèche		Nursery/crèche management Nursery/crèche equipment maintenance		
General management		ICF Management agency	Senior/strategic management Salaries Benefits Overheads Contract management (in-house and outsourced contracts) Expenses Equipment	Blue-collar workers Task supervisors

Category	Subcategories (contract bundle items)	Principal examples/ cost elements	Items included in costs	Items excluded
			Course fees Professional subscriptions	
Professional services		Health and safety management Project management Estate management Space management Conference/meeting room management	Facilities consultants	Blue-collar workers Task supervisors

All categories to include outsourced contracts and in-house staff.

In-house staff costs to include benefits and overhead allowances.

All costs to be exclusive of VAT.

Costs of space occupied by category to be excluded but may be shown separately.

Depreciation and amortisation costs are not included within the categories but may be shown separately.

AV: audiovisual; CCTV: closed circuit television; DECT: digital enhanced cordless telecommunications; DTP: desktop publishing; ICF: intelligent client function; ISDN: integrated services digital network; ISP: Internet service provider; IT: information technology; MIS: management information system; PAT: Portable Appliance Testing; PC: personal computer; PPM: Planned Preventative Maintenance; UPS: uninterruptible power supply.

Index